A Concept of Mathematical Physics

MODELS FOR SPACE-TIME

A Concept
of Mathematical Physics

MODELS
FOR SPACE-TIME

by

TAMÁS MATOLCSI

Eötvös Loránd University
Budapest

1828

Akadémiai Kiadó, Budapest 1984

ISBN 963 05 3245 X

Printed in Hungary

ACKNOWLEDGEMENTS

I should like to express my gratitude to all those who have helped me in writing this book. In particular, to Professor B. Sz.-Nagy for his valuable assistance, to Professor A. Kósa for giving the opportunity to research, to Professors Gy. Soós and J. Erdős for proposing the publication and for advising some useful modifications, to my colleague J. Kristóf with whom I have discussed all the topics and to whom most of the exercises are due, and to my student T. Gruber for having read the manuscript and corrected a number of errors.

One should keep the need for a sound mathematical basis dominating one's search for a new theory. Any physical or phylosophical ideas that one has must be adjusted to fit the mathematics. Not the way round.

Too many physicists are inclined to start from some preconceived physical ideas and then to try to develop them and find a mathematical scheme that incorporates them. Such a line of attack is unlikely to lead to success. One runs into difficulties and finds no reasonable way out of them. One ought then to realize that one's whole line of approach is wrong and to seek a new starting point with a sound mathematical basis.

P. A. M. Dirac

(in Mathematical Foundations of Quantum Theory,
edited by A. R. Marlow, Ac. Press, New York 1978)

CONTENTS

PART TWO
MATHEMATICAL TOOLS

PREFACE

1. Mathematics reached a crisis at the end of the last century when a number of paradoxes came to light. Mathematicians surmounted the difficulties by revealing the origin of the troubles: the obscure notions, the inexact definitions; then the modern mathematical exactness was created and all the earlier notions and results were reappraised. After this great work nowadays mathematics is firmly based upon its exactness.

Theoretical physics—in quantum field theory—reached its own crisis in the last decades ([10], [13]). The reason of the troubles is the same. Earlier physics has treated common, visible and palpable phenomena, everything has been obvious. On the other hand, modern physics deals with phenomena of the microworld where nothing is common, nothing is visible, nothing is obvious. Most of the notions applied to describe phenomena of the microworld are the old ones and in the new framework they are necessarily confused.

It is quite evident, we have to follow a way similar to that followed by mathematicians to create *mathematical physics*, a firm theory based on mathematical exactness; having mathematical exactness as a guiding principle, we must reappraise physics, its most common, most visible and most palpable notions as well. Doing so we can hope we shall be able to overcome the difficulties.

2. According to the concept of the present book, *mathematical physics* should be a *mathematical theory* of the *whole physics*, a mathematical theory based on mathematical exactness, a mathematical theory in which only *mathematically defined notions* appear and in which *all the notions used in physics* are defined in a mathematically exact way.

We can illustrate this by the following example. In present axiomatic quantum field theory one defines that the observables of a physical system constitute a C^*-algebra. By doing so one defines what an observable is without defining what a physical system is. Furthermore, one speaks about degrees of freedom, reference frames etc. without an exact mathematical definition. Hence, axiomatic field theory in its present state cannot be a part of mathematical physics.

To build up the new theory, mathematical physics, we must start with the simplest, most common notions of physics; we cannot start with quantum field theory but we hope that we can end up with it.

3. The fundamental notion of mathematical physics is that of *models*. Our aim is to construct mathematical models for physical phenomena. The modelling procedure has two sides of equal importance: the mathematical model and the modelled part of physical reality. We shall sharply distinguish these two sides. The physical reality is *independent* of our mind, it is such as it is. A mathematical model *depends* on our mind, it is such as it is constructed by us. The confusion of the physical reality and its models led to heavy misunderstandings in connection with quantum mechanics.

A mathematical model is constructed on results of experiments and theoretical considerations and conclusions of the model are controlled by experiments. This relation between models and modelled parts of Nature is illuminated below.

i) A mathematical model is a *system of mathematical axioms* which we require to be exact and consistent. This requirement can be formulated *universally in an exact way.*

On the other hand, nothing can be required of the physical reality, of course.

ii) A mathematical model *must not contradict* to general principles of physics (such as "conservation laws"). This requirement can be *exactly formulated in each model.*

iii) A mathematical model is expected to well *reflect* some properties of the modelled piece of reality. However, it lies outside the model to answer what and how it reflects and to decide in what sense it is good or bad. To answer these questions, we have to go *beyond the exact framework* of the model.

4. The whole world is an undivisible unity. However, to make physics, we are forced by our limited biological, mental etc. capacity to divide it in theory.

Today's physics suggests the arrangement of physical phenomena in three groups; the corresponding three entities can be called Space-time, Matter and Field.

The phenomena of these three entities interact among themselves and they determine each other mutually. At present it is impossible to give a good description of the complicated situation in which everything interacts with everything, which can be illustrated graphically as follows:

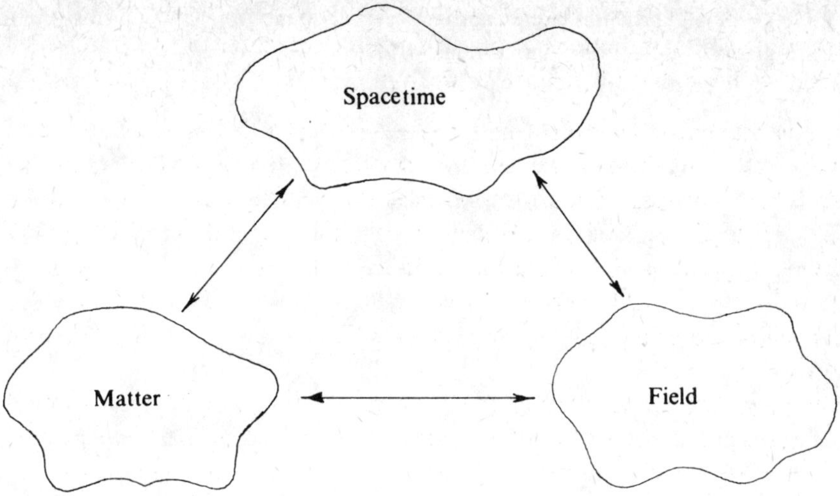

5. Fortunately, a great number of phenomena allows us to neglect some aspects of the interactions. More precisely, we can construct a good theory if we can *replace interaction by action*, i.e. we can consider as if the phenomena of two of the entities above were given, known, prefixed, "stiff" and only the phenomena of the third one were "flexible", unknown, and looked for. Then the stiff phenomena of the two entities are supposed to act upon and even to determine the phenomena of the third one which do not react. We obtain different theories according to which two entities are considered as prefixed, namely:

Mechanics (classical and quantal) if space-time and field phenomena are given to determine phenomena of matter:

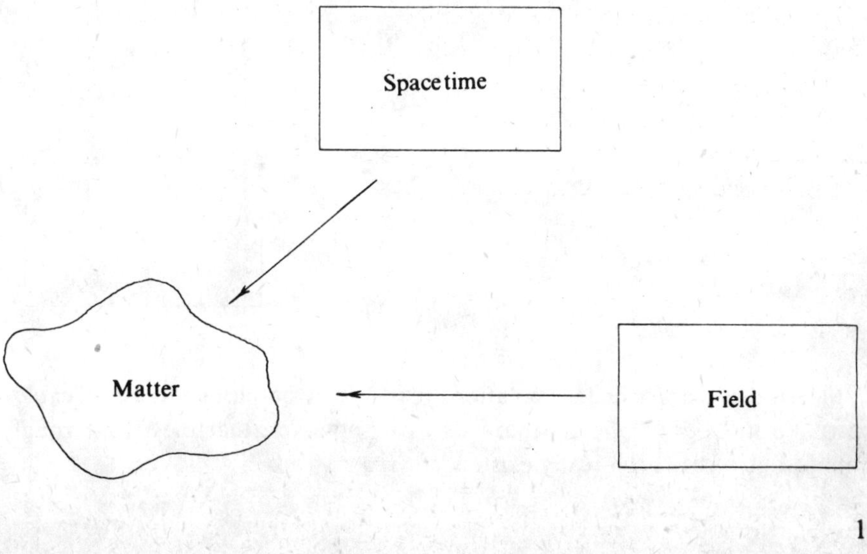

Field theory (classical, i.e. electrodynamics) if space-time and matter phenomena are given to determine phenomena of field:

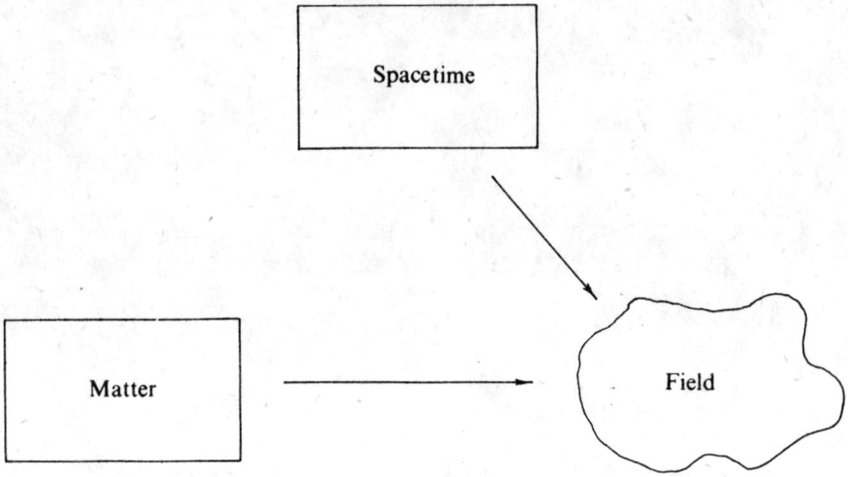

Gravitation theory if matter and field phenomena are given to determine space-time:

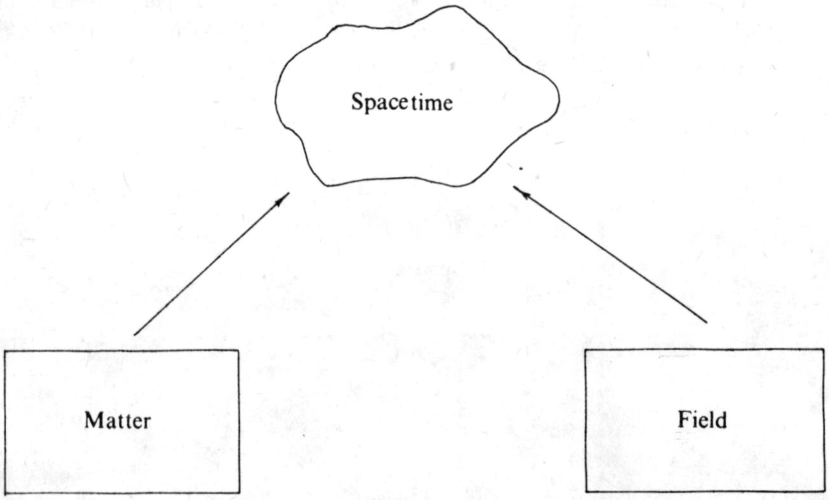

These theories in usual formulation are relatively simple and well applicable to describe a number of phenomena; it is clear, however, that they draw roughly simplified pictures of the really existing physical world.

6. Difficulties arise when we want to describe complicated situations in which only one of the three entities listed can be considered as known and interactions occur among the phenomena of the other two entities. The following graphically delineated possibilities exist:

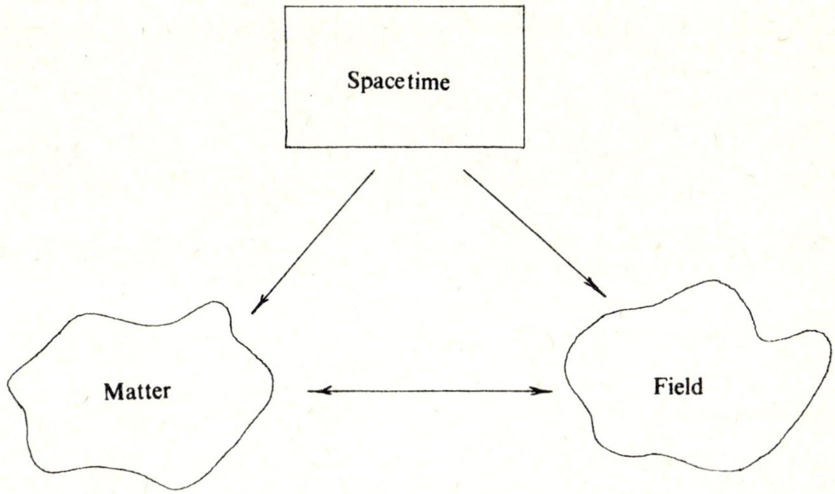

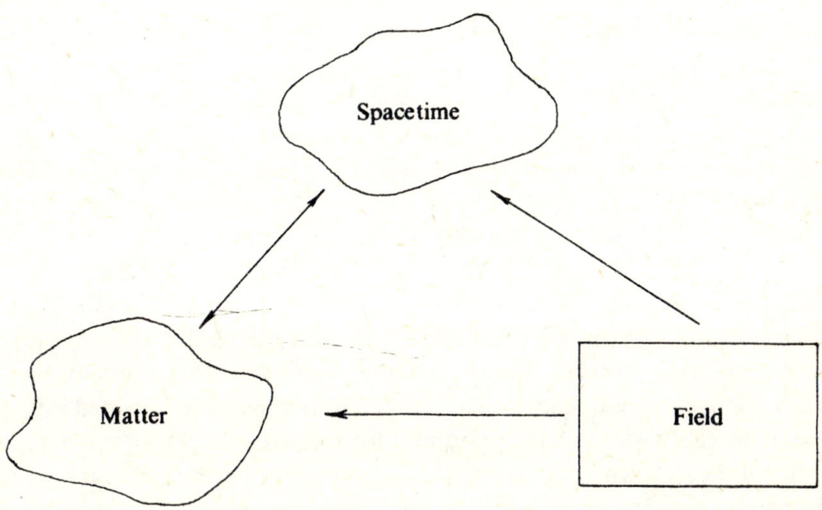

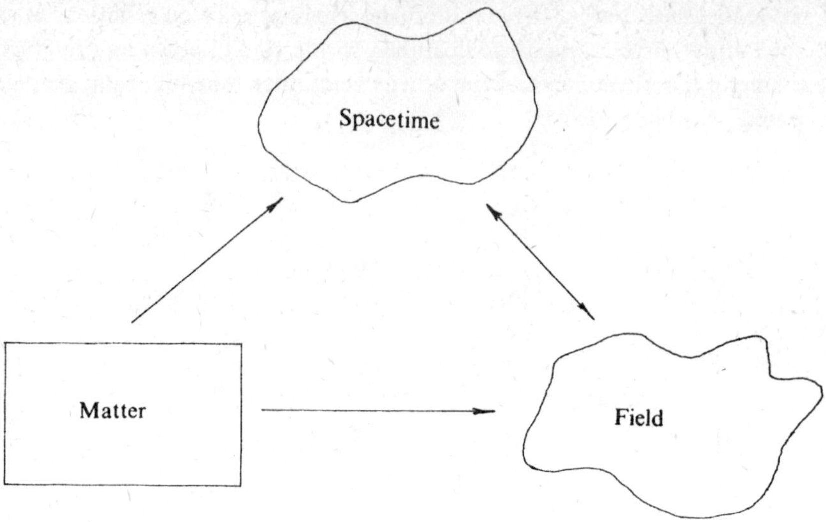

The third one is of no practical interest, so far. On the contrary, the other two are very important and we are forced to deal with them. They represent qualitatively new problems and they cannot be reduced to the previous well-known theories, except the special case treated in the next paragraph.

Electromagnetic radiation of microparticles is a phenomenon, for instance, which needs such a theory. Usual quantum electrodynamics has the purpose of serving as a theory for its description, and in general usual quantum field theory is destined to describe the interaction of field and matter.

As it is well-known, usual quantum field theory has failed in its aim. One might suspect the reason of the failure is that usual quantum field theory was created in such a way that the notions and formulae of *mechanics* were mixed with those of *field theory*. This way leads to nowhere: in *mechanics* the field phenomena are fixed, in *field theory* the matter phenomena are fixed, the corresponding notions "stiff" on one side cannot be fused correctly to produce notions "flexible" on both sides.

The complicated mathematics of quantum electrodynamics prevents us from exhibiting a simple example to illustrate the foregoings, whereas classical electrodynamics offers an excellent example. The electromagnetic field of a point charge moving on a prescribed path is obtained by the Lienard-Wiechert potential which allows us to calculate the force acting upon the charge because of the electromagnetic radiation. Then the Newtonian equation is supplemented with this radiation reacting force—which is deduced for a point charge *moving on a given path*—to get the so called Lorentz–Dirac equation which is used *to give the motion* of a point charge put in an electromagnetic field. One must not wonder that the result is a nonsense ("runaway solutions", see [3]).

Electromagnetic radiation is an irreversible process; in fact every process in Nature is irreversible. *The description of interaction must reflect irreversibility.*

Mechanics (Newtonian equation, Schrödinger equation) and electrodynamics (Maxwell equations), i.e. the theories dealing with action instead of interaction, do not know irreversibility. Evidently, no amalgamation of these theories can describe interaction and irreversibility.

7. There is a special case in which interaction can be reduced to some combination of actions yielding a good approximation. Assume matter phenomena can be divided into two parts, a "big" one and a "small" one. The big one and field (or space-time) are considered given and supposed to produce space-time (or field) which in turn act on the small matter phenomena to determine it. The situations can be illustrated graphically as follows

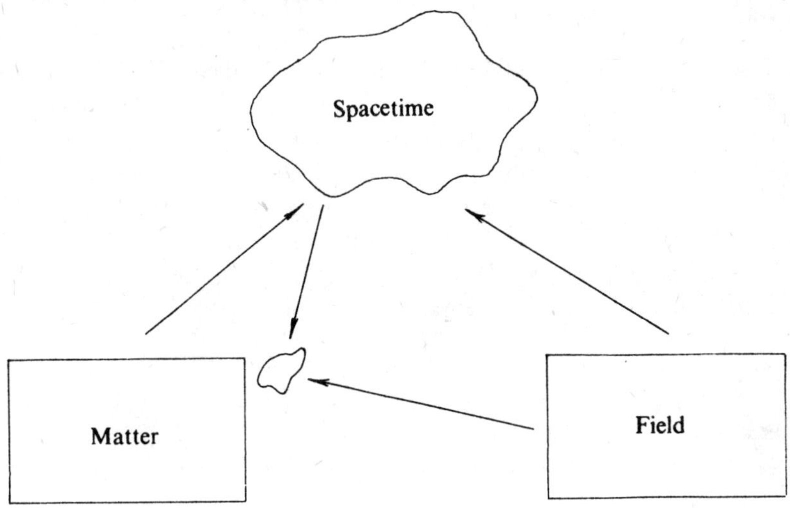

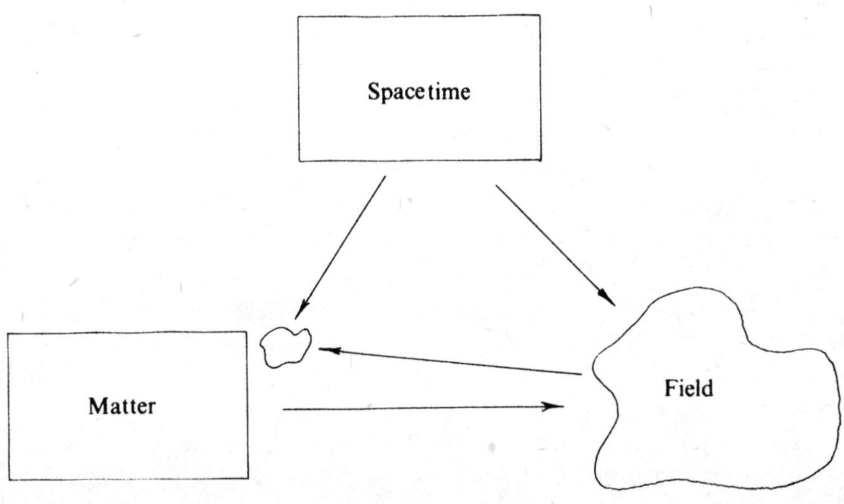

An example for the application of this trick is the description of planetary motions in general relativity, more closely, the advance of the perihelion of Mercury. Then field is supposed absent, the big Sun produces space-time and this space-time determines the motion of the small Mercury. Note that doing so we neglect that space-time is influenced by the Mercury and the motion of the Sun is influenced by space-time as well, i.e. we neglect interaction.

The second example is similar. A given space-time and a heavy point charge are supposed to produce an electromagnetic field and this electromagnetic field determines the motion of a light point charge. Doing so we neglect that the electromagnetic field is influenced by the light point charge and the motion of the heavy point charge is influenced by the electromagnetic field as well, i.e. we neglect interactions.

8. It seems, an entirely new theory should be established for the description of matter-field interaction. The present state of physics indicates that the successful construction of such a theory needs an exact mathematical physics.

The present work is an attempt to formulate the primary theories (i.e. those dealing with action instead of interaction) in the language of mathematical physics. According to our aim, we strictly distinguish the exact mathematical notions, relations, formulae and the physical motivations, interpretations. Since the former ones constitute the essence of this work, the latter ones are put between the marks ⌜ ⌟.

PART ONE

SPACE-TIME MODELS

INTRODUCTION

1. ⌈The notions of *space* and *time* are fundamental in physics; we conceive that Natural phenomena proceed in space and time.

Space is constituted by material objects. If I am in a room, then the walls and the furniture realize space for me. If I am in a flying aeroplane then its body and the seats realize space for me. If I sit near a road, the road, the trees, the buildings constitute the physical space for me. Consider someone who travels on the road in a car; then the steering-wheel, the dashboard, the seats constitute the physical space for him. The road is not a part of his space and the car is not a part of my space.

Time, too, is constituted by material objects; this is a less obvious assertion. However, time is nothing else than the processes of phenomena connected with the material objects in question.

We shall call a set of material objects an *observer*. So we can say that an observer realizes space and time for itself. Our simplest experiences show very well that, in general, different observers realize different spaces (see for instance the road and the moving car). Thus *space does not exist absolutely*, it depends on observers. On the other hand, our simplest experiences *do not answer* the question whether time exists absolutely or does not, i.e. whether times constituted by different observers are different or not.

Anyhow, if we want to establish a *relation between spaces* of two observers moving relatively to each other, we must *involve their times*, too.

This suggests the idea that instead of space and time there is an *absolute unity: space-time;* only observers perceive this absolute unity separately as space and time.

Since our first rough experiments do not indicate whether time is absolute, yes or no, first we set up the so called *nonrelativistic space-time model* based on the assumption that time is absolute. More delicate experiments show that not only space but also time depends on observers, whence we construct the diverse *relativistic space-time models.*⌋

2. ⌈Constructing models for physical phenomena we always meet the problem to *describe the magnitudes* of physical quantities. One usually considers the magnitudes as real numbers though they evidently are not real numbers; only if we choose a magnitude for unit then the other magnitudes of values of the corresponding physical quantity become real numbers: the ones showing the rates of the magnitudes relative to the unit. If the unit is changed, the numbers giving the magnitudes change as well.

Usually the magnitudes of values of a physical quantity form either a one dimensional real vector space (e.g. electric charge) or a one dimensional cone (e.g. space distance), i.e. a set in which the multiplication by non-negative real numbers is defined and there is an element such that all others are its non-negative multiples. Such a cone can be canonically embedded in an oriented one dimensional real vector space in such a way that the set of non-negative elements of the vector space coincides with the cone. If C denotes a cone then we define $-C$ as the set of ordered pairs $(-1, a)$ where $a \in C$ and the set $C \cup -C$ can be endowed with a vector space structure and orientation trivially (see IV. 4.1.).

Hence we find convenient to consider *the set of magnitudes of values of a physical quantity* as *a one dimensional oriented real vector space.*

We often meet physical quantities whose magnitudes can be expressed as products or quotients of magnitudes of other quantities. For instance, the magnitudes of the velocity are quotients of space lengths by time lengths. The vector space of such product and quotient magnitudes can be correctly defined as *tensor products* and *tensor quotients* (see Sections IV. 1. and IV. 2.). ⌋

I. NONRELATIVISTIC SPACE-TIME MODEL

1. Fundamentals

Necessary mathematical tools:

Tensor products and quotients (Chapter IV.)
Affine spaces and affine maps (Section VI. 1.)
Euclidean vector spaces (Section V. 3.)

1.1. ⌈We have seen in Introduction 1 that in any case we are forced to construct space-time models in which space and time do not exist separately, space-time is an absolute unity.

The basic assumption of a nonrelativistic space-time model is that absolute time exists.

Furthermore, according to our simplest experiences, spaces relative to observers (the Earth or the car, see Introduction 1) have the properties that a "vector" can be assigned to each ordered pair of space points, and these vectors, considered as freely removable, obey the usual rules of multiplication by real numbers and addition; the length of a vector and the angle between two vectors can be defined; there are only three mutually orthogonal vectors, and right and left oriented orthogonal systems can be distinguished. The "parity breaking" property of the so called weak interactions shows that the two orientations are not physically equivalent. In other words, we find convenient to describe space relative to an observer as a three dimensional oriented Euclidean affine space. Similary, we describe time as a one dimensional oriented affine space; the orientation reflects that past and future have different physical roles, they are distinguished by the irreversibility of Natural phenomena.

As a consequence, we assume that space-time has an affine structure, an orientation, it is of four dimensions, there are an absolute time and some Euclidean structure on the set of simultaneous space-time points. Since the set of space lengths is considered as a one dimensional oriented real vector space (see Introduction 2.), we take the Euclidean structure in the form given in Section V. 3.⌋

1.2. Definition. A *nonrelativistic space-time model* is a quintuplet $(M, I, \tau, \mathbf{D}, \gamma)$ where

M is an oriented four dimensional real affine space
over the vector space $\mathbf{M}$,

I is an oriented one dimensional real affine space over the vector space $\mathbf{I}$,
$\tau: M \longrightarrow I$ is an affine surjection,
$\mathbf{D}$ is an oriented one dimensional real vector space,
$\gamma: (\operatorname{Ker} \mathrm{D}\tau) \times (\operatorname{Ker} \mathrm{D}\tau) \longrightarrow \mathbf{D} \otimes \mathbf{D}$ is a positive definite symmetric bilinear map; in other words, $(\operatorname{Ker} \mathrm{D}\tau, \mathbf{D}, \gamma)$ is a Euclidean vector space.

We assume in the sequel that a nonrelativistic space-time model $(M, I, \tau, \mathbf{D}, \gamma)$ *is given.*

1.3. We shall use the notations

$$\boldsymbol{\tau} := \mathrm{D}\tau, \qquad \mathbf{E} := \operatorname{Ker} \boldsymbol{\tau}.$$

M and I are called *space-time* and (absolute) *time*, respectively. τ assigns the corresponding time points to space-time points, i.e. $\tau(x)$ is the time point of the space-time point x.

For $t \in I$ the set $\overset{-1}{\tau}\{t\}$ is a three dimensional affine subspace over $\mathbf{E}$ of M which is the set of (absolutely) *simultaneous* space-time points and can be viewed as the *instantaneous space at the time point t* (but there is no space!). $\mathbf{D}$ is the vector space of *space lengths* and γ is the Euclidean structure on $\mathbf{E}$, i.e.

$$|\mathbf{q}| := \sqrt[\otimes]{\gamma(\mathbf{q}, \mathbf{q})} \in \mathbf{D}$$

(notation: IV. 3. 9.) is the *length* of $\mathbf{q} \in \mathbf{E}$ and the real number

$$\frac{\gamma(\mathbf{q}_1, \mathbf{q}_2)}{|\mathbf{q}_1| \otimes |\mathbf{q}_2|}$$

(notation: IV. 2.2., see also IV. 2.3. and V. 3.2.) is the *cosine of the angle* between the non-zero elements $\mathbf{q}_1$ and $\mathbf{q}_2$ of $\mathbf{E}$.

1.4. The correspondence $t \mapsto \overset{-1}{\tau}\{t\}$ is a bijection from I onto the set of affine subspaces over $\mathbf{E}$ of M. Using this correspondence we shall find convenient to *identify I with the set of affine subspaces over* $\mathbf{E}$ *of M.*

1.5. Notice the interesting and important fact that there is time but there is no space in our model; space-time is not a Cartesian product of time and space. A curious complementarity of $\mathbf{I}$ ("time translations") and $\mathbf{E}$ ("instantaneous space translations") is reflected by the fact that there is a linear surjection from $\mathbf{M}$ onto $\mathbf{I}$, namely $\boldsymbol{\tau}$, and there is an injection from $\mathbf{E}$ into $\mathbf{M}$, namely the canonical embedding $\mathbf{i}$, and the kernel of the surjection is the image of the injection; in other words, the sequence of maps

$$0 \longrightarrow \mathbf{E} \xrightarrow{\ \mathbf{i}\ } \mathbf{M} \xrightarrow{\ \boldsymbol{\tau}\ } \mathbf{I} \longrightarrow 0$$

is exact. Hence the sequence of maps

$$0 \longrightarrow \mathbf{I}^* \xrightarrow{\tau^*} \mathbf{M}^* \xrightarrow{\mathbf{i}^*} \mathbf{E}^* \longrightarrow 0$$

is exact as well.

Observe that this allows us to *consider* $\mathbf{I}^*$, *identified with* $\tau^*[\mathbf{I}^*]$, *as a linear subspace of* $\mathbf{M}^*$ and that $\mathbf{i}^*(\mathbf{k})$ is the restriction of the linear form $\mathbf{k} \in \mathbf{M}^*$ to the subspace $\mathbf{E}$ of $\mathbf{M}$.

1.6. Definition. Let $\mathbf{A}$ be a vector space constructed of $\mathbf{I}$, $\mathbf{I}^*$, $\mathbf{D}$ and $\mathbf{D}^*$ by tensor products and tensor quotients. Then the elements of

$\mathbf{A}$	are called	*scalars of type* $\mathbf{A}$,
$\mathbf{A}\otimes\mathbf{M}$	are called	*vectors of type* $\mathbf{A}$,
$\dfrac{\mathbf{M}}{\mathbf{A}}$	are called	*vectors of cotype* $\mathbf{A}$,
$\mathbf{A}\otimes\mathbf{M}^* = \left(\dfrac{\mathbf{M}}{\mathbf{A}}\right)^*$	are called	*covectors of type* $\mathbf{A}$.

Tensors and cotensors, tensor fields and cotensor fields of type and of cotype $\mathbf{A}$ are defined as in VI. 3.1.

In the case $\mathbf{A} = \mathbb{R}$ we omit the terms "of type $\mathbb{R}$" and "of cotype $\mathbb{R}$".

1.7. Definition. An element $\mathbf{x}$ of $\mathbf{M}$ is called *spacelike* or *timelike* if $\mathbf{x}$ is or is not in the subspace $\mathbf{E}$, respectively. An element $\mathbf{k}$ of $\mathbf{M}^*$ is called *timelike* or *spacelike* if $\mathbf{k}$ is or is not in the subspace $\mathbf{I}^*$ (see 1.5.), respectively. Similar names are used for vectors and covectors of type $\mathbf{A}$ as well. ∎

Evidently, a vector $\mathbf{x}$ is spacelike if and only if $\tau(\mathbf{x}) = \mathbf{0}$ and a covector $\mathbf{k}$ is timelike if and only if $\mathbf{i}^*(\mathbf{k}) = \mathbf{k}|_{\mathbf{E}} = \mathbf{0}$.

Notice that a timelike covector, as an element of $\mathbf{I}^*$, is actually a scalar.

1.8. The set of timelike vectors has two connected components, namely

$$T^{\rightarrow} := \{\mathbf{x} \in \mathbf{M} \,\vdots\, \tau(\mathbf{x}) > \mathbf{0}\}, \qquad T^{\leftarrow} := \{\mathbf{x} \in \mathbf{M} \,\vdots\, \tau(\mathbf{x}) < \mathbf{0}\}.$$

The elements of $T^{\rightarrow}$ and $T^{\leftarrow}$ are called *positive* and *negative*, respectively.

1.9. $\mathbf{M}^*$, $\mathbf{I}^*$ and $\mathbf{D}^*$ are considered to be oriented, as duals of oriented vector spaces (see IV. 3.1.). Moreover, the orientation of $\mathbf{M}$ and $\mathbf{I}$ and the surjection τ define an orientation of $\mathbf{E}$:

Definition. A basis $\mathbf{e}_1, \mathbf{e}_2, \mathbf{e}_3$ in $\mathbf{E}$ is called *positively oriented* if for some (hence for all) $\mathbf{x} \in T^{\rightarrow}$ the basis $\mathbf{x}, \mathbf{e}_1, \mathbf{e}_2, \mathbf{e}_3$ in $\mathbf{M}$ is positively oriented.

1.10. ⌈A nonrelativistic space-time model is defined as a mathematical structure. The questions arise: what relations can be established between such structures? how many different structures of this kind exist? They are interesting mathematical questions and have a deep physical significance: as many different nonrelativistic space-time models so many different nonrelativistic physics. To answer these questions first we need an exact definition what the "difference" or the "sameness" of two nonrelativistic space-time models mean. ⌋

Definition. Two nonrelativistic space-time models $(M, I, \tau, \mathbf{D}, \gamma)$ and $(M', I', \tau', \mathbf{D}', \gamma')$ are *isomorphic* if there are

(i) an orientation preserving affine bijection $F: M \longrightarrow M'$,

(ii) an orientation preserving affine bijection $B: I \longrightarrow I'$,

(iii) an orientation preserving linear bijection $Z: \mathbf{D} \longrightarrow \mathbf{D}'$

in such a way that

(a) $$\tau' \circ F = B \circ \tau,$$

(b) $$\gamma' \circ (\mathrm{D}F \times \mathrm{D}F) = (Z \otimes Z) \circ \gamma.$$

(Notice that (a) implies $\mathrm{D}\tau' \circ \mathrm{D}F = \mathrm{D}B \circ \mathrm{D}\tau$ involving $\mathrm{D}F[\mathrm{Ker}\,\mathrm{D}\tau] = \mathrm{Ker}\,\mathrm{D}\tau'$, thus the left hand side of (b) makes sense and (b) says that $\mathrm{D}F$ is a γ-γ'-orthogonal map, [see II. V. 2.10.]).

The triplet (F, B, Z) is an *isomorphism* between the two nonrelativistic space-time models. If the two models coincide, an isomorphism is called an *automorphism*. An automorphism (F, B, Z) is called *strict* if $\mathrm{D}B = \mathrm{id}_{\mathbf{I}}$ and $Z = \mathrm{id}_{\mathbf{D}}$.

1.11. To advance further, we need some auxiliary results concerning a nonrelativistic space-time model $(M, I, \tau, \mathbf{D}, \gamma)$. First we introduce the notations

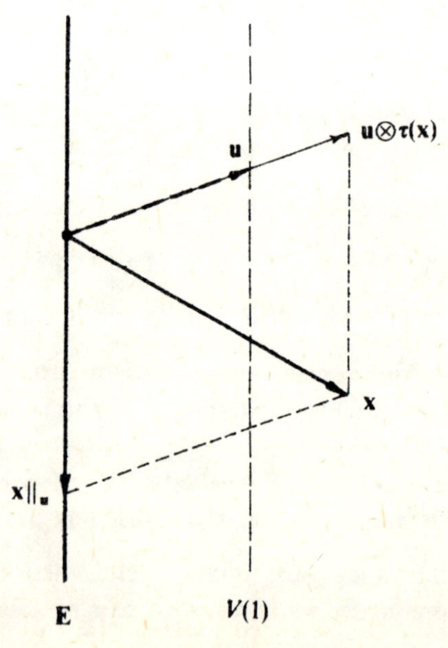

$$V(1) := \left\{ \mathbf{u} \in \frac{\mathbf{M}}{\mathbf{I}} : \tau_{\mathbf{I}}(\mathbf{u}) = 1 \right\}$$

(notation: IV. 2.10. (ii)),

$$\mathbf{x}\|_{\mathbf{u}} := \mathbf{x} - \mathbf{u} \otimes \tau(\mathbf{x}) \qquad (\mathbf{x} \in \mathbf{M},\ \mathbf{u} \in V(1)).$$

The $\frac{\mathbf{D}\otimes\mathbf{D}}{\mathbf{I}\otimes\mathbf{I}}$ valued bilinear map on $\frac{\mathbf{E}}{\mathbf{I}}$ arising from γ is denoted by $[\ |\ ]$:

$$[\mathbf{v}|\mathbf{v}'] := \frac{\gamma(\mathbf{v}\otimes\mathbf{t}, \mathbf{v}'\otimes\mathbf{t})}{\mathbf{t}\otimes\mathbf{t}}, \qquad \left(\mathbf{v}, \mathbf{v}' \in \frac{\mathbf{E}}{\mathbf{I}},\ \mathbf{0} \neq \mathbf{t} \in \mathbf{I}\right).$$

1.12. Proposition. $V(1)$ is an affine hyperplane, over $\frac{\mathbf{E}}{\mathbf{I}}$, of $\frac{\mathbf{M}}{\mathbf{I}}$. $\left(V(1), \frac{\mathbf{D}}{\mathbf{I}}, [\ |\ ]\right)$ is a three dimensional Euclidean affine space.

1.13. Let us orient $\mathbf{E}$ according to Definition 1.9. and attach to $\mathbf{I}\times\mathbf{E}$ the orientation determined by the orientations of $\mathbf{I}$ and $\mathbf{E}$.

Proposition. Let $\mathbf{c} \in V(1)$; then the map

$$\mathbf{s}_{\mathbf{c}} : \mathbf{M} \longrightarrow \mathbf{I}\times\mathbf{E}, \qquad \mathbf{x} \mapsto (\tau(\mathbf{x}), \mathbf{x}\|_{\mathbf{c}})$$

is an orientation preserving linear bijection whose inverse is

$$(\mathbf{t}, \mathbf{q}) \mapsto \mathbf{c}\otimes\mathbf{t} + \mathbf{q} \qquad ((\mathbf{t}, \mathbf{q}) \in \mathbf{I}\times\mathbf{E}).$$

Furthermore, if $o \in M$, the map

$$s_{\mathbf{c},o} := \mathbf{s}_{\mathbf{c}} \circ O_o : M \longrightarrow \mathbf{I}\times\mathbf{E}, \quad x \mapsto (\tau(x-o), (x-o)\|_{\mathbf{c}})$$

is an orientation preserving affine bijection whose inverse is

$$(\mathbf{t}, \mathbf{q}) \mapsto o + \mathbf{c}\otimes\mathbf{t} + \mathbf{q} \qquad ((\mathbf{t}, \mathbf{q}) \in \mathbf{I}\times\mathbf{E}),$$

and the derivative of $s_{\mathbf{c},o}$ equals $\mathbf{s}_{\mathbf{c}}$.

1.14. As a consequence of the previous evident results, we deduce immediately the following one.

Proposition. Let $(M, I, \tau, \mathbf{D}, \gamma)$ be a nonrelativistic space-time model. Then $(\mathbf{I}\times\mathbf{E}, \mathbf{I}, \mathrm{pr}_{\mathbf{I}}, \mathbf{D}, \gamma)$ is a nonrelativistic space-time model and if $(\mathbf{c}, o) \in V(1)\times M$ then $(s_{\mathbf{c},o}, O_{\tau(o)}, \mathrm{id}_{\mathbf{D}})$ is an isomorphism between the two nonrelativistic space-time models, where

$$O_{\tau(o)}(t) = t - \tau(o) \qquad (t \in I).$$

1.15. Proposition. $(\mathbb{R}\times\mathbb{R}^3, \mathbb{R}, \mathrm{pr}_{\mathbb{R}}, \mathbb{R}, \langle\ ,\ \rangle)$ is a nonrelativistic space-time model (where $\langle\ ,\ \rangle$ denotes, as usually, the inner product on $\mathbb{R}^3$, $\mathbb{R}$ and $\mathbb{R}^3$ are oriented by their standard bases) and every nonrelativistic space-time model is isomorphic to it.

Proof. Let $(M, I, \tau, \mathbf{D}, \gamma)$ be a nonrelativistic space-time model. Take a positive element $\mathbf{e}$ of $\mathbf{I}$, a positive element $\mathbf{d}$ of $\mathbf{D}$, a positively oriented γ-orthogonal basis $\mathbf{e}_1$, $\mathbf{e}_2$, $\mathbf{e}_3$ in $\mathbf{E}$, normed to $\mathbf{d}$. Then the linear maps

$$\mathbf{I} \times \mathbf{E} \longrightarrow \mathbb{R} \times \mathbb{R}^3, \quad (\mathbf{t}, \mathbf{q}) \mapsto \left(\frac{\mathbf{t}}{\mathbf{e}}, \left\{ \frac{\gamma(\mathbf{e}_k, \mathbf{q})}{\mathbf{d} \otimes \mathbf{d}} \right\}_{k=1,2,3} \right),$$

$$\mathbf{I} \longrightarrow \mathbb{R}, \qquad \mathbf{t} \mapsto \frac{\mathbf{t}}{\mathbf{e}},$$

$$\mathbf{D} \longrightarrow \mathbb{R}, \qquad \mathbf{z} \mapsto \frac{\mathbf{z}}{\mathbf{d}}$$

constitute an isomorphism between $(\mathbf{I} \times \mathbf{E}, \mathbf{I}, \mathrm{pr}_{\mathbf{I}}, \mathbf{D}, \gamma)$ and $(\mathbb{R} \times \mathbb{R}^3, \mathbb{R}, \mathrm{pr}_{\mathbb{R}}, \mathbb{R}, \langle\ ,\ \rangle)$. Applying Proposition 1.14. we get the desired result.

1.16. The previous assertion has the following very important corollary.
Theorem. Any two nonrelativistic space-time models are isomorphic.

1.17. Now we see that the nonrelativistic space-time model as a mathematical structure is unique. For investigations and applications of the nonrelativistic space-time model we can use an arbitrarily given one. However, an actual model such as $(\mathbf{I} \times \mathbf{E}, \mathbf{I}, \mathrm{pr}_{\mathbf{I}}, \mathbf{D}, \gamma)$ or $(\mathbb{R} \times \mathbb{R}^3, \mathbb{R}, \mathrm{pr}_{\mathbb{R}}, \mathbb{R}, \langle\ ,\ \rangle)$ has a number of additional structures, too. For instance, space-time and time are vector spaces, time is canonically embedded in space-time as $\mathbf{I} \times \{\mathbf{0}\}$ or as $\mathbb{R} \times \{\mathbf{0}\}$. These models tempt us to consider $\mathbf{E}$ or $\mathbb{R}^3$ as space (but space does not exist!) and space-time as a Cartesian product of time and space. In the latter model time lengths and space lengths have a canonical unit, namely the number one in $\mathbb{R}$.

To avoid such confusions we should keep away from similar specially constructed models; we can and will use only the structures listed in Definition 1.2. for the investigation and application of the nonrelativistic space-time model.

2. World lines

Necessary mathematical tools:

Tensor products and quotients (Chapter IV.)
Differentiation in affine spaces (Sections VI. 2., VI. 3.)

2.1. In usual classical mechanics the *motion* of a masspoint is determined by the *forces* acting upon the masspoint and is described by the *Newtonian equation.* However, all the usual notions are related to observers. Now we are to define the corresponding absolute notions connected with masspoints. Masspoint itself is not defined here; volume Models in Mechanics has the task to give the rather complicated definition of a masspoint.⌋

2.2. Definition. A piecewise twice differentiable function (see VI. 2.8.) $r : I \rightarrowtail M$ having the properties

(i) Dom r is connected,

(ii) $\tau \circ r \subset \mathrm{id}_I$

is called a *world line function.* A subset of M is called a *world line* if it is the range of a world line function.

A world line which is a straight line segment is called *intertial.*

A world line function and the corresponding world line are called *global* if r is everywhere defined, i.e. $\tau \circ r = \mathrm{id}_I$.

⌈A world line function r is the model of the history of a masspoint in the time interval Dom r; $r(t)$ is the space-time point where the masspoint exists in the time-point t; evidently $\tau(r(t)) = t$ must hold.⌋

2.3. Proposition.

(i) The world line function corresponding to a world line is uniquely determined.

(ii) If r is a world line function then for all t in Dom r we have

$$\boldsymbol{\tau}_{\mathbf{I}}(\mathrm{D}^+ r(t)) = \boldsymbol{\tau}_{\mathbf{I}}(\mathrm{D}^- r(t)) = 1\,,$$

$$\boldsymbol{\tau}_{\mathbf{I}\otimes\mathbf{I}}((\mathrm{D}^+)^2 r(t)) = \boldsymbol{\tau}_{\mathbf{I}\otimes\mathbf{I}}((\mathrm{D}^-)^2 r(t)) = 0$$

(notation: IV. 2.10. (ii)).

Remark. If r is twice differentiable at t then the above formulae can be written in the form

$$\boldsymbol{\tau}_{\mathbf{I}}(\dot{r}(t)) = 1\,, \qquad \boldsymbol{\tau}_{\mathbf{I}\otimes\mathbf{I}}(\ddot{r}(t)) = 0\,,$$

or in other words,

$$\dot{r}(t) \in V(1)\,, \qquad \ddot{r}(t) \in \frac{\mathbf{E}}{\mathbf{I}\otimes\mathbf{I}}\,.$$

2.4. ⌈For a twice differentiable world line function r, $\dot{r}(t)$ and $\ddot{r}(t)$ can be interpreted as the value of the (absolute) velocity and the (absolute) acceleration at $t \in \mathrm{Dom}\, r$, respectively, of a masspoint having the world line function r as its history.⌋

That is why we call the elements of $V(1)$ *velocities* and the elements of $\frac{\mathbf{E}}{\mathbf{I}\otimes\mathbf{I}}$ *accelerations.* Notice the very important facts that

(i) — a velocity is a positive timelike vector of cotype **I**,

— an acceleration is a spacelike vector of cotype **I⊗I**;

(ii) — the set of velocities is an affine space over $\frac{\mathbf{E}}{\mathbf{I}}$, i.e. there is no distinguished velocity, all velocities play equal role,
— the set of accelerations is a vector space, hence there is a distinguished acceleration, namely the zero one;

(iii) — both $V(1)$ and $\frac{\mathbf{E}}{\mathbf{I}\otimes\mathbf{I}}$ have a Euclidean structure deriving from the positive definite symmetric bilinear map γ, but
— since $V(1)$ is an affine space, only the *relative velocities*, i.e. the differences of velocities have magnitudes; a velocity cannot be big or small compared with a given one, only the differences of two velocities can be compared in such a way,
— the accelerations themselves have magnitudes.

2.5. The magnitude of a relative velocity $\mathbf{v}\in\frac{\mathbf{E}}{\mathbf{I}}$ is the element

$$|\mathbf{v}|:=\sqrt[\otimes]{[\mathbf{v}|\mathbf{v}]} \quad \text{of} \quad \frac{\mathbf{D}}{\mathbf{I}}$$

(notations: 1.11. and IV. 3.9.). Similarly, the magnitude of an acceleration is an element of $\frac{\mathbf{D}}{\mathbf{I}\otimes\mathbf{I}}$.

⌈**D** and **I** are the vector spaces of space lengths and time lengths, respectively. Choosing a positive element in **D** and in **I** we fix the unit of space lengths and time lengths, respectively; for instance, cm $\in\mathbf{D}$ and sec $\in\mathbf{I}$. Then the unit of relative velocity magnitudes is usually taken as "cm divided by sec".

Its exact meaning can be defined in a mathematically correct way as the tensorial quotient $\frac{\text{cm}}{\text{sec}}$, an element of $\frac{\mathbf{D}}{\mathbf{I}}$, as it must be. Similarly, the chosen units of space lengths and time lengths fixes the unit of acceleration magnitudes as $\frac{\text{cm}}{\text{sec}\otimes\text{sec}}$.

The mass of a masspoint will be considered as a positive element of $\frac{\mathbf{I}}{\mathbf{D}\otimes\mathbf{D}}$. Thus, the units of space lengths and time lengths above determine the unit of the mass as $\frac{\text{sec}}{\text{cm}\otimes\text{cm}}$. This corresponds to the choice of units of measurements in which the Planck constant is a real number (dimensionless). A justification for this choice will appear in volume Models in Mechanics.

As a consequence, the vector space of force magnitudes is $\frac{\mathbf{I}}{\mathbf{D}\otimes\mathbf{D}}\otimes\frac{\mathbf{D}}{\mathbf{I}\otimes\mathbf{I}}=$ $=\frac{\mathbb{R}}{\mathbf{I}\otimes\mathbf{D}}$ which is reflected in the following definitions, too.⌋

2.6. Definition. A twice differentiable world line function r and the corresponding world line is called *rotation free* if $\ddot{r}(t)$ is parallel with $\ddot{r}(s)$ for all $t, s \in \operatorname{Dom} r$. ∎

Evidently, r is rotation free if and only if $\frac{\ddot{r}(t)}{|\ddot{r}(t)|}$ is the same for all $t \in \operatorname{Dom} r$, $\ddot{r}(t) \neq 0$ (where $|\ddot{r}(t)|$ denotes the magnitude of $\ddot{r}(t)$ (see 3.5.)); in other words, $\frac{\ddot{r}}{|\ddot{r}|}$ is a constant map.

2.7. Definition. (i) A twice differentiable covector field $\mathbf{K}: M \rightarrowtail \mathbf{M}^*$ is called a *potential.*

The potential $\mathbf{K}$ is *static* if there are a map $\mathbf{Z}: \mathbf{E} \rightarrowtail \mathbf{M}^*$, a $\mathbf{c} \in V(1)$ and an $o \in M$ such that

$$\mathbf{K}(x) = \mathbf{Z}((x-o)\|_{\mathbf{c}}) \qquad (x \in \operatorname{Dom} \mathbf{K}).$$

(ii) A differentiable map

$$\mathbf{f}: M \times V(1) \rightarrowtail \frac{\mathbf{E}}{\mathbf{I} \otimes \mathbf{D} \otimes \mathbf{D}} = \frac{\mathbf{E}^*}{\mathbf{I}}$$

(see V. 3.1.) is called a *force field.* The force field $\mathbf{f}$ is said to have a potential if there is a potential $\mathbf{K}$ such that

$$\mathbf{f}(x, \mathbf{u}) = \mathbf{i}_{\mathbf{I}}^*(\mathrm{d}\mathbf{K}[x]_{\mathbf{I}}\mathbf{u}) = (\mathbf{i}^* \circ \mathrm{d}\mathbf{K}[x])_{\mathbf{I}}(\mathbf{u}) \qquad ((x, \mathbf{u}) \in \operatorname{Dom} \mathbf{f})$$

where $\mathrm{d}\mathbf{K}$ is the exterior derivative of $\mathbf{K}$ (see V. 5.2.) and $\mathbf{i}^*: \mathbf{M}^* \longrightarrow \mathbf{E}^*$ is the adjoint of the canonical embedding $\mathbf{i}: \mathbf{E} \longrightarrow \mathbf{M}$ (see 1.5.).

(iii) A positive element of $\frac{\mathbf{I}}{\mathbf{D} \otimes \mathbf{D}}$ is called a *mass.*

2.8. Definition. A twice differentiable world line function r is called *Newtonian* if there are

(i) a force field $\mathbf{f}$

(ii) a mass m

such that r satisfies the second order differential equation (see VI. 3.2. and VI. 3.9.)

$$m \otimes \ddot{x} = \mathbf{f} \circ (x, \dot{x})$$

or

$$\dot{x} = \mathbf{u}, \qquad \dot{\mathbf{u}} = \frac{\mathbf{f} \circ (x, \mathbf{u})}{m}.$$

2.9. Let r be a world line function. It is very important to know, both from theoretical and practical point of view, whether the acceleration $\ddot{r}$ is the non-trivial part of $\ddot{\mathbf{r}}$ or $\nabla_{\dot{r}}\dot{r}$ (see VII. 9.9.).

Simple applications (use of curvilinear coordinates and usual physical considerations) as well as theoretical arguments (general relativity) show that $\ddot{r}$ represents $\nabla_{\dot{r}}\dot{r}$.

3. Observers

Necessary mathematical tools:

Tensor products and quotients (Chapter IV.)
Affine spaces, differentiation (Sections VI. 1., VI. 2.)

The essence of Propositions 3.3 and 3.4 can be grasped without the involved knowledge on smooth manifolds (foliations).

3.1. ⌈The notion of observers is fundamental in physics. In 1.1. and 2.1. we spoke about observers in the usual intuitive sense.

An observer is a collection of masspoints—see the man and car which appeared in Introduction 1—which has a spatial extent and a temporal duration, i.e. occupies a region of space-time.

The world lines of continuously distributed masspoints can be well given as integral curves of a vector field. That is why we accept the following definition.⌋

3.2. Definition. An *observer* is a smooth map $\mathbf{U}: M \rightarrowtail V(1) \subset \frac{\mathbf{M}}{\mathbf{I}}$ (i.e. a vector field of cotype $\mathbf{I}$); depending on whether $\operatorname{Dom} \mathbf{U} \neq M$ or $\operatorname{Dom} \mathbf{U} = M$, the observer is called *local* or *global*, respectively.

The maximal integral curves of $\mathbf{U}$ are called $\mathbf{U}$-*lines.*

The $\mathbf{U}$-line passing through $x \in \operatorname{Dom} \mathbf{U}$ is denoted by $C_{\mathbf{U}}(x)$, and $E_{\mathbf{U}}$ stands for the set of $\mathbf{U}$-lines.

Remark. The integral curves of an observer are evidently world lines. Note that the maximal integral curves of a global observer need not be global world lines (see Exercise 10.).

⌈On the base of 2.4. we can imagine that an observer is a *velocity field.* Furthermore, $D\mathbf{U}[\cdot]_{\mathbf{I}} \circ \mathbf{U}(\cdot): M \rightarrowtail \frac{\mathbf{E}}{\mathbf{I} \otimes \mathbf{I}}$ can be considered as the *acceleration field* of the observer $\mathbf{U}$, since if $r: I \rightarrowtail M$ gives an integral curve of $\mathbf{U}$ then $\dot{r} = \mathbf{U} \circ r$ involves $\ddot{r}(t) = D\mathbf{U}[r(t)]_{\mathbf{I}}\dot{r}(t) = D\mathbf{U}[r(t)]_{\mathbf{I}}\mathbf{U}(r(t))$ for $t \in \operatorname{Dom} r$.

We emphasize that we are dealing with mathematical models; an observer as it is defined is a mathematical model for a physical object. To underline this fact we might use the term "observer model" instead of "observer" but we want to avoid ponderousness.⌋

3.3. Proposition. Let $\mathbf{U}$ be an observer. Then $E_{\mathbf{U}}$ is a manifold with the smooth structure described in VII. 7.7. The tangent space of $E_{\mathbf{U}}$ at every point can be identified with $\mathbf{E}$ and then the derivative of the smooth map $C_{\mathbf{U}}: \operatorname{Dom} \mathbf{U} \longrightarrow E_{\mathbf{U}}$,

$x \mapsto C_{\mathbf{U}}(x)$ is given by

$$DC_{\mathbf{U}}[x]\mathbf{x} = \mathbf{x}\|_{\mathbf{U}(x)} \qquad (x \in \mathrm{Dom}\,\mathbf{U},\ \mathbf{x} \in \mathbf{M}).$$

(notation: 1.11.).

Proof. Since the **U**-lines are world lines it is trivial that the intersection of $t \in I$ (see 1.4.) with an **U**-line is either void or a singleton. It is evident that the hyperplanes directed by **E** are submanifolds transversal to $E_{\mathbf{U}}$. Hence we can apply VII. 7.7. to endow $E_{\mathbf{U}}$ with a smooth structure. The assertion concerning the tangent spaces is an immediate consequence of VII. 7.10.

3.4. Theorem. Let **U** be an observer. Then the map

$$S_{\mathbf{U}}: \mathrm{Dom}\,\mathbf{U} \longrightarrow I \times E_{\mathbf{U}}, \qquad x \mapsto (\tau(x),\ C_{\mathbf{U}}(x))$$

is a little diffeomorphism whose inverse has the form

$$(t, q) \mapsto t \cap q \qquad ((t, q) \in \mathrm{Ran}\,S_{\mathbf{U}} \subset I \times E_{\mathbf{U}})$$

(we denote the unique element of the singleton $t \cap q$ by $t \cap q$ as well).

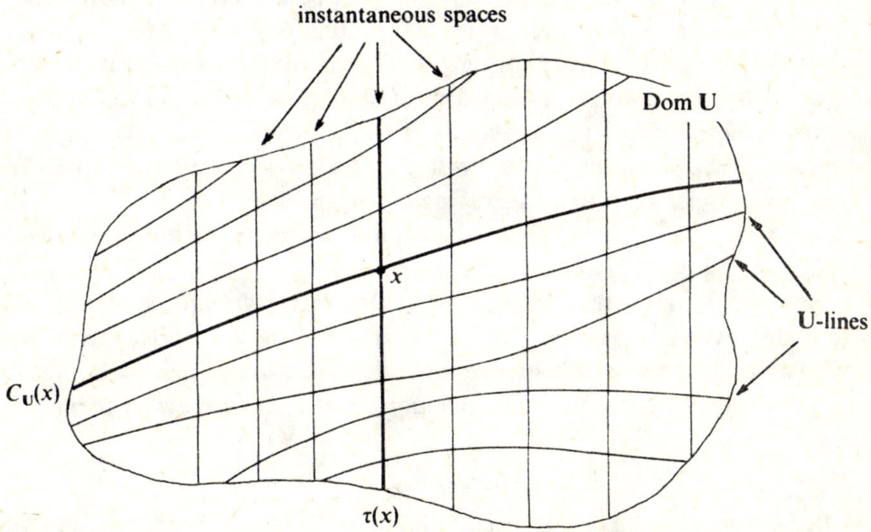

Furthermore,

$$DS_{\mathbf{U}}[x] = \mathbf{s}_{\mathbf{U}(x)} := (\boldsymbol{\tau}, (\,\cdot\,)\|_{\mathbf{U}(x)}) \qquad (x \in \mathrm{Dom}\,\mathbf{U}).$$

Proof. The assertion of the theorem is a reformulation of VII. 7.8. for the present case.

3.5. ⌈The material points of a physical observer constitute the space of the observer, as we pointed out in Introduction 1. A material point of a physical observer modelled by **U** is modelled by a **U**-line which describes the history of the material point. Thus a space point of the observer is modelled by a **U**-line and the space corresponding to the observer must be modelled by the set of **U**-lines.

Consider a **U**-line q; the space-time position of the corresponding material point at the time point t is $t \frown q$; it can be considered as the instantaneous value of q at t. ⌋

Definition. Let **U** be an observer. $E_{\mathbf{U}}$ is called **U**-*space* or *space relative* to **U** and $S_{\mathbf{U}}$ is the *splitting* of (a part of) space-time into time and **U**-space.

If q is a **U**-space point and $t \in I$ then $t \frown q$ is the *instantaneous value* of q at t; the set $\{t \frown q : q \in E_{\mathbf{U}}\}$ is the *instantaneous U-space* at the time point t.

The *distance* of the **U**-space points q and q' at the time point t makes sense if neither of $t \frown q$ and $t \frown q'$ is void and then

$$d_t(q, q') := \sqrt{\gamma(t \frown q - t \frown q', t \frown q - t \frown q')} .$$

3.6. $E_{\mathbf{U}}$ can be a very complicated and "bad" object. It is not, in general, a metric space, even the instantaneous distance d_t need not be defined on all $E_{\mathbf{U}} \times E_{\mathbf{U}}$. Moreover, $E_{\mathbf{U}}$ need not be either a Riemannian manifold.

⌈The space of a physical observer can be very bad. For instance, if the observer is a waving and whirling water, the distance of space points (which are the molecules of the water) and even the direction of their mutual positions vary with time. A physical observer with a space having better properties is a rock: it is rigid, the distance of its space points is independent of time. However, it can rotate, i.e. the directions of mutual positions of space points vary with time. Rigid and rotation free observers are nearly the best ones. Note that an observer can be rotation free without being rigid: for instance a waving but not whirling water.

Our next definition aims to input these notions in mathematical physics, i.e. to formulate the mathematical models of rigidity etc.

Lastly we mention that physical observers, in reality, are never rigid and rotation free: at least molecular motion contradict to these properties. Besides, a physical observer is never global, i.e. cannot fill all space-time. All these notions, as all models, are idealizations, extrapolations for a convenient mathematical description. ⌋

3.7. Definition. An observer **U** is called

(i) *rigid* if the distances of **U**-space points at given time points are independent of the time points;

(ii) *quasi rotation free* if for all **U**-space points (**U**-lines) q_1 and q_2 and for all $t, s \in I$ such that $t \frown q_1$, $t \frown q_2$, $s \frown q_1$, $s \frown q_2$ are not void

$$(t \frown q_1) - (t \frown q_2) \quad \text{is parallel to} \quad (s \frown q_1) - (s \frown q_2);$$

rotation free if it is quasi rotation free and all **U**-lines are rotation free;

(iii) *inertial* if **U** is a constant map or, equivalently, the **U**-lines are parallel straight line segments. ∎

Evidently, inertial observers are rigid and rotation free.

3.8. ⌈The notion of time and space arose for mankind in connection with the Earth; the Earth (and a man resting on it) is an observer splitting space-time into time and space. In everyday language space means the space related to the Earth, i.e. Earth-space. Our first experience mentioned in 1.1 that space can be viewed as a Euclidean affine space concerns Earth-space. However, the famous Foucault experiment shows that Earth-space is not affine; we can roughly say that in a short time interval, i.e. instantaneously, Earth-space is a Euclidean affine space but for a long time is not. The rotation of the Earth is relatively slow that is why we need a long time period to demonstrate the non-affine feature of Earth-space.

The following assertions illuminate the situation in our exact framework.⌋

Proposition. Let **U** be a global observer. Then

(i) the instantaneous **U**-space at the time point t is a three dimensional affine subspace of M, over **E**, for all $t \in I$;

(ii) **U**-space is an affine space over **E** with the addition

$$q+\mathbf{q}:=\{a+\mathbf{q}\colon\ a\in q\} \qquad (q\in E_{\mathbf{U}},\ \mathbf{q}\in \mathbf{E})$$

if and only if **U** is rigid and quasi rotation free, and all **U**-lines are global.

Proof. Assertion (i) is trivial since the instantaneous **U**-space at the time point t coincides with t if **U** is global.

(ii) Assume $E_{\mathbf{U}}$ to be affine with the above addition. Then **U** is obviously rigid and quasi rotation free. Furthermore suppose that there is a $q_0 \in E_{\mathbf{U}}$ which is not global, i.e. $t \cap q_0$ is void for some $t \in I$. Since $E_{\mathbf{U}}$ is affine, all its elements are of the form $q_0+\mathbf{q}$ where $\mathbf{q}$ is in **E**. As a consequence, $t\cap q$ is void for all $q \in E_{\mathbf{U}}$ but this is impossible because of the globality of **U**. Thus all **U**-lines are global.

Assume now that **U** is rigid, quasi rotation free and all **U**-lines are global. Take an arbitrary $q_0 \in E_{\mathbf{U}}$. A simple consideration shows that

$$t\cap C_{\mathbf{U}}(a+\mathbf{q})=t\cap C_{\mathbf{U}}(a)+\mathbf{q}$$

for all $t \in I$, $a \in q_0$, $\mathbf{q} \in \mathbf{E}$. This implies that $q_0+\mathbf{q}$ is a **U**-line as well, and all **U**-lines are of this form because of their globality. It is left to the reader to check the validity of axioms of affine spaces for the given addition.

Remark. In particular, the **U**-space of a global inertial observer **U** having constant value $\mathbf{c} \in V(1)$ is the affine space, over **E**, of the lines directed by $\mathbf{c}\otimes \mathbf{I}$, and

$$(x+\mathbf{c}\otimes\mathbf{I})+\mathbf{q}=(x+\mathbf{q})+\mathbf{c}\otimes\mathbf{I} \qquad (x\in M,\ \mathbf{q}\in\mathbf{E}).$$

(see 3.10.)

3.9. ⌈Now we have the exact definition of observers and we see how an observer **U** splits space-time into time and **U**-space. In general **U**-space is a complicated object and even if it is simple, e.g. an affine space, one likes to speak about a "position vector", i.e. to treat the space of an observer as a vector space. This is achieved by choosing a space point for the "origin".⌋

Definition. An *observer with origin* is a pair $(\mathbf{U}, o)$ where $\mathbf{U}$ is an observer and o is an element of Dom $\mathbf{U}$. The map

$$S_o^{\mathbf{U}} : I \times E_{\mathbf{U}} \rightarrowtail \mathbf{I} \times \mathbf{E}, \qquad (t, q) \mapsto (t - \tau(o), t \cap q - t \cap C_{\mathbf{U}}(o))$$

is the *vectorization* of (a part of) time and $\mathbf{U}$-space. ∎

Note that $S_o^{\mathbf{U}}$ need not be everywhere defined even if $\mathbf{U}$ is a global observer (see Exercise 10.).

Proposition. $S_o^{\mathbf{U}}$ is a little diffeomorphism.

Proof. $S_o^{\mathbf{U}}$ is evidently injective. It is smooth because it is the composition of the maps

$$I \times E_{\mathbf{U}} \rightarrowtail M, \qquad (t, q) \mapsto t \cap q$$

and

$$M \longrightarrow \mathbf{I} \times \mathbf{E}, \qquad x \mapsto (\tau(x) - \tau(o), x - \tau(x) \cap C_{\mathbf{U}}(o)).$$

The first map is smooth by 3.4. and the second one is smooth because of the fact that the world line functions of $\mathbf{U}$ are smooth, in particular, $I \longrightarrow M$, $t \mapsto t \cap C_{\mathbf{U}}(o)$ is smooth.

3.10. To deal with observers, we must leave in general the nice affine structure for the more complicated smooth structure. Even if the space relative to an observer is an affine space, the splitting of space-time need not be an affine map. Exceptions are the global inertial observers which preserve the affine structure.

Proposition. Let $\mathbf{U}$ be a global inertial observer and assume $\mathbf{c} \in V(1)$ is the constant value of $\mathbf{U}$. Then $E_{\mathbf{U}}$ is the set of affine lines directed by $\mathbf{c} \otimes \mathbf{I}$ and it is an affine space over $\mathbf{E}$ with the subtraction

$$q' - q := (x' - x)\|_{\mathbf{c}} \qquad (q', q \in E_{\mathbf{U}}, x' \in q', \; x \in q).$$

Moreover, $S_{\mathbf{U}}$ is an affine bijection,

$$S_{\mathbf{U}}(x) = (\tau(x), x + \mathbf{c} \otimes \mathbf{I}) = (x + \mathbf{E}, x + \mathbf{c} \otimes \mathbf{I}) \qquad (x \in M)$$

(see. 1.4.), and $DS_{\mathbf{U}} = \mathbf{s}_{\mathbf{c}}$. ∎

3.11. Proposition. Let $(\mathbf{U}, o)$ be a global inertial observer with origin and assume $\mathbf{c} \in V(1)$ is the constant value of $\mathbf{U}$. Then $S_o^{\mathbf{U}}$ is an affine bijection, and $DS_o^{\mathbf{U}} = \mathrm{id}_{\mathbf{I} \times \mathbf{E}}$, $S_o^{\mathbf{U}} \circ S_{\mathbf{U}} = s_{\mathbf{c}, o}$.

4. Projecting of world line functions

Necessary mathematical tools:

Tensor products and quotients (Chapter IV.)
Differentiation in affine spaces (Section VI. 2.)

4.1. ⌈An experimenter observes the movement of a masspoint; he describes it as a function assigning a space point to each time point of the observation. In our framework, the history of a masspoint, a world line function, appears to an observer **U** as a function involving time and **U**-space, or, in other words, the world line is projected into such a function.

Let **U** be a global inertial observer with velocity **u** and suppose r is an inertial world line function, i.e. $\dot{r}$ is a constant, say $\mathbf{u}'$. Then the position of the masspoint with world line function r at the time point t in **U**-space is $r_{\mathbf{U}}(t) := = r(t) + \mathbf{u} \otimes \mathbf{I}$. The velocity of the masspoint relative to the observer is the derivative of the function $I \longrightarrow E_{\mathbf{U}}, t \mapsto r_{\mathbf{U}}(t)$. The subtraction defined in $E_{\mathbf{U}}$ (see 3.10.) involves that

$$\lim_{\mathbf{h} \to 0} \frac{r_{\mathbf{U}}(t+\mathbf{h}) - r_{\mathbf{U}}(t)}{\mathbf{h}} = \mathbf{u}' - \mathbf{u} \qquad (t \in I).$$

This suggests to introduce the following notions.⌋

4.2. Definition. Let $\mathbf{u}, \mathbf{u}'$ be elements of $V(1)$. Then we call

$$\mathbf{v}_{\mathbf{u}'\mathbf{u}} := \mathbf{u}' - \mathbf{u}$$

the *relative velocity* of $\mathbf{u}'$ *with respect to* $\mathbf{u}$.

Proposition. Suppose $\mathbf{u}, \mathbf{u}'$ and $\mathbf{u}''$ are elements of $V(1)$. Then

(i) $\mathbf{v}_{\mathbf{u}'\mathbf{u}}$ is an element of $\dfrac{\mathbf{E}}{\mathbf{I}}$,

(ii) $\mathbf{v}_{\mathbf{u}'\mathbf{u}} = -\mathbf{v}_{\mathbf{u}\mathbf{u}'}$,

(iii) $\mathbf{v}_{\mathbf{u}''\mathbf{u}} = \mathbf{v}_{\mathbf{u}''\mathbf{u}'} + \mathbf{v}_{\mathbf{u}'\mathbf{u}}$.

4.3. Definition. Let **U** be an observer and let r be a twice differentiable world line function, $\operatorname{Ran} r \subset \operatorname{Dom} \mathbf{U}$. The maps

$$r_{\mathbf{U}} : I \rightarrowtail E_{\mathbf{U}}, \qquad t \mapsto C_{\mathbf{U}}(r(t)),$$

$$\mathbf{v}_{\mathbf{U}} : I \rightarrowtail \frac{\mathbf{E}}{\mathbf{I}}, \qquad t \mapsto \dot{r}(t) - \mathbf{U}(r(t)) = \mathbf{v}_{\dot{r}(t)\mathbf{U}(r(t))}$$

$$(t \in \operatorname{Dom} r)$$

are called the **U**-*projections* of r and $\dot{r}$, respectively.

If $(\mathbf{U}, o)$ is an observer with origin then

$$\mathbf{r}_{\mathbf{U}}: I \rightarrowtail \mathbf{E}, \qquad t \mapsto r(t) - t \cap C_{\mathbf{U}}(o)$$

is called the *vectorized* **U**-*projection* of r.

Remarks. (i) The **U**-projections are non-trivial parts of **U**-splittings. Indeed, a world line function r is split by an observer **U** into the map $I \rightarrowtail I \times E_{\mathbf{U}}$, $t \mapsto (t, r_{\mathbf{U}}(t))$, whose first component is trivial. Similarly, $\dot{r}$ is split into $(1, \mathbf{v}_{\mathbf{U}})$ (cf. 5.2.).

(ii) $r_{\mathbf{U}}$ is, in general, a complicated function because of the complicatedness of $E_{\mathbf{U}}$. That is why one mostly uses a vectorization to produce $\mathbf{r}_{\mathbf{U}}$.

(iii) Using Proposition 3.3. we easily find that $\dot{\mathbf{r}}_{\mathbf{U}} = \mathbf{v}_{\mathbf{U}}$.

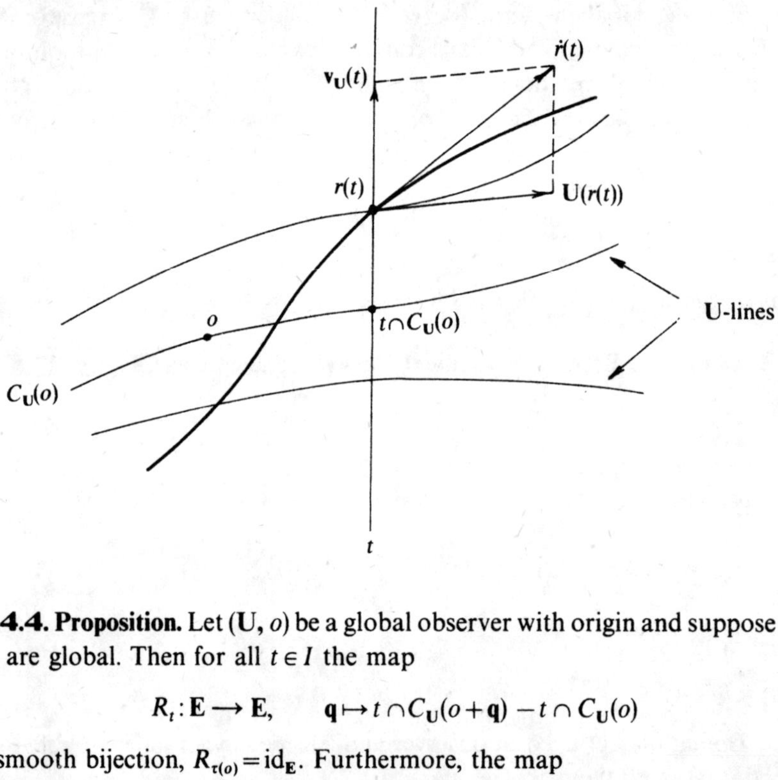

4.4. Proposition. Let $(\mathbf{U}, o)$ be a global observer with origin and suppose all **U**-lines are global. Then for all $t \in I$ the map

$$R_t: \mathbf{E} \longrightarrow \mathbf{E}, \qquad \mathbf{q} \mapsto t \cap C_{\mathbf{U}}(o + \mathbf{q}) - t \cap C_{\mathbf{U}}(o)$$

is a smooth bijection, $R_{\tau(o)} = \mathrm{id}_{\mathbf{E}}$. Furthermore, the map

$$\dot{R}_t: \mathbf{E} \longrightarrow \frac{\mathbf{E}}{\mathbf{I}}, \qquad \mathbf{q} \mapsto \frac{\mathrm{d}(R_t(\mathbf{q}))}{\mathrm{d}t} = \mathbf{U}(t \cap C_U(o + \mathbf{q})) - \mathbf{U}(t \cap C_{\mathbf{U}}(o))$$

is smooth.

Proof. R_t is smooth because the solutions of a smooth differential equation depend smoothly on the initial values, it is injective because of the uniqueness of solutions with given initial values and it is surjective because of the globality of **U**. $\dot{R}_t$ is evidently smooth.

4.5. Proposition. Let $(\mathbf{U}, o)$ be as before and introduce the notation

$$\Omega(t) := \dot{R}_t \circ R_t^{-1} : \mathbf{E} \longrightarrow \frac{\mathbf{E}}{\mathbf{I}} \qquad (t \in I).$$

Suppose r is a twice differentiable world line function. Then

$$\dot{\mathbf{r}}_{\mathbf{U}}(t) = \mathbf{v}_{\mathbf{U}}(t) + \Omega(t)(\mathbf{r}_{\mathbf{U}}(t)) \qquad (t \in \operatorname{Dom} r).$$

Proof. The definitions of R_t and $\dot{R}_t$ imply that

$$\dot{R}_t(\mathbf{q}) = \mathbf{U}(t \cap C_{\mathbf{U}}(o) + R_t(\mathbf{q})) - \mathbf{U}(t \cap C_{\mathbf{U}}(o))$$

for all $\mathbf{q} \in \mathbf{E}$. We can substitute $\mathbf{q}$ for $R_t(\mathbf{q})$ to get

$$\Omega(t)(\mathbf{q}) = \mathbf{U}(t \cap C_{\mathbf{U}}(o) + \mathbf{q}) - \mathbf{U}(t \cap C_{\mathbf{U}}(o)) \qquad (t \in I,\ \mathbf{q} \in \mathbf{E}).$$

This equality and the following trivial ones yield the desired result:

$$\dot{\mathbf{r}}_{\mathbf{U}}(t) = \dot{r}(t) - \mathbf{U}(t \cap C_{\mathbf{U}}(o)) = \dot{r}(t) - \mathbf{U}(r(t)) + \mathbf{U}(r(t)) - \mathbf{U}(t \cap C_{\mathbf{U}}(o)) =$$
$$= \mathbf{v}_{\mathbf{U}}(t) + \mathbf{U}(t \cap C_{\mathbf{U}}(o) + \mathbf{r}_{\mathbf{U}}(t)) - \mathbf{U}(t \cap C_{\mathbf{U}}(o)).$$

Remark. The relation between $\dot{\mathbf{r}}_{\mathbf{U}}$ and $\mathbf{v}_{\mathbf{U}}$ is well known in the literature for rigid observers; then Ω is the angular velocity (see Exercise 8.).

4.6. We should like to have a relation concerning the relative acceleration as well. However, the very definition of the relative acceleration is not so easy. One can give somehow a covariant derivation $\nabla^{\mathbf{U}}$ corresponding to an observer $\mathbf{U}$ and then the acceleration of a world line function r relative to $\mathbf{U}$ is defined as $(\nabla^{\mathbf{U}})_{\dot{r}}\dot{r}$ (cf. 2.9.). However, the construction of $\nabla^{\mathbf{U}}$ is not straightforward. The interested reader can consult [1].

5. Splitting of tensor fields

Necessary mathematical tools:

Tensor products and quotients (Chapter IV.)

The formulae of splittings given in 5.7. and 5.8. can be understood without the involved knowledge on smooth manifolds.

5.1. An observer splits not only the corresponding space-time region but $\mathbf{M}$ and $\mathbf{M}^*$ and their various tensor products, too. This splitting, however, depends on the points of the domain of the observer, or better to say, on "the instantaneous velocities" of the observer, i.e. on the values of the vector field that the observer is.

After some preparations we give the splitting due to an observer in 5.7.

5.2. We recall that for $\mathbf{u} \in V(1)$ the map

$$\mathbf{s_u} : \mathbf{M} \longrightarrow \mathbf{I} \times \mathbf{E}, \qquad \mathbf{x} \mapsto (\tau(\mathbf{x}), \mathbf{x}\|_{\mathbf{u}}) = (\tau(\mathbf{x}), \mathbf{x} - \mathbf{u} \otimes \tau(\mathbf{x}))$$

is a linear bijection whose inverse is

$$(\mathbf{t}, \mathbf{q}) \mapsto \mathbf{u} \otimes \mathbf{t} + \mathbf{q} \qquad ((\mathbf{t}, \mathbf{q}) \in \mathbf{I} \times \mathbf{E})$$

(Proposition 1.14.). Observe that $\mathbf{u} \otimes \tau(\mathbf{x})$ and $\mathbf{x}\|_{\mathbf{u}}$ are projections of $\mathbf{x}$ parallel to the complementary subspaces $\mathbf{E}$ and $\mathbf{u} \otimes \mathbf{I}$, respectively.

Definition. $\tau(\mathbf{x})$ and $\mathbf{x}\|_{\mathbf{u}}$ are called the **u**-*time component* and the **u**-*space component* of $\mathbf{x} \in \mathbf{M}$, respectively. $\mathbf{s_u}$ is the *splitting* of vectors into time and space components corresponding to **u**.

Remark. If $\mathbf{A}$ is a space of scalars then $\frac{\mathbf{M}}{\mathbf{A}}$ is split into $\frac{\mathbf{I}}{\mathbf{A}} \times \frac{\mathbf{E}}{\mathbf{A}}$ by $(\mathbf{s_u})_{\mathbf{A}} := \frac{\mathbf{s_u}}{\mathrm{id}_{\mathbf{A}}}$ (see IV. 2.10.). In particular, $(\mathbf{s_u})_{\mathbf{I}}$ maps $\frac{\mathbf{M}}{\mathbf{I}}$ into $\mathbb{R} \times \frac{\mathbf{E}}{\mathbf{I}}$ and for all $\mathbf{u}' \in V(1)$

$$(\mathbf{s_u})_{\mathbf{I}}(\mathbf{u}') = (1, \mathbf{u}' - \mathbf{u}) = (1, \mathbf{v}_{\mathbf{u}'\mathbf{u}}).$$

In other words, if **u** and **u**′ are velocities then the **u**-space component of **u**′ is the relative velocity of **u**′ with respect to **u** (see 4.2.).

5.3. The **u**-time component of a vector is the same for all $\mathbf{u} \in V(1)$. On the contrary, the **u**-space components vary with **u**, except when the vector is space-like; then the **u**-time component is zero and the **u**-space component coincides with the vector itself. The so called transformation law that shows how the **u**-space components vary with **u** can be well seen from the following.

Proposition. Let $\mathbf{u}, \mathbf{u}' \in V(1)$. Then

$$(\mathbf{s}_{\mathbf{u}'} \circ \mathbf{s}_{\mathbf{u}}^{-1})(\mathbf{t}, \mathbf{q}) = (\mathbf{t}, -\mathbf{v}_{\mathbf{u}'\mathbf{u}} \otimes \mathbf{t} + \mathbf{q}) \qquad ((\mathbf{t}, \mathbf{q}) \in \mathbf{I} \times \mathbf{E}).$$

5.4. Now we turn to the splitting of $\mathbf{M}^*$.

Proposition. Let **u** be an element of $V(1)$. Then the map

$$\mathbf{r_u} := (\mathbf{s}_{\mathbf{u}}^{-1})^* : \mathbf{M}^* \longrightarrow (\mathbf{I} \times \mathbf{E})^*$$

is a linear bijection and if $(\mathbf{I} \times \mathbf{E})^*$ is identified with $\mathbf{I}^* \times \mathbf{E}^*$ by

$$((\mathbf{a}, \mathbf{p}) | (\mathbf{t}, \mathbf{q})) := (\mathbf{a} | \mathbf{t}) + (\mathbf{p} | \mathbf{q}) \qquad \begin{pmatrix} (\mathbf{a}, \mathbf{p}) \in \mathbf{I}^* \times \mathbf{E}^* \\ (\mathbf{t}, \mathbf{q}) \in \mathbf{I} \times \mathbf{E} \end{pmatrix}$$

then

$$\mathbf{r_u}(\mathbf{k}) = (\mathbf{k}_{\mathbf{I}}(\mathbf{u}), \mathbf{k}|_{\mathbf{E}}) = (\mathbf{k}_{\mathbf{I}}(\mathbf{u}), \mathbf{i}^*\mathbf{k}) \qquad (\mathbf{k} \in \mathbf{M}^*).$$

Proof. Let $\mathbf{t} \in \mathbf{I}$, $\mathbf{q} \in \mathbf{E}$. Then

$$(\mathbf{r_u}(\mathbf{k}) | (\mathbf{t}, \mathbf{q})) = ((\mathbf{s}_{\mathbf{u}}^{-1})^* (\mathbf{k}) | (\mathbf{t}, \mathbf{q})) = (\mathbf{k} | \mathbf{s}_{\mathbf{u}}^{-1}(\mathbf{t}, \mathbf{q})) =$$
$$= (\mathbf{k} | \mathbf{u} \otimes \mathbf{t} + \mathbf{q}) = \mathbf{k}_{\mathbf{I}}(\mathbf{u}) \otimes \mathbf{t} + (\mathbf{k} | \mathbf{q}).$$

Since $\mathbf{k_I(u)}$ is an element of $\frac{\mathbb{R}}{\mathbf{I}} = \mathbf{I}^*$, we can write $\mathbf{k_I(u)} \otimes \mathbf{t} = (\mathbf{k_I(u)} | \mathbf{t})$ (see IV. 2.8. (ii).).

Definition. $\mathbf{k_I(u)}$ and $\mathbf{k}|_{\mathbf{E}}$ are called the **u**-*time component* and the **u**-*space component* of $\mathbf{k} \in \mathbf{M}^*$, respectively. $\mathbf{r_u}$ is the *splitting* of covectors into time and space components corresponding to **u**.

5.5. The **u**-space component of a covector is the same for all $\mathbf{u} \in V(1)$. On the contrary, the **u**-time components vary with **u**, except when the covector is timelike (i.e. a scalar in fact, see 1.7.); then the **u**-space component is zero and the **u**-time component coincides with the covector itself considered as an element of $\mathbf{I}^*$. The transformation law which shows how the **u**-time components vary with **u** can be well seen from the following.

Proposition. Let $\mathbf{u}, \mathbf{u}' \in V(1)$. Then

$$(\mathbf{r_{u'}} \circ \mathbf{r_u}^{-1})(\mathbf{a}, \mathbf{p}) = (\mathbf{a} + \mathbf{p_I}(\mathbf{v_{u'u}}), \mathbf{p}) \qquad ((\mathbf{a}, \mathbf{p}) \in \mathbf{I}^* \times \mathbf{E}^*).$$

Proof. $\mathbf{r_{u'}}(\mathbf{k}) - \mathbf{r_u}(\mathbf{k}) = (\mathbf{k_I}(\mathbf{u}') - \mathbf{k_I}(\mathbf{u}), \mathbf{0})$ for all $\mathbf{k} \in \mathbf{M}^*$. Put $\mathbf{k} := \mathbf{r_u}^{-1}(\mathbf{a}, \mathbf{p})$ to get

$$(\mathbf{r_{u'}} \circ \mathbf{r_u}^{-1})(\mathbf{a}, \mathbf{p}) = (\mathbf{a}, \mathbf{p}) + (\mathbf{p_I}(\mathbf{u}' - \mathbf{u}), \mathbf{0}),$$

because $\mathbf{u}' - \mathbf{u} \in \frac{\mathbf{E}}{\mathbf{I}}$ and the restriction of $\mathbf{k_I}$ onto $\frac{\mathbf{E}}{\mathbf{I}}$ is $\mathbf{p_I}$ by definition.

5.6. Remark. Formulae of 5.3. and 5.5. reflect well that vectors and covectors are split in a different way. This is so because **M** and $\mathbf{M}^*$ cannot be identified, in contradistinction to the relativistic case (see II. 5.6.).

5.7. Let U be an observer. The splitting of (Dom U) × **M** and (Dom U) × $\mathbf{M}^*$ corresponding to U can be obtained by the maps appearing in the following proposition whose proof is evident from the smoothness of U and S_U.

Proposition. The maps

$$(\mathrm{Dom}\, U) \times \mathbf{M} \longrightarrow (I \times E_U) \times (\mathbf{I} \times \mathbf{E}),$$

$$(x, \mathbf{x}) \mapsto (S_U(x), \mathbf{s}_{U(x)}(\mathbf{x})) = (\tau(x), C_U(x), \boldsymbol{\tau}(\mathbf{x}), \mathbf{x}\|_{U(x)});$$

$$(\mathrm{Dom}\, U) \times \mathbf{M}^* \longrightarrow (I \times E_U) \times (\mathbf{I}^* \times \mathbf{E}^*),$$

$$(x, \mathbf{k}) \mapsto (S_U(x), \mathbf{r}_{U(x)}(\mathbf{k})) = (\tau(x), C_U(x), \mathbf{k_I}(U(x)), \mathbf{k}|_{\mathbf{E}})$$

are smooth injections.

Remark. (Dom U) × **M** and (Dom U) × $\mathbf{M}^*$ are the tangent bundle and the cotangent bundle of Dom U, respectively (see VII. 2.12., VII. 2.17.).

The tangent bundle of E_U is $E_U \times \mathbf{E}$ (see 3.3.) and we easily find that the above given splitting of (Dom U) × **M** is DS_U, the derivative of S_U.

Similarly, the splitting of (Dom U) × $\mathbf{M}^*$ is DS_U^{*-1} (see VII. 2.20.).

5.8. Vector fields and covector fields are split by an observer **U** as follows Let $\mathbf{X}: \mathrm{Dom}\,\mathbf{U} \rightarrowtail \mathbf{M}$ be a vector field and $\mathbf{K}: \mathrm{Dom}\,\mathbf{U} \rightarrowtail \mathbf{M}^*$ a covector field They are transferred by the observer **U** into the maps

$$I \times E_{\mathbf{U}} \rightarrowtail \mathbf{I} \times \mathbf{E}, \quad (t, q) \mapsto (\tau(\mathbf{X}(t \cap q)), \mathbf{X}(t \cap q)\|_{\mathbf{U}(t \cap q)}),$$

$$I \times E_{\mathbf{U}} \rightarrowtail \mathbf{I}^* \times \mathbf{E}^*, \quad (t, q) \mapsto (\mathbf{K}(t \cap q)_{\mathbf{I}}(\mathbf{U}(t \cap q)), \mathbf{K}(t \cap q)|_{\mathbf{E}}),$$

respectively. The following diagrams help to see what is done:

$$I \times E_{\mathbf{U}} \xrightarrow{S_{\mathbf{U}}^{-1}} M \xrightarrow{\mathbf{X}} \mathbf{M} \xrightarrow{\mathbf{s}_{\mathbf{U}(\cdot)}} \mathbf{I} \times \mathbf{E}$$

$$I \times E_{\mathbf{U}} \xrightarrow{S_{\mathbf{U}}^{-1}} M \xrightarrow{\mathbf{K}} \mathbf{M}^* \xrightarrow{\mathbf{r}_{\mathbf{U}(\cdot)}} \mathbf{I}^* \times \mathbf{E}^*$$

In terms of the theory of smooth manifolds we can say that the split form of **X** and **K** by **U** are $(S_{\mathbf{U}})_* \mathbf{X}$ and $(S_{\mathbf{U}}^{-1})^* \mathbf{K}$, respectively (see VII. 3.9. and VII. 4.8.).

5.9. An observer with origin $(\mathbf{U}, o)$ transfers vector fields and covector fields into maps $\mathbf{I} \times \mathbf{E} \rightarrowtail \mathbf{I} \times \mathbf{E}$ and $\mathbf{I} \times \mathbf{E} \rightarrowtail \mathbf{I}^* \times \mathbf{E}^*$, respectively; it is evident, how. We write down the formulae only for a global inertial observer with origin; let us denote the constant value of **U** by **c**.

Then the vector field **X** becomes

$$\mathbf{I} \times \mathbf{E} \rightarrowtail \mathbf{I} \times \mathbf{E}, \quad (\mathbf{t}, \mathbf{q}) \mapsto (\tau(\mathbf{X}(o + \mathbf{c} \otimes \mathbf{t} + \mathbf{q})), \mathbf{X}(o + \mathbf{c} \otimes \mathbf{t} + \mathbf{q})\|_{\mathbf{c}}),$$

and the covector field **K** becomes

$$\mathbf{I} \times \mathbf{E} \rightarrowtail \mathbf{I}^* \times \mathbf{E}^*, \quad (\mathbf{t}, \mathbf{q}) \mapsto (\mathbf{K}(o + \mathbf{c} \otimes \mathbf{t} + \mathbf{q})_{\mathbf{I}}(\mathbf{c}), \mathbf{K}(o + \mathbf{c} \otimes \mathbf{t} + \mathbf{q})|_{\mathbf{E}}).$$

5.10. Observers manipulate tensor fields and cotensor fields, too. We shall detail only the splitting of 2-tensors and 2-cotensors, the general formulae can be derived from it.

Let $\mathbf{u} \in V(1)$. Then

$$\mathbf{s}_{\mathbf{u}} \otimes \mathbf{s}_{\mathbf{u}}: \mathbf{M} \otimes \mathbf{M} \longrightarrow (\mathbf{I} \times \mathbf{E}) \otimes (\mathbf{I} \times \mathbf{E}) = (\mathbf{I} \otimes \mathbf{I}) \times (\mathbf{I} \otimes \mathbf{E}) \times (\mathbf{E} \otimes \mathbf{I}) \times (\mathbf{E} \otimes \mathbf{E}),$$

$$\mathbf{r}_{\mathbf{u}} \otimes \mathbf{r}_{\mathbf{u}}: \mathbf{M}^* \otimes \mathbf{M}^* \longrightarrow (\mathbf{I}^* \times \mathbf{E}^*) \otimes (\mathbf{I}^* \times \mathbf{E}^*) =$$

$$= (\mathbf{I}^* \otimes \mathbf{I}^*) \times (\mathbf{I}^* \otimes \mathbf{E}^*) \times (\mathbf{E}^* \otimes \mathbf{I}^*) \times (\mathbf{E}^* \otimes \mathbf{E}^*)$$

are the splittings of 2-tensors and 2-cotensors according to **u**.

Since $(\mathbf{I} \times \mathbf{E}) \otimes (\mathbf{I} \times \mathbf{E})$ equals the vector space of bilinear forms over $(\mathbf{I} \times \mathbf{E})^* = \mathbf{I}^* \times \mathbf{E}^*$, we shall find convenient to write the elements of $(\mathbf{I} \times \mathbf{E}) \otimes (\mathbf{I} \times \mathbf{E})$

in the form

$$\begin{pmatrix} \mathbf{h} & \mathbf{b} \\ \mathbf{a} & \mathbf{A} \end{pmatrix}$$

where $\mathbf{h} \in \mathbf{I} \otimes \mathbf{I}$, $\mathbf{b} \in \mathbf{I} \otimes \mathbf{E}$, $\mathbf{a} \in \mathbf{E} \otimes \mathbf{I}$, $\mathbf{A} \in \mathbf{E} \otimes \mathbf{E}$. Doing so we can easily calculate the values of the bilinear maps in the customary way (see V. 1.2.). We use a similar form for the elements of $(\mathbf{I}^* \times \mathbf{E}^*) \otimes (\mathbf{I}^* \times \mathbf{E}^*)$.

If the element above is antisymmetric then $\mathbf{h} = \mathbf{0}$, $\mathbf{b} = -\mathbf{a}^*$ (notation: IV. 1.5. (ii)), $\mathbf{A} \in \mathbf{E} \wedge \mathbf{E}$. Then we use the notation

$$(\mathbf{a}, \mathbf{A}) := \begin{pmatrix} \mathbf{0} & -\mathbf{a}^* \\ \mathbf{a} & \mathbf{A} \end{pmatrix}$$

and similarly for antisymmetric elements of $(\mathbf{I}^* \times \mathbf{E}^*) \otimes (\mathbf{I}^* \times \mathbf{E}^*)$

$$(\mathbf{z}, \mathbf{Z}) := \begin{pmatrix} \mathbf{0} & -\mathbf{z}^* \\ \mathbf{z} & \mathbf{Z} \end{pmatrix}$$

where $\mathbf{z} \in \mathbf{E}^* \otimes \mathbf{I}^*$, $\mathbf{Z} \in \mathbf{E}^* \wedge \mathbf{E}^*$.

5.11. An element F of $\mathbf{M}^* \wedge \mathbf{M}^*$ can be considered either as a linear map $\mathbf{M} \longrightarrow \mathbf{M}^*$ or as a bilinear map $\mathbf{M} \times \mathbf{M} \longrightarrow \mathbb{R}$. According to these possibilities we have $(\mathbf{r_u} \otimes \mathbf{r_u})(\mathrm{F}) = \mathbf{r_u} \circ \mathrm{F} \circ \mathbf{s_u}^{-1}$ or $(\mathbf{r_u} \otimes \mathbf{r_u})(\mathrm{F}) = \mathrm{F} \circ (\mathbf{s_u}^{-1} \times \mathbf{s_u}^{-1})$. Correspondingly, the elements of $\mathbf{E}^* \otimes \mathbf{I}^*$ are considered either as linear maps $\mathbf{I} \longrightarrow \mathbf{E}^*$ or bilinear maps $\mathbf{E} \times \mathbf{I} \longrightarrow \mathbb{R}$, and the elements of $\mathbf{E}^* \wedge \mathbf{E}^*$ are considered either as linear maps $\mathbf{E} \longrightarrow \mathbf{E}^*$ or as bilinear maps $\mathbf{E} \times \mathbf{E} \longrightarrow \mathbb{R}$.

Now the demonstration of the following proposition is straightforward for cotensors; for tensors take $T := \mathbf{x} \otimes \mathbf{y} - \mathbf{y} \otimes \mathbf{x}$ to have the desired result.

Proposition. Let T and F be antisymmetric 2-tensor and 2-cotensor, respectively. Then for $\mathbf{u} \in V(1)$ we have

$$(\mathbf{s_u} \otimes \mathbf{s_u})(T) = ((\mathrm{id}_M \otimes \tau)(T), T - (\mathrm{id}_M \otimes \tau)(T) \wedge \mathbf{u})$$

(cf. IV. 2.8. (iii)) and

$$(\mathbf{r_u} \otimes \mathbf{r_u})(F) = ((\mathbf{i}^* \circ F)_{\mathbf{I}}(\mathbf{u}), (\mathbf{i}^* \circ F)|_{\mathbf{E}}) = (F|_{\mathbf{E} \times \mathbf{u} \otimes \mathbf{I}}, F|_{\mathbf{E} \times \mathbf{E}}).$$

Definition. We shall employ the following names:

$(\mathrm{id}_M \otimes \tau)(T)$	is the **u**-space-time component of T,
$T - (\mathrm{id}_M \otimes \tau)(T) \wedge \mathbf{u}$	is the **u**-space-space component of T,
$(\mathbf{i}^* \circ F)_{\mathbf{I}}(\mathbf{u})$ or $F\|_{\mathbf{E} \times \mathbf{u} \otimes \mathbf{I}}$	is the **u**-space-time component of F,
$(\mathbf{i}^* \circ F)\|_{\mathbf{E}}$ or $F\|_{\mathbf{E} \times \mathbf{E}}$	is the **u**-space-space component of F.

5.12. The **u**-space-time components of $T \in \mathbf{M} \wedge \mathbf{M}$ are the same for all $\mathbf{u} \in V(1)$ while the **u**-space-space components vary with **u**, except when T is in $\mathbf{E} \wedge \mathbf{E}$; then the **u**-space-time component is zero and the **u**-space-space component is T itself.

The **u**-space-space components of $F \in \mathbf{M}^* \wedge \mathbf{M}^*$ are the same for all $\mathbf{u} \in V(1)$ whereas the **u**-space-time components vary with **u**, except when F is in $\mathbf{M}^* \wedge \mathbf{I}^*$ (which is the subspace of $\mathbf{M}^* \wedge \mathbf{M}^*$ generated by antisymmetric tensor products of elements from $\mathbf{M}^*$ and from $\mathbf{I}^*$ viewed as a subspace of $\mathbf{M}^*$).

The transformation law which shows how the diverse components vary with the elements of $V(1)$ can be well seen from the following.

Proposition. Let $\mathbf{u}, \mathbf{u}' \in V(1)$. Then

$$((\mathbf{s}_{\mathbf{u}'} \otimes \mathbf{s}_{\mathbf{u}'}) \circ (\mathbf{s}_{\mathbf{u}} \otimes \mathbf{s}_{\mathbf{u}})^{-1})(\mathbf{a}, \mathbf{A}) = (\mathbf{a}, \mathbf{v}_{\mathbf{u}'\mathbf{u}} \wedge \mathbf{a} + \mathbf{A})$$
$$(\mathbf{a} \in \mathbf{E} \otimes \mathbf{I},\ \mathbf{A} \in \mathbf{E} \wedge \mathbf{E}),$$

$$((\mathbf{r}_{\mathbf{u}'} \otimes \mathbf{r}_{\mathbf{u}'}) \circ (\mathbf{r}_{\mathbf{u}} \otimes \mathbf{r}_{\mathbf{u}})^{-1})(\mathbf{z}, \mathbf{Z}) = (\mathbf{z} + \mathbf{Z}_{\mathbf{I}}(\mathbf{v}_{\mathbf{u}'\mathbf{u}}), \mathbf{Z})$$
$$(\mathbf{z} \in \mathbf{E}^* \otimes \mathbf{I}^*,\quad \mathbf{Z} \in \mathbf{E}^* \wedge \mathbf{E}^*),$$

where $\mathbf{E}^* \otimes \mathbf{I}^*$ is identified with $\frac{\mathbf{E}^*}{\mathbf{I}}$ and $\mathbf{Z}$ is considered as a linear map $\mathbf{E} \longrightarrow \mathbf{E}^*$.

Proof. We easily deduce that

$$(\mathbf{s}_{\mathbf{u}} \otimes \mathbf{s}_{\mathbf{u}})^{-1}(\mathbf{a}, \mathbf{A}) = \mathbf{a} \wedge \mathbf{u} + \mathbf{A}$$

and that $(\mathrm{id}_{\mathbf{M}} \otimes \tau)(\mathbf{a} \wedge \mathbf{u} + \mathbf{A}) = \mathbf{a}$, which give the first desired result.

Furthermore, if we consider $(\mathbf{z}, \mathbf{Z})$ as a linear map $\mathbf{I} \times \mathbf{E} \longrightarrow \mathbf{I}^* \times \mathbf{E}^*$ then

$$(\mathbf{r}_{\mathbf{u}'} \otimes \mathbf{r}_{\mathbf{u}'})(F) - (\mathbf{r}_{\mathbf{u}} \otimes \mathbf{r}_{\mathbf{u}})(F) = ((\mathbf{i}^* \circ F)_{\mathbf{I}}(\mathbf{u}' - \mathbf{u}), \mathbf{0})$$

for all $F \in \mathbf{M}^* \wedge \mathbf{M}^*$; put $F := (\mathbf{r}_{\mathbf{u}} \otimes \mathbf{r}_{\mathbf{u}})^{-1}(\mathbf{z}, \mathbf{Z})$ to get the second desired result.

5.13. The values of a force field (see Definition 2.7.) are spacelike vectors of cotype $\mathbf{I} \otimes \mathbf{D} \otimes \mathbf{D}$, hence their **u**-time components are zero and **u**-space components are the same for all $\mathbf{u} \in V(1)$. ⌈Thus a force appears to all observer the same.⌋

Let us consider a potential, i.e. a covector field $\mathbf{K}$ (see 2.7.). Then it is split by a global inertial observer $\mathbf{U}$ with constant velocity **u** into an **u**-time component $-V_{\mathbf{u}} : I \times E_{\mathrm{U}} \longrightarrow \mathbf{I}^*$ and **u**-space component $\mathbf{A}_{\mathbf{u}} : I \times E_{\mathrm{U}} \longrightarrow \mathbf{E}^*$ (see 5.8.), called in usual language scalar potential and vector potential, respectively. The vector potential is the same for all global inertial observers in the sense that

$$\mathbf{A}_{\mathbf{u}'} \circ S_{\mathrm{U}'} = \mathbf{A}_{\mathbf{u}} \circ S_{\mathrm{U}} \qquad (\mathbf{u}, \mathbf{u}' \in V(1)).$$

The scalar potential depends in general on observers:

$$V_{\mathbf{u}'} \circ S_{\mathrm{U}'} = V_{\mathbf{u}} \circ S_{\mathrm{U}} - (\mathbf{A}_{\mathbf{u}} \circ S_{\mathrm{U}})_{\mathbf{I}} \mathbf{v}_{\mathbf{u}'\mathbf{u}}, \qquad (\mathbf{u}, \mathbf{u}' \in V(1)).$$

However, if $\mathbf{K}$ is timelike then it is actually a scalar field (see 1.7.) which is reflected by the fact that the vector potential is zero and the scalar potential is the same for all global inertial observers.

The existence of potentials (i.e. covector fields) *that are in fact scalar fields is a peculiar feature of our nonrelativistic space-time model,* in contradistinction to relativistic space-time models (cf. II. 5.12.). ⌈Gravitational fields, electrostatic fields and elastic fields (harmonic oscillator) are modelled by such timelike potentials, whereas electromagnetic fields are not.⌋

Introducing the notations

$$V := V_{\mathbf{u}} \circ S_{\mathbf{U}}, \qquad V' := V_{\mathbf{u}'} \circ S_{\mathbf{U}'},$$

$$\mathbf{A} := \mathbf{A}_{\mathbf{u}} \circ S_{\mathbf{U}}, \qquad \mathbf{A}' := \mathbf{A}_{\mathbf{u}'} \circ S_{\mathbf{U}'},$$

$$\mathbf{v} := \mathbf{v}_{\mathbf{u}'\mathbf{u}},$$

we get the above transformation formulae in the more familiar form

$$V' = V - \mathbf{A} \cdot \mathbf{v}, \qquad \mathbf{A}' = \mathbf{A}$$

where $\mathbf{A} \cdot \mathbf{v} := \mathbf{A}_{\mathbf{I}}(\mathbf{v})$.

Similar observations can be made on an antisymmetric 2-cotensor field F. By a definition similar to the one accepted for potentials we denote by E and E' the **u**-space-time and **u**′-space-time components of F, and by **B** and **B**′ the **u**-space-space and **u**′-space-space components of F, respectively. Then we have the familiar transformation formulae

$$E' = E + \mathbf{B}\mathbf{v}, \qquad \mathbf{B}' = \mathbf{B}$$

where $\mathbf{B}\mathbf{v} := \mathbf{B}_{\mathbf{I}}(\mathbf{v})$.

⌈This nonrelativistic transformation law is well known for the electromagnetic field where E is the electric field and **B** is the magnetic field (which is considered usually as an $\mathbf{E}^* \otimes \mathbf{D}^* = \frac{\mathbf{E}^*}{\mathbf{D}} = \frac{\mathbf{E}}{\mathbf{D} \otimes \mathbf{D} \otimes \mathbf{D}}$ valued map, due to the identification described in V. 3.6., V. 3.7.). Notice that electric field and magnetic field do not exist absolutely. Electromagnetic field is absolute, it is an antisymmetric 2-tensor field, and an observer splits electromagnetic field into electric field and magnetic field, and the splitting depends on observers.⌋

6. Reference systems

Necessary mathematical tools:

Tensor products and quotients (Chapter IV.)
Affine spaces, differentiation (Sections VI. 1., VI. 2.)

The notion of Galilean transformations appear here; the Galilean group is defined and investigated in the next section.

6.1. ⌈To describe physical phenomena, an observer coordinatizes his own time and space. The coordinatization is realized physically by measuring time with a clock (having a dial) and by introducing numbered reference lines in space. Then a time point is represented by a number and a space point is represented by a triplet of numbers.⌋

6.2. Definition. A *reference system* is a pair $(\mathbf{U}, K)$ where $\mathbf{U}$ is an observer and $K: \mathrm{Dom}\,\mathbf{U} \rightarrowtail \mathbb{R}^4$ is a *coordinatization*, i.e. $\mathrm{Dom}\,K$ and $\mathrm{Ran}\,K$ are open subsets and K is a diffeomorphism between $\mathrm{Dom}\,K$ and $\mathrm{Ran}\,K$.

The reference system $(\mathbf{U}, K)$ is called

(i) *natural* if there are maps $T: I \rightarrowtail \mathbb{R}$, $S: E_{\mathbf{U}} \rightarrowtail \mathbb{R}^3$ such that

$$K = (T \times S) \circ S_{\mathbf{U}},$$

i.e. time and $\mathbf{U}$-space are coordinatized separately: in this case we say that the coordinatization K is *adapted* to the observer $\mathbf{U}$;

(ii) *global* if $\mathbf{U}$ is global and $\mathrm{Dom}\,K = M$;

(iii) *rigid*, *rotation free* and *inertial* if $\mathbf{U}$ is rigid, rotation free and inertial, respectively.

6.3. Definition. A *reference system with origin* is a triplet $(\mathbf{U}, o, K)$ where $(\mathbf{U}, o)$ is an observer with origin and $(\mathbf{U}, K)$ is a reference system such that o is an accumulation point of $\mathrm{Dom}\,K$.

$(\mathbf{U}, o, K)$ is called

(i) *natural* if there are maps $T: \mathbf{I} \rightarrowtail \mathbb{R}$, $S: \mathbf{E} \rightarrowtail \mathbb{R}^3$ such that

$$K = (T \times S) \circ S_o^{\mathbf{U}} \circ S_{\mathbf{U}}$$

(ii) *global* etc if $(\mathbf{U}, K)$ is global etc.

6.4. We have seen that global inertial observers are of the nicest kind. Reference systems connected with such observers may have bad properties if the coordinatization is unpleasant. The adapted affine coordinatizations have a great importance because they are the simplest ones.

Definition. An *affine reference system* is a global, inertial natural reference system whose coordinatization is an affine map. If, moreover, the derivative of space coordinatization is a $\gamma - \langle\ ,\ \rangle$-orthogonal map $\mathbf{E} \longrightarrow \mathbb{R}^3$ ($\langle\ ,\ \rangle$ is the standard inner product on $\mathbb{R}^3$), the reference system is called *Galilean*.

6.5. Proposition. Let $(\mathbf{U}, K)$ be an affine reference system and suppose $\mathbf{c} \in V(1)$ is the constant value of $\mathbf{U}$. Then there are an $o \in M$, a non-zero $\mathbf{e} \in \mathbf{I}$, linearly independent vectors $\mathbf{e}_1, \mathbf{e}_2, \mathbf{e}_3 \in \mathbf{E}$, all these elements are uniquely determined, and

$$K(x) = \left(\frac{\tau(x-o)}{\mathbf{e}}, (\mathbf{e}^{k*} | (x-o)\|_{\mathbf{c}})_{k=1,2,3} \right) \qquad (x \in M),$$

where $\{\mathbf{e}^{k*} : k = 1, 2, 3\}$ is the basis in $\mathbf{E}^*$ dual to $\{\mathbf{e}_k : k = 1, 2, 3\}$.

Proof. Let o be the unique element of M that $K(o)=0$ holds for. Then $K(x)=\mathrm{D}K(x-o)$ $(x\in M)$.

K is adapted to $\mathbf{U}$, hence it has the form $K=(T\times S)\circ S_{\mathbf{U}}$. As a consequence, $\mathrm{D}K=(\mathrm{D}T\times \mathrm{D}S)\circ \mathbf{s}_{\mathbf{c}}$ (see Proposition 4.11.). Let $\mathbf{e}$ be the element of $\mathbf{I}$ that $\mathrm{D}T(\mathbf{e})=1$ holds for, and let $\mathbf{e}_k$ $(k=1,2,3)$ be the elements of $\mathbf{E}$ whose images by $\mathrm{D}S$ are the standard basis elements of $\mathbb{R}^3$.

Now it is an easy task to check that K has the form stated in our proposition.

Definition. Using the above-given notations we call $|\mathbf{e}|\in\mathbf{I}^+$ (see IV. 3.6.) the *time unit* of the affine reference system, and the unique element $\mathbf{d}\in\mathbf{D}^+$ that

$$\det\left(\left(\frac{\gamma(\mathbf{e}_i,\mathbf{e}_k)}{\mathbf{d}\otimes\mathbf{d}}\right)_{i,k=1,2,3}\right)=1$$

holds for is called the *space unit* of the affine reference system. Furthermore, we call $\mathbf{c}\in V(1), o\in M, \mathbf{e}\in\mathbf{I}, \mathbf{e}_1,\mathbf{e}_2,\mathbf{e}_3\in\mathbf{E}$ the *velocity*, the *origin*, the *time basis* and the *space basis* of the affine reference system, respectively.

6.6. Proposition. Use the previous notations; the affine reference system is Galilean if and only if $\mathbf{e}_1$, $\mathbf{e}_2$, $\mathbf{e}_3$ is a γ-orthogonal basis normed to $\mathbf{d}$, and then

$$K(x)=\left(\frac{\tau(x-o)}{\mathbf{e}},\left(\frac{\gamma(\mathbf{e}_k,(x-o)\|_{\mathbf{c}})}{\mathbf{d}\otimes\mathbf{d}}\right)_{k=1,2,3}\right)\qquad (x\in M).$$

6.7. Proposition. If $(\mathbf{U}, K)$ is a Galilean reference system and L is a Galilean transformation then $((\mathrm{ar}\, L)\mathrm{D}L_{\mathbf{I}}\circ\mathbf{U}, K\circ L^{-1})$ is a Galilean reference system, too. The time units and space units of the two systems are equal, respectively. If L is a proper Galilean transformation then the time bases and the space bases of the two systems are equally oriented, respectively.

Proof. It is trivial that $(\mathrm{ar}\, L)\mathrm{D}L_{\mathbf{I}}(\mathbf{u})\in V(1)$ for $\mathbf{u}\in V(1)$, hence $(\mathrm{ar}\, L)\mathrm{D}L_{\mathbf{I}}\circ\mathbf{U}$ is a global inertial observer.

Let $\mathbf{c}$, o, $\mathbf{e}$, $\mathbf{e}_1$, $\mathbf{e}_2$, $\mathbf{e}_3$ be as before. Since for $x\in M$

$$\tau(L^{-1}(x)-o)=(\mathrm{ar}\, L)\tau(x-L(o)),$$

$$\mathrm{D}L(L^{-1}(x)-o-\mathbf{c}\otimes\tau(L^{-1}(x)-o))=x-L(o)-(\mathrm{ar}\, L)\mathrm{D}L_I(\mathbf{c})\otimes\tau(x-L(o)),$$

we check at once that $K\circ L^{-1}$ is given by the origin $L(o)$ and bases $(\mathrm{ar}\, L)\mathbf{e}$, $\mathrm{D}L(\mathbf{e}_k)$ $(k=1,2,3)$, ending the proof.

6.8. Proposition. If $(\mathbf{U}, K)$ and $(\mathbf{U}', K')$ are Galilean reference systems with equal time units and equal space units then there exists a unique Galilean transformation L such that $U'=(\mathrm{ar}\, L)\mathrm{D}L_{\mathbf{I}}\circ\mathbf{U}$, $K'=K\circ L^{-1}$. If the time bases and the space bases of the two systems, respectively, are equally oriented then L is a proper Galilean transformation.

Proof. Let $\mathbf{e} \in \mathbf{I}$, $\mathbf{e}_1, \mathbf{e}_2, \mathbf{e}_3 \in \mathbf{E}$ and $\mathbf{e}' \in \mathbf{I}$, $\mathbf{e}'_1, \mathbf{e}'_2, \mathbf{e}'_3 \in \mathbf{E}$ be the corresponding bases. Then $|\mathbf{e}'| = |\mathbf{e}|$ and $\gamma(\mathbf{e}'_i, \mathbf{e}'_k) = \gamma(\mathbf{e}_i, \mathbf{e}_k)$ $(i, k = 1, 2, 3)$.

Let R be the element of $\mathrm{O}(\gamma)$ defined by $R\mathbf{e}_k := \mathbf{e}'_k$ $(k = 1, 2, 3)$. If $\mathbf{c}$ and $\mathbf{c}'$ denote the constant value of $\mathbf{U}$ and $\mathbf{U}'$, respectively, o and o' are the corresponding origins, then

$$L(x) := o' + R((x-o)\|_{\mathbf{c}}) + \frac{\mathbf{e}}{\mathbf{e}'}\mathbf{c}' \otimes \tau(x-o) \qquad (x \in M)$$

is a Galilean transformation, $\operatorname{ar} L = \dfrac{\mathbf{e}}{\mathbf{e}'}$ and

$$\mathrm{D}L(\mathbf{x}) = R(\mathbf{x}\|_{\mathbf{c}}) + \frac{\mathbf{e}}{\mathbf{e}'}\mathbf{c}' \otimes \tau(\mathbf{x}) \qquad (\mathbf{x} \in \mathbf{M}).$$

Hence $o' = L(o)$, $\mathrm{D}L_1(\mathbf{c}) = (\operatorname{ar} L)\mathbf{c}'$ and an easy calculation based on the formulae of the preceding proof shows that $K' = K \circ L^{-1}$.

If N is another Galilean transformation such that $K' = K \circ N^{-1} = K \circ L^{-1}$ then composing the equality by K^{-1} from the left we get that $N = L$.

6.9. The previous theorem and proposition give some connection between some different reference systems. It is an important task to introduce a general notion of the comparison of reference systems because of the following reason.

⌈Consider two physical reference systems, i.e. physical observers with clocks and numbered reference lines, measuring rods. We have a physical method to decide whether the two systems are similar or not. For instance, they are similar if both observers are inertial, have clocks and measuring rods of the same kind and the reference lines are straight.

Furthermore, we have a physical method to decide whether the two systems are equivalent or not from a physical point of view. We consider them equivalent if all experiments prepared by the same procedure give the same result. The following examples can serve as an illustration.

Let us consider two Galilean reference systems with the same time units and space units and with the same orientation of time and space. Then according to our experience, all equally prepared experiments proceed equally in the two systems. On the contrary, for instance, if the two Galilean systems have different time units, then the same bullet hitting with unit velocity against the same sheet of glass breaks up the glass in one of the systems but does not break up in the other one. Hence these reference systems are not equivalent.

Similarly, two reference systems with uniformly accelerating observers, with the same magnitude of acceleration, with the same time units and space units, with equally oriented time and space, with orthogonal and equally normed bases in space chosen in such a way that the first axes have the direction of the accelerations, respectively, are equivalent. On the contrary, for instance, if in one of the systems the first axis is not directed parallel to the acceleration, the two systems are not equivalent: "leave a masspoint alone with zero initial velocity then it moves along the first axis" is true in one of the systems and is not true in the other one.

We draw from experience that reference systems connected by proper Galilean transformations are equivalent. ⌋

6.10. We need some notions which are customary in the theory of smooth manifolds; we define them here in the special context to spare the reader some troubles.

A *little diffeomorphism* of M is an injective smooth map $M \rightarrowtail M$ whose inverse is smooth as well.

If $\mathbf{X}: M \rightarrowtail \frac{\mathbf{M}}{\mathbf{I}}$ is a vector field of cotype $\mathbf{I}$ and $F: M \rightarrowtail M$ is a little diffeomorphism and $\operatorname{Dom} \mathbf{X} \cap \operatorname{Dom} F \neq \emptyset$, we define a vector field of cotype $\mathbf{I}$ by

$$(F_* \mathbf{X})(x) := \mathrm{D}F(x)_{\mathbf{I}} \mathbf{X}(F^{-1}(x)) \qquad (x \in \operatorname{Dom}(\mathbf{X} \circ F^{-1})).$$

6.11. Definition. Let $(\mathbf{U}, K)$ and $(\mathbf{U}', K')$ be two reference systems. A *morphism* from $(\mathbf{U}, K)$ into $(\mathbf{U}', K')$ is a pair (F, λ) where

(i) $F: \operatorname{Dom} \mathbf{U} \longrightarrow \operatorname{Dom} \mathbf{U}'$ is a little diffeomorphism,

(ii) $\lambda: F[\operatorname{Dom} \mathbf{U}] \longrightarrow \mathbb{R} \setminus \{0\}$ is a smooth map such that

(a) $F_* \mathbf{U} \subset \lambda \mathbf{U}'$,

(b) $K \circ F^{-1} \subset K'$.

If F is surjective (i.e. F is a diffeomorphism between $\operatorname{Dom} \mathbf{U}$ and $\operatorname{Dom} \mathbf{U}'$) and equality holds in (b) then (F, λ) is called an *isomorphism* and the two reference systems are *isomorphic* by (F, λ). ∎

Note that (b) implies F maps $\operatorname{Dom} K$ into $\operatorname{Dom} K'$.

6.12. Definition. Let $(\mathbf{U}, o, K)$ and $(\mathbf{U}', o', K')$ be two reference systems with origin. A *morphism* from $(\mathbf{U}, o, K)$ into $(\mathbf{U}', o', K')$ is a morphism (F, λ) from $(\mathbf{U}, K)$ into $(\mathbf{U}', K')$ with the supplementary property $F(o) = o'$.

6.13. Proposition. If L is a Galilean transformation and (L, λ) is a morphism of reference systems then λ is a constant map, $\lambda = \operatorname{ar} L$.

Proof. Suppose $\mathbf{U}$ and $\mathbf{U}'$ are the observers corresponding to the reference systems. Then

$$\mathrm{D}L_{\mathbf{I}} \mathbf{U}(L^{-1}(x)) = \lambda(x) \mathbf{U}'(x) \qquad (x \in L[\operatorname{Dom} \mathbf{U}] \subset \operatorname{Dom} \mathbf{U}').$$

Applying $\boldsymbol{\tau}_{\mathbf{I}}$ to both sides and using $\boldsymbol{\tau} \circ \mathrm{D}L = (\operatorname{ar} L)\boldsymbol{\tau}$, $\boldsymbol{\tau}_{\mathbf{I}} \circ \mathbf{U} = \boldsymbol{\tau}_{\mathbf{I}} \circ \mathbf{U}' = 1$, we obtain the desired result. ∎

As a consequence, in such a case it is convenient to consider the Galilean transformation L as a morphism instead of $(L, \operatorname{ar} L)$.

6.14. Definition. Two reference systems or two reference systems with origin being isomorphic by a proper Galilean transformation are called *equivalent*.

7. The Galilean group

Necessary mathematical tools:

Tensor products and quotients (Chapter IV.)
Elements of group theory
Elements of Lie group theory (Section VIII. 1.)
Some knowledge on certain Lie groups (Section VIII. 4.)

7.1. ⌈Galilean trasformations play a fundamental role in nonrelativistic physics; they are considered as "symmetries" of space-time. Unfortunately, one usually takes $\mathbb{R}\times\mathbb{R}^3$ as space-time, thus Galilean transformations are given as transformations of $\mathbb{R}\times\mathbb{R}^3$. Consequently, the usually defined transformation group of $\mathbb{R}\times\mathbb{R}^3$ ([12], pp. 135 and 171)—let us call it the *arithmetic Galilean group*—has a number of additional structures that the Galilean group defined by means of a nonrelativistic space-time model has not.

We emphasize that using a nonrelativistic model we must define the Galilean group only in terms of the structure given in Definition 1.2.⌋

7.2. In the sequel we consider a nonrelativistic space-time model $(M, I, \tau, \mathbf{D}, \gamma)$ and we use the notations introduced in 1.3.

The orthogonal group $\mathrm{O}(\gamma)$ is a three dimensional Lie group whose Lie algebra is $\mathbf{A}(\gamma)$ (see VIII. 4.11. and VIII. 4.12.); $\mathrm{O}(\gamma)$ has two connected components denoted by $\mathrm{O}^+(\gamma)$ and $\mathrm{O}^-(\gamma)$; the identity component $\mathrm{O}^+(\gamma)$ is usually denoted by $\mathrm{SO}(\gamma)$. The elements of $\mathrm{O}^+(\gamma)$ and $\mathrm{O}^-(\gamma)$ preserve and reverse the orientation of $\mathbf{E}$, respectively.

7.3. Definition. The *Noether group* $\mathcal{N}$ is the subgroup of the bijective linear transformations of $\mathbf{M}$ defined by

$$\mathcal{N} := \{F \in \mathrm{GL}(\mathbf{M}) \,\vdots\, \boldsymbol{\tau}\circ F = \pm\boldsymbol{\tau}, \quad F|_{\mathbf{E}} \in \mathrm{O}(\gamma)\}.$$

7.4. Definition. For $F\in\mathcal{N}$ we put

$$\operatorname{ar} F := \begin{cases} +1 & \text{if} \quad \boldsymbol{\tau}\circ F = \boldsymbol{\tau}, \\ -1 & \text{if} \quad \boldsymbol{\tau}\circ F = -\boldsymbol{\tau}, \end{cases}$$

$$\operatorname{sign} F := \begin{cases} +1 & \text{if} \quad F|_{\mathbf{E}} \in \mathrm{O}^+(\gamma) \\ -1 & \text{if} \quad F|_{\mathbf{E}} \in \mathrm{O}^-(\gamma) \end{cases}$$

and we introduce

$$\mathcal{N}^{+\rightarrow} := \{F\in\mathcal{N} \,\vdots\, \operatorname{sign} F = \operatorname{ar} F = 1\},$$

$$\mathcal{N}^{+\leftarrow} := \{F\in\mathcal{N} \,\vdots\, \operatorname{sign} F = -\operatorname{ar} F = 1\},$$

$$\mathcal{N}^{-\rightarrow} := \{F \in \mathcal{N} \mid \text{sign } F = -\text{ar } F = -1\},$$

$$\mathcal{N}^{-\leftarrow} := \{F \in \mathcal{N} \mid \text{sign } F = \text{ar } F = -1\}.$$

$\mathcal{N}^{+\rightarrow}$ is called the *proper Noether group.*
$\mathcal{N}^{\rightarrow} := \mathcal{N}^{+\rightarrow} \cup \mathcal{N}^{-\rightarrow}$ is the *orthocronous Noether group.*

7.5. Proposition. The Noether group is a six dimensional Lie group having four connected components which are the above listed subsets. The identity component is the proper Noether group. The elements of $\mathcal{N}^{+\rightarrow}$ and $\mathcal{N}^{-\leftarrow}$ preserve the orientation of **M** whereas the elements of $\mathcal{N}^{+\leftarrow}$ and $\mathcal{N}^{-\rightarrow}$ do not.

Proof. The identity component of $\mathcal{N}$ is evidently in $\mathcal{N}^{+\rightarrow}$. We know that the elements of $O^+(\gamma)$ are orientation preserving transformations of **E** whereas the elements of $O^-(\gamma)$ are not. Hence the desired result on the orientation preserving feature of Noether transformations follows at once from Definitions 7.4. and 1.9.

The Noether group is a Lie group as a closed subgroup of the Lie group GL(**M**). The other assertions can be proved by establishing a Lie group isomorphism from $\mathcal{N}$ onto an other Lie group. This is done in 7.20.

7.6. Proposition. The Lie algebra of the Noether group is

$$\mathbf{L}(\mathcal{N}) = \{A : \mathbf{M} \longrightarrow \mathbf{M} \mid A \text{ is linear}, \quad \boldsymbol{\tau} \circ A = \mathbf{0}, \quad A|_{\mathbf{E}} \in \mathbf{A}(\gamma)\}$$

equipped with the commutator

$$[A, B] = AB - BA.$$

Proof. We know that $\mathbf{L}(\mathcal{N})$ is a Lie subalgebra of the linear maps $\mathbf{M} \longrightarrow \mathbf{M}$ equipped with the above commutator (VIII. 3.3., VIII. 4.2.). The definition of the Noether group and that of the Lie algebra of a Lie group imply immediately the conditions imposed on the linear maps to be elements of $\mathbf{L}(\mathcal{N})$.

7.7. Definition. The *Galiliean group* $\mathcal{G}$ is the subgroup of the bijective affine transformations of M defined by

$$\mathcal{G} := \{L \in \text{GA}(M) \mid \text{D}L \in \mathcal{N}\}$$

(cf. VIII. 4.10. (iii).).

7.8. Definition. For $L \in \mathcal{G}$ we put

$$\text{ar } L := \text{ar D}L, \qquad \text{sign } L := \text{sign D}L$$

and we introduce $\mathcal{G}^{+\rightarrow}$, $\mathcal{G}^{+\leftarrow}$, $\mathcal{G}^{-\rightarrow}$, $\mathcal{G}^{-\leftarrow}$ as the subsets of $\mathcal{G}$ consisting of the elements whose derivative is in $\mathcal{N}^{+\rightarrow}$, $\mathcal{N}^{+\leftarrow}$, $\mathcal{N}^{-\rightarrow}$, $\mathcal{N}^{-\leftarrow}$, respectively.

$\mathcal{G}^{+\rightarrow}$ is called the *proper Galilean group.*

$\mathcal{G}^{\rightarrow} := \mathcal{G}^{+\rightarrow} \cup \mathcal{G}^{-\rightarrow}$ is the *orthocronous Galilean group.*

7.9. Theorem. The Galilean group is a ten dimensional Lie group having four connected components which are the above listed subsets. The identity component is the proper Galilean group. The elements of $\mathcal{G}^{+\rightarrow}$ and $\mathcal{G}^{-\leftarrow}$ preserve the orientation of M whereas the elements of $\mathcal{G}^{+\leftarrow}$ and $\mathcal{G}^{-\rightarrow}$ do not.

Proof. Combine Proposition 7.5. with VIII. 4.10. (ii)–(iii).

7.10. Theorem. The Lie algebra of the Galilean group is

$$\mathsf{L}(\mathcal{G})=\{X:M\longrightarrow \mathbf{M} \,\vdots\, X \text{ is affine}, \quad \mathrm{D}X\in \mathsf{L}(\mathcal{N})\},$$

equipped with the commutator

$$[X, Y]=(\mathrm{D}X)\circ Y-(\mathrm{D}Y)\circ X.$$

Proof. See VIII. 4.8. and VIII. 4.10. (iii).

Remark. If $X\in \mathsf{L}(\mathcal{G})$ then $\tau\circ X$ is constant.

7.11. Proposition. Make the identification given in 1.4., i.e. consider $t\in I$ as an affine subspace over $\mathbf{E}$ of M. Then $L[t]$ is again an element of I for every Galilean transformation $L$and the map

$$\mathrm{ti}\, L: I\longrightarrow I, \qquad t\mapsto L[t]$$

is affine.

Proof. If $x, y\in t$, i.e. $\tau(x)=\tau(y)=t$ then

$$\tau(L(x))-\tau(L(y))=\boldsymbol{\tau}(\mathrm{D}L(x-y))=(\mathrm{ar}\, L)\boldsymbol{\tau}(x-y)=\mathbf{0},$$

hence $L[t]$ is indeed an affine subspace over $\mathbf{E}$.

Notice that we can write

$$L[t]=\tau(L(x)) \qquad (x\in t)$$

and we easily deduce that

$$(\mathrm{ti}\, L)(t)-(\mathrm{ti}\, L)(s)=(\mathrm{ar}\, L)(t-s) \qquad (t,\ s\in I)$$

which shows that $\mathrm{ti}\, L$ is an affine map having derivative $(\mathrm{ar}\, L)\,\mathrm{id}_{\mathbf{I}}$.

7.12. Proposition. The proper Galilean group can be considered as the subgroup of the automorphism group of $(M, I, \tau, \mathbf{D}, \gamma)$ consisting of the strict automorphisms (see Definition 1.10.).

Proof. If $L\in\mathcal{G}^{+\rightarrow}$ then $(L, \mathrm{ti}\, L, \mathrm{id}_{\mathbf{D}})$ is a strict automorphism.

Conversely, if $(F, B, \mathrm{id}_{\mathbf{D}})$ is a strict automorphism then $\mathrm{D}B=\mathrm{id}_{\mathbf{D}}$, hence (a) of Definition 1.10. implies that F is a Galilean transformation, $\mathrm{ar}\, F=1$, and F is orientation preserving by definition, consequently F is in $\mathcal{G}^{+\rightarrow}$. Furthermore we easily derive that $B=\mathrm{ti}\, F$.

7.13. We have a Galilean group which, at first sight, does not resemble the usually defined arithmetic Galilean group. One might conjecture that this last group is the "Galilean group of the nonrelativistic space-time model $(\mathbb{R}\times\mathbb{R}^3, \mathbb{R}, \mathrm{pr}_{\mathbb{R}}, \mathbb{R}, \langle\ ,\ \rangle)$". It is indeed so, as we shall see. Since the properties of the arithmetic Galilean group are well known and the arithmetic Galilean group is used up to now in physics, we should like to have a relation between the Galilean group and the arithmetic Galilean group. A relation can be exhibited by the isomorphisms given in Propositions 1.14. and 1.15. The first isomorphism is more interesting; the second one is very simple. That is why we concentrate on intermediate steps between the Galilean group and the arithmetic Galilean group, on the so called vectorial Galilean group and split Galilean group; the latter one is very similar to the arithmetic Galilean group and there is a trivial connection between them.

7.14. Definition. The *vectorial Galilean group* $\mathcal{G}_V$ is $\mathbf{M} \times \mathcal{N}$ equipped with the multiplication law

$$(\mathbf{x}', F')(\mathbf{x}, F) := (\mathbf{x}' + F'\mathbf{x}, F'F).$$

Remarks. (i) The vectorial Galilean group is the semi-direct product of Tn ($\mathbf{M}$) (the additive group of $\mathbf{M}$) and the Noether group $\mathcal{N}$ (see VIII. 4.5.).

(ii) The elements of the vectorial Galilean group can be considered in a natural way as invertible affine transformations of $\mathbf{M}$:

$$\hat{\mathbf{x}} \mapsto (\mathbf{x}, F)(\hat{\mathbf{x}}) := F\hat{\mathbf{x}} + \mathbf{x}.$$

Then the group multiplication coincides with the composition.

(iii) The Lie algebra of $\mathcal{G}_V$ is $\mathbf{M} \times \mathsf{L}(\mathcal{N})$ equipped with the commutator

$$[(\mathbf{x}', A'), (\mathbf{x}, A)] = (A'\mathbf{x} - A\mathbf{x}', A'A - AA')$$

(see VIII. 4.4. and VIII. 4.5.); the elements of $\mathsf{L}(\mathcal{G}_V)$ can be considered as affine transformations of $\mathbf{M}$.

(iv) We can introduce the subsets $\mathcal{G}_V^{+\rightarrow}$, $\mathcal{G}_V^{+\leftarrow}$, $\mathcal{G}_V^{-\rightarrow}$, $\mathcal{G}_V^{-\leftarrow}$ by an evident definition based on Definition 7.4.

7.15. We know that given $o \in M$, we can establish isomorphisms z_o: $\mathcal{G} \longrightarrow \mathcal{G}_V$ and $w_o : \mathsf{L}(\mathcal{G}) \longrightarrow \mathsf{L}(\mathcal{G}_V)$; for different o and o' the isomorphisms are different as it is shown in VIII. 4.7. and VIII. 4.8. (i).

The topological properties of $\mathcal{N}$ (implying the topological properties of $\mathcal{G}$) can be seen by "splitting" the proper Noether group, i.e. by giving a Lie group isomorphism between $\mathcal{N}$ and a semi-direct product of Lie groups. We shall realize the splitting of the Galilean group because this one is required by applications; the desired results for the Noether group (to prove Proposition 7.5.) can be extracted from the corresponding formulae.

7.16. Definition. The *proper split Galilean group* $\mathcal{G}_0$ is the set $\mathbf{I} \times \mathbf{E} \times \frac{\mathbf{E}}{\mathbf{I}} \times \mathrm{SO}(\gamma)$ equipped with the multiplication law

$$(\mathbf{t}', \mathbf{q}', \mathbf{v}', R')(\mathbf{t}, \mathbf{q}, \mathbf{v}, R) := (\mathbf{t}' + \mathbf{t}, \mathbf{q}' + R'\mathbf{q} + \mathbf{v}' \otimes \mathbf{t}, \mathbf{v}' + R'_{\mathbf{I}}\mathbf{v}, R'R)$$

(notation: IV. 2.10. (ii)).

Remarks. (i) The elements of the proper split Galilean group can be considered in a natural way as affine transformations of $\mathbf{I}\times\mathbf{E}$:

$$(\hat{\mathbf{t}},\hat{\mathbf{q}})\mapsto(\mathbf{t},\mathbf{q},\mathbf{v},R)(\hat{\mathbf{t}},\hat{\mathbf{q}}):=(\hat{\mathbf{t}}+\mathbf{t},\mathbf{v}\otimes\hat{\mathbf{t}}+R\hat{\mathbf{q}}+\mathbf{q}).$$

Then the group multiplication coincides with the composition.

(ii) $\mathbf{I}\times\{\mathbf{0}\}\times\{\mathbf{0}\}\times\{\mathrm{id}_{\mathbf{E}}\}$,

$\{\mathbf{0}\}\times\mathbf{E}\times\{\mathbf{0}\}\times\{\mathrm{id}_{\mathbf{E}}\}$,

$\{\mathbf{0}\}\times\{\mathbf{0}\}\times\frac{\mathbf{E}}{\mathbf{I}}\times\{\mathrm{id}_{\mathbf{E}}\}$,

$\{\mathbf{0}\}\times\{\mathbf{0}\}\times\{\mathbf{0}\}\times\mathrm{SO}(\gamma)$

are Lie subgroups of $\mathscr{G}_0$ to which we refer as the *fundamental subgroups.*

7.17. Definition. The *split Galilean group* $\mathscr{G}_s$ is the transformation group of $\mathbf{I}\times\mathbf{E}$ generated by $\mathscr{G}_0$ and by the two inversions

$$\begin{aligned}T(\hat{\mathbf{t}},\hat{\mathbf{q}})&:=(-\hat{\mathbf{t}},\hat{\mathbf{q}})\\ S(\hat{\mathbf{t}},\hat{\mathbf{q}})&:=(\hat{\mathbf{t}},-\hat{\mathbf{q}})\end{aligned}\qquad((\hat{\mathbf{t}},\hat{\mathbf{q}})\in\mathbf{I}\times\mathbf{E}).$$

Remark. S can be incorporated in the given form of proper split Galilean transformations, namely $S:=(\mathbf{0},\mathbf{0},\mathbf{0},-\mathrm{id}_{\mathbf{E}})$; in other words, $\mathscr{G}_0$ extended by S is $\mathbf{I}\times\mathbf{E}\times\frac{\mathbf{E}}{\mathbf{I}}\times\mathrm{O}(\gamma)$.

On the other hand, T cannot be handled in a similar way.

Every element of $\mathscr{G}_s$ can be written in the form

$$(\mathbf{t},\mathbf{q},\mathbf{v},R)\,T(\alpha)$$

where $(\mathbf{t},\mathbf{q},\mathbf{v},R)\in\mathbf{I}\times\mathbf{E}\times\frac{\mathbf{E}}{\mathbf{I}}\times\mathrm{O}(\gamma)$, $\alpha=\pm1$ and

$$T(\alpha):=\begin{cases}\mathrm{id}_{\mathbf{I}\times\mathbf{E}} & \text{if}\quad \alpha=1\\ T & \text{if}\quad \alpha=-1.\end{cases}$$

7.18. Proposition. The split Galilean group is a ten dimensional Lie group having four connected components, namely

$$\mathscr{G}_0,\qquad T\mathscr{G}_0,\qquad S\mathscr{G}_0,\qquad TS\mathscr{G}_0.$$

Proof. $\mathscr{G}_0$ is evidently ten dimensional and connected; the other three subsets, as images of $\mathscr{G}_0$ by continuous maps, are connected as well. We end the proof by noting the obvious fact that the subsets in question are disjoint and their union is the whole group.

7.19. Proposition. The Lie algebra $\mathbf{L}(\mathcal{G}_s)$ of $\mathcal{G}_s$ is $\mathbf{I}\times\mathbf{E}\times\frac{\mathbf{E}}{\mathbf{I}}\times\mathbf{A}(\gamma)$ equipped with the commutator

$$[(\mathbf{t}',\mathbf{q}',\mathbf{v}',B'),(\mathbf{t},\mathbf{q},\mathbf{v},B)]=$$
$$=(0,\mathbf{v}'\otimes\mathbf{t}-\mathbf{v}\otimes\mathbf{t}'+B'\mathbf{q}-B\mathbf{q}',B'_{\mathbf{I}}\mathbf{v}-B_{\mathbf{I}}\mathbf{v}',B'B-BB').$$

Proof. The underlying set of $\mathbf{L}(\mathcal{G}_s)$ is $\mathbf{I}\times\mathbf{E}\times\frac{\mathbf{E}}{\mathbf{I}}\times\mathbf{A}(\gamma)$ because the tangent space at a point of a Cartesian product of manifolds is the Cartesian product of the corresponding tangent spaces.

Remark. The elements of $\mathbf{L}(\mathcal{G}_s)$ can be considered as affine transformations of $\mathbf{I}\times\mathbf{E}$, as follows:

$$(\hat{\mathbf{t}},\hat{\mathbf{q}})\mapsto(\mathbf{t},\mathbf{q},\mathbf{v},B)(\hat{\mathbf{t}},\hat{\mathbf{q}}):=(\mathbf{t},\mathbf{v}\otimes\hat{\mathbf{t}}+B\hat{\mathbf{q}}+\mathbf{q}).$$

Then the commutator has the form (see VIII. 4.4.)

$$[(\mathbf{t}',\mathbf{q}',\mathbf{v}',B'),(\mathbf{t},\mathbf{q},\mathbf{v},B)]=$$
$$=\mathrm{D}(\mathbf{t}',\mathbf{q}',\mathbf{v}',B')\circ(\mathbf{t},\mathbf{q},\mathbf{v},B)-\mathrm{D}(\mathbf{t},\mathbf{q},\mathbf{v},B)\circ(\mathbf{t}',\mathbf{q}',\mathbf{v}',B')$$

where $\mathrm{D}(\mathbf{t},\mathbf{q},\mathbf{v},B)$ is the derivative of the affine map $(\mathbf{t},\mathbf{q},\mathbf{v},B)$ and equals $(\mathbf{0},\mathbf{0},\mathbf{v},B)$.

7.20. A straightforward but somewhat laborious calculation leads us to the following important results.

Theorem. Let $\mathbf{c}\in V(1)$. Then

$$\mathbf{z_c}:\mathcal{G}_V\longrightarrow\mathcal{G}_s,\qquad(\mathbf{x},F)\mapsto\mathbf{s_c}\circ(\mathbf{x},F)\circ\mathbf{s_c}^{-1}$$

(see 1.13.) is a Lie group isomorphism, in detail

$$\mathbf{z_c}(\mathbf{x},F)=(\tau(\mathbf{x}),\mathbf{x}\|_{\mathbf{c}},(\operatorname{ar}F)F_{\mathbf{I}}\mathbf{c}-\mathbf{c},F|_{\mathbf{E}})\,T(\operatorname{ar}F)\qquad\circledast$$

for $(\mathbf{x},F)\in\mathcal{G}_V$, and if $(\mathbf{t},\mathbf{q},\mathbf{v},R)\,T(\alpha)$ is a split Galilean transformation then its image by $\mathbf{z_c}^{-1}$ is the vectorial Galilean transformation given by

$$\mathbf{x}\mapsto R(\mathbf{x}\|_{\mathbf{c}})+\alpha(\mathbf{c}+\mathbf{v})\otimes\tau(\mathbf{x})+\mathbf{c}\otimes\mathbf{t}+\mathbf{q}.$$

Furthermore we have

$$\mathbf{z_c}[\mathcal{G}_V^{+\rightarrow}]=\mathcal{G}_0,\qquad\mathbf{z_c}[\mathcal{G}_V^{+\leftarrow}]=T\mathcal{G}_0,$$
$$\mathbf{z_c}[\mathcal{G}_V^{-\rightarrow}]=S\mathcal{G}_0,\qquad\mathbf{z_c}[\mathcal{G}_V^{-\leftarrow}]=TS\mathcal{G}_0.$$

As a consequence, for $o\in M$,

$$z_{\mathbf{c},o}:=\mathbf{z_c}\circ z_o:\mathcal{G}\longrightarrow\mathcal{G}_s$$

(see. 7.15.) is a Lie group isomorphism as well. We can obtain $z_{\mathbf{c},o}(L)$ for $L\in\mathcal{G}$ from the right hand side of $\circledast$ by substituting $L(o)-o$ for $\mathbf{x}$ and $\mathrm{D}L$ for F.

Furthermore we have

$$z_{\mathbf{c},o}[\mathcal{G}^{+\rightarrow}]=\mathcal{G}_0, \qquad z_{\mathbf{c},o}[\mathcal{G}^{+\leftarrow}]=T\mathcal{G}_0,$$

$$z_{\mathbf{c},o}[\mathcal{G}^{-\rightarrow}]=S\mathcal{G}_0, \qquad z_{\mathbf{c},o}[\mathcal{G}^{-\leftarrow}]=TS\mathcal{G}_0.$$

Remarks. (i) Our theorem furnishes a proof of Theorem 7.9.

(ii) Since the Noether group in the form $\{(\mathbf{0}, F) : F \in \mathcal{N}\}$ is a subgroup of the vectorial Galilean group, a simple consideration shows how the Noether group is mapped by $\mathbf{z}_\mathbf{c}$ onto the group generated by $\{\mathbf{0}\} \times \{\mathbf{0}\} \times \frac{\mathbf{E}}{\mathbf{I}} \times \mathrm{SO}(\gamma)$ and S and T, providing a proof of Proposition 7.5.

7.21. Consider the elements of $\mathsf{L}(\mathcal{G}_V)$ and $\mathsf{L}(\mathcal{G}_s)$ as affine transformations of $\mathbf{M}$ and $\mathbf{I} \times \mathbf{E}$, respectively (see 7.14. and 7.19.) to derive, by a calculation similar to the previous one, the following important results.

Theorem. Let $\mathbf{c} \in V(1)$. Then

$$\mathbf{w}_\mathbf{c} : \mathsf{L}(\mathcal{G}_V) \longrightarrow \mathsf{L}(\mathcal{G}_s), \qquad (\mathbf{x}, A) \mapsto \mathbf{s}_\mathbf{c} \circ (\mathbf{x}, A) \circ \mathbf{s}_\mathbf{c}^{-1}$$

(see 1.13.) is a Lie algebra isomorphism, in detail

$$\mathbf{w}_\mathbf{c}(\mathbf{x}, A) = (\boldsymbol{\tau}(\mathbf{x}), \mathbf{x}\|_\mathbf{c}, A_\mathbf{I}\mathbf{c}, A|_\mathbf{E}) \qquad \circledast\circledast$$

for $(\mathbf{x}, A) \in \mathsf{L}(\mathcal{G}_V)$ and if $(\mathbf{t}, \mathbf{q}, \mathbf{v}, B)$ is an element of $\mathsf{L}(\mathcal{G}_s)$ then its image by $\mathbf{w}_\mathbf{c}^{-1}$ is the affine map $\mathbf{M} \longrightarrow \mathbf{M}$,

$$\mathbf{x} \mapsto B(\mathbf{x}\|_\mathbf{c}) + \mathbf{v} \otimes \boldsymbol{\tau}(\mathbf{x}) + \mathbf{c} \otimes \mathbf{t} + \mathbf{q}.$$

As a consequence, for $o \in M$,

$$w_{\mathbf{c},o} := \mathbf{w}_\mathbf{c} \circ w_o : \mathsf{L}(\mathcal{G}) \longrightarrow \mathsf{L}(\mathcal{G}_s)$$

(see 7.15) is a Lie algebra isomorphism as well. We can obtain $w_{\mathbf{c},o}(X)$ for $X \in \mathsf{L}(\mathcal{G})$ from the right hand side of $\circledast\circledast$ by substituting $X(o)$ for $\mathbf{x}$ and $\mathrm{D}X$ for A.

7.22. The isomorphisms constructed in Theorems 7.20. and 7.21. depend heavily on the elements of $V(1)$ and M. If $\mathbf{c}, \mathbf{c}' \in V(1)$ and $o, o' \in M$ then

$$z_{\mathbf{c}',o'}(L) = m_{\mathbf{cc}'oo'}^{-1} \circ z_{\mathbf{c},o}(L) \circ m_{\mathbf{cc}'oo'} \qquad (L \in \mathcal{G}),$$

$$w_{\mathbf{c}',o'}(X) = m_{\mathbf{cc}'}^{-1} \circ w_{\mathbf{c},o}(X) \circ m_{\mathbf{cc}'oo'} \qquad (X \in \mathsf{L}(\mathcal{G})),$$

where

$$m_{\mathbf{cc}'} := \mathbf{s}_\mathbf{c} \circ \mathbf{s}_{\mathbf{c}'}^{-1}, \qquad m_{\mathbf{cc}'oo'} := s_{\mathbf{c},o} \circ s_{\mathbf{c}',o'}^{-1}.$$

The following easily provable formulae reflect the dependence of the isomorphisms under consideration on the elements of $V(1) \times M$.

Proposition. $m_{\mathbf{cc}'}$ and $m_{\mathbf{cc}'oo'}$ are proper split Galilean transformations, namely

$$m_{\mathbf{cc}'} = (\mathbf{0}, \mathbf{0}, \mathbf{c}' - \mathbf{c}, \mathrm{id}_\mathbf{E}),$$

$$m_{\mathbf{cc}'oo'} = (\boldsymbol{\tau}(o' - o), (o' - o)\|_\mathbf{c}, \mathbf{c}' - \mathbf{c}, \mathrm{id}_\mathbf{E}).$$

Furthermore,

$$m^{-1}_{cc'oo'}(\mathbf{t}, \mathbf{q}, \mathbf{v}, R)\, m_{cc'oo'} =$$

$$=\big(\mathbf{t}, \mathbf{q}+R((o'-o)\|_{\mathbf{c}})-(o'-o)\|_{\mathbf{c}}+\mathbf{v}\otimes\tau(o'-o)-(\mathbf{c}'-\mathbf{c})\otimes\mathbf{t},$$
$$\mathbf{v}+R_{\mathbf{I}}(\mathbf{c}'-\mathbf{c})-(\mathbf{c}'-\mathbf{c}), R\big)$$

for all proper split Galilean transformations $(\mathbf{t}, \mathbf{q}, \mathbf{v}, R)$ and

$$m^{-1}_{cc'}(\mathbf{t}, \mathbf{q}, \mathbf{v}, B)\, m_{cc'oo'} =$$

$$=\big(\mathbf{t}, \mathbf{q}+B((o'-o)\|_{\mathbf{c}})+\mathbf{v}\otimes\tau(o'-o)-(\mathbf{c}'-\mathbf{c})\otimes\mathbf{t}, \mathbf{v}+B_{\mathbf{I}}(\mathbf{c}'-\mathbf{c}), B\big)$$

for all elements $(\mathbf{t}, \mathbf{q}, \mathbf{v}, B)$ of the Lie algebra of the split Galilean group.

7.23. The Galilean group is isomorphic as a Lie group to the split Galilean group, however, there is *no canonical isomorphism* between the two groups. Given $(\mathbf{c}, o) \in V(1) \times M$, we can establish an isomorphism but for different $(\mathbf{c}, o)$ and $(\mathbf{c}', o')$ the isomorphisms are different, and all elements of $V(1) \times M$ have the "same right" to generate an isomorphism.

One can prove that the split Galilean group is "the Galilean group of the nonrelativistic space-time model $(\mathbf{I} \times \mathbf{E}, \mathbf{I}, \mathrm{pr}_{\mathbf{I}}, \mathbf{D}, \gamma)$". Due to the additional structures of this special model (see 1.17.), the split Galilean group has a number of additional structures that the Galilean group has not. The form of split Galilean transformations, in accordance with the usual terminology for arithmetic Galilean transformations, tempts us to consider elements

$(\mathbf{t}, \mathbf{0}, \mathbf{0}, \mathrm{id}_{\mathbf{E}})$	as a time translation,
$(\mathbf{0}, \mathbf{q}, \mathbf{0}, \mathrm{id}_{\mathbf{E}})$	as a space translation,
$(\mathbf{0}, \mathbf{0}, \mathbf{v}, \mathrm{id}_{\mathbf{E}})$	as a special Galilean transformation,
$(\mathbf{0}, \mathbf{0}, \mathbf{0}, R)$	as a space rotation,
T	as time inversion,
S	as space inversion.

However, the split Galilean transformations are not transformations of our space-time; the above notions arise from the additional structures of $(\mathbf{I} \times \mathbf{E}, \mathbf{I}, \mathrm{pr}_{\mathbf{I}}, \mathbf{D}, \gamma)$ as we shall see in detail by examining the Galilean group further.

7.24. We remind of the definitions of the Noether group $\mathcal{N}$ and the Galilean group $\mathcal{G}$ and we introduce the following groups.

Definition.

(i) The *special vectorial Galilean group* is

$$\mathcal{V} := \{F \in \mathcal{N}^{\to} \,\vdots\, F|_{\mathbf{E}} = \mathrm{id}_{\mathbf{E}}\};$$

(ii) the *time translation group* $\mathrm{Tn}(\mathbf{I})$ is the additive group of $\mathbf{I}$;

(iii) the *space-time translation group* $\mathrm{Tn}(\mathbf{M})$ is the additive group of $\mathbf{M}$;

(iv) the *space translation group* Tn(**E**) is the additive group of **E**;
(v) the *instantaneous Galilean group* is

$$\mathcal{G}_i := \{L \in \mathcal{G} : \tau \circ L = \tau\};$$

(vi) the *Neumann group* is

$$\mathcal{C} := \{L \in \mathcal{G}_i : DL|_{\mathbf{E}} = \mathrm{id}_{\mathbf{E}}\}.$$

7.25. Proposition.

(i) Tn(**I**), O(γ), $\mathcal{V}$ and $\mathcal{N}$ are not subgroups of $\mathcal{G}$.

(ii) Tn(**E**) is a subgroup of Tn(**M**) and $\mathcal{C}$, $\mathcal{C}$ is a subgroup of $\mathcal{G}_i$, Tn(**M**) and $\mathcal{G}_i$ are subgroups of $\mathcal{G}$.

(iii) We have the following exact sequences of Lie group homomorphisms:

$$e \longrightarrow \mathrm{Tn}(\mathbf{E}) \longrightarrow \mathrm{Tn}(\mathbf{M}) \longrightarrow \mathcal{G} \xrightarrow[L \mapsto DL]{} \mathcal{N} \longrightarrow e$$

$$e \longrightarrow \mathcal{V} \longrightarrow \mathcal{N} \xrightarrow[F \mapsto F|_{\mathbf{E}}]{} \mathrm{O}(\gamma) \longrightarrow e$$

$$e \longrightarrow \mathrm{Tn}(\mathbf{E}) \longrightarrow \mathcal{C} \longrightarrow \mathcal{G}_i \xrightarrow[L \mapsto DL]{} \mathcal{V} \longrightarrow e$$

$$e \longrightarrow \mathcal{G}_i \longrightarrow \mathcal{G}^{\to} \xrightarrow[L \mapsto \mathrm{ti}\, L]{} \mathrm{Tn}(\mathbf{I}) \longrightarrow e$$

where e is the trivial group.

(iv)

$$\mathrm{Tn}\left(\frac{\mathbf{E}}{\mathbf{I}}\right) \longrightarrow \mathcal{V}, \qquad \mathbf{v} \mapsto F_{\mathbf{v}}$$

defined by

$$F_{\mathbf{v}}(\mathbf{x}) := \mathbf{x} + \mathbf{v} \otimes \tau(\mathbf{x}) \qquad (\mathbf{x} \in \mathbf{M})$$

is a Lie group isomorphism where $\mathrm{Tn}\left(\frac{\mathbf{E}}{\mathbf{I}}\right)$ is the additive group of $\frac{\mathbf{E}}{\mathbf{I}}$.

Proof. Only (iv) is not trivial. The given map is clearly an injective Lie group homomorphism.

Let F be an element of $\mathcal{V}$. Then for all $\mathbf{u} \in V(1)$, $\mathbf{v}_F := F_{\mathbf{I}}(\mathbf{u}) - \mathbf{u}$ is the same element of $\frac{\mathbf{E}}{\mathbf{I}}$, for

$$(F_{\mathbf{I}}(\mathbf{u}) - \mathbf{u}) - (F_{\mathbf{I}}(\mathbf{u}') - \mathbf{u}') = (F_{\mathbf{I}}(\mathbf{u}) - F_{\mathbf{I}}(\mathbf{u}')) - (\mathbf{u} - \mathbf{u}') =$$

$$= DF_{\mathbf{I}}(\mathbf{u} - \mathbf{u}') - (\mathbf{u} - \mathbf{u}') = \mathbf{0}.$$

It is trivial then that $F \mapsto \mathbf{v}_F$ is the inverse of $\mathbf{v} \mapsto F_{\mathbf{v}}$ from which we easily deduce that the latter is a Lie group isomorphism.

7.26. Summarizing our results we emphasize: the *space rotation-inversion group* O(γ), *the special vectorial Galilean group* $\mathscr{V}$ and the *time translation group* Tn(**I**) *are not subgroups of the Galilean group* $\mathscr{G}$. ⌈In customary nonrelativistic physics one uses these subgroups of the arithmetic Galilean group and one attaches to them a physical meaning. We shall see in volume Models in Mechanics the physical significance of the non-existence of these subgroups of the Galilean group.⌋

Furthermore we see there are *no time inversion, no space inversion and no space-time inversion* in the Galilean group. This is obvious because time is not a subset of space-time, there is no space, and neither time nor space-time have a distinguished point, i.e. they have no origin to which they could be inverted.

The Noether group contains an element, namely $-\mathrm{id}_{\mathbf{M}}$, which can be considered as "vectorial space-time inversion" but there are no "vectorial time inversion" and "vectorial space inversion" either in the Noether group.

On the base of 2.20. we can say symbolically that the Galilean transformations

in $\mathscr{G}^{+\rightarrow}$ do not invert either time or space,

in $\mathscr{G}^{+\leftarrow}$ invert time and do not invert space,

in $\mathscr{G}^{-\rightarrow}$ do not invert time and invert space,

in $\mathscr{G}^{-\leftarrow}$ invert both time and space,

but there are no elements that one could call time inversion and space inversion.

The reader is advised to consult the exercises for a thorough understanding of the situation.

7.27. Proposition. A Galilean transformation L is instantaneous, i.e. it is in $\mathscr{G}_i$ if and only if for all $t \in I$ the instantaneous space at time point t, i.e. the hyperplane t (see 1.4.) is invariant for L.

Proof. The assertion is a simple reformulation of the definition of $\mathscr{G}_i$.

7.28. Proposition. The Neumann group is a six dimensional commutative normal subgroup of the Galilean group.

Proof. Given $(\mathbf{c}, o) \in V(1) \times M$, we immediately derive from 7.20. that $z_{\mathbf{c},o}[\mathscr{C}]$ equals $\{\mathbf{0}\} \times \mathbf{E} \times \frac{\mathbf{E}}{\mathbf{I}} \times \{\mathrm{id}_{\mathbf{E}}\}$ which is a six dimensional commutative normal subgroup of the split Galilean group.

Exercises

1. Show that a global observer $\mathbf{U}$ is rigid if and only if

$$\gamma([\mathbf{U}(x)-\mathbf{U}(y)]\otimes\mathbf{t}, x-y)=\mathbf{0}$$

for some (hence for all) non-zero $\mathbf{t}\in\mathbf{I}$, for all $x, y\in M, \tau(x)=\tau(y)$, or, equivalently,

$$\gamma([\mathbf{U}(x+\mathbf{q})-\mathbf{U}(x)]\otimes\mathbf{t}, \mathbf{q})=\mathbf{0}$$

for all $x\in M, \mathbf{q}\in\mathbf{E}$.

2. Let $\mathbf{U}$ be a global observer and suppose the $\mathbf{U}$-lines are global. Demonstrate that $\mathbf{U}$ is quasi rotation free if and only if there is a map α: $I\times M\times\mathbf{E}\longrightarrow\mathbb{R}$ such that

(i)

$$t\cap C_{\mathbf{U}}(x+\mathbf{q})=t\cap C_{\mathbf{U}}(x)+\alpha(t, x, \mathbf{q})\mathbf{q}$$

for all $(t, x, \mathbf{q})\in I\times M\times\mathbf{E}$;

(ii)

$$\alpha(\tau(x), x, \mathbf{q})=1 \qquad (x\in M,\ \mathbf{q}\in\mathbf{E});$$

(ii)

$$\alpha(t, x, \mathbf{0})=1 \qquad (t\in I,\ x\in M).$$

3. Let $\mathbf{U}$ be a global observer and suppose the $\mathbf{U}$-lines are global. Prove that the following three statements are equivalent:

(i) $\mathbf{U}$ is rigid and quasi rotation free;
(ii) $\mathbf{U}(x+\mathbf{q})=\mathbf{U}(x) \quad (x\in M, \mathbf{q}\in\mathbf{E})$;
(iii) there is a smooth map $\mathbf{V}: I\longrightarrow V(1)$ such that

$$\mathbf{U}=\mathbf{V}\circ\tau.$$

(see Proposition 7.8.)

4. Since a global observer is a map $M\longrightarrow V(1)$ and both M and $V(1)$ are affine spaces, it makes sense to say that a global observer is *affine*. If $\mathbf{U}$ is an affine observer then $D\mathbf{U}: \mathbf{M}\longrightarrow \frac{\mathbf{E}}{\mathbf{I}}$ is a linear map. As a consequence of Exercise 1, $\mathbf{U}$ is rigid if and only if $(\mathbf{t}\otimes D\mathbf{U})|_{\mathbf{E}}$ (notation: IV. 1.9.) is antisymmetric with respect to γ for all $\mathbf{t}\in\mathbf{I}$.

We distinguish four cases for a rigid affine observer $\mathbf{U}$:

(i) the general case,

$$D\mathbf{U}|_{\mathbf{E}}\neq\mathbf{0}, \quad D\mathbf{U}_{\mathbf{I}}(\mathbf{u})\neq\mathbf{0} \quad \text{for all} \quad \mathbf{u}\in V(1);$$

(ii) the *uniformly rotating* observer,

$$D\mathbf{U}|_{\mathbf{E}}\neq\mathbf{0}, \quad D\mathbf{U}_{\mathbf{I}}(\mathbf{u})=\mathbf{0} \quad \text{for some} \quad \mathbf{u}\in V(1);$$

(iii) the *uniformly accelerating* observer,

$$\mathbf{DU}|_{\mathbf{E}}=\mathbf{0},\quad \mathbf{DU}_{\mathbf{I}}(\mathbf{u})\neq\mathbf{0}\quad \text{for all}\quad \mathbf{u}\in V(1);$$

(iv) the *inertial* observer,

$$\mathbf{DU}|_{\mathbf{E}}=\mathbf{0},\quad \mathbf{DU}_{\mathbf{I}}(\mathbf{u})=\mathbf{0}\quad \text{for some (hence for all)}\quad \mathbf{u}\in V(1),$$

in other words, $\mathbf{DU}=\mathbf{0}$.

5. Let $\mathbf{U}$ be a uniformly rotating observer (see the previous exercise). Prove the following assertions.

(i) The kernel of $\Omega:=\mathbf{DU}|_{\mathbf{E}}$ is one-dimensional, it is spanned, say, by $\mathbf{e}\neq\mathbf{0}$; if $\mathbf{u}\in V(1)$ for which $\mathbf{DU}_{\mathbf{I}}(\mathbf{u})=\mathbf{0}$ then

$$\operatorname{Ker}\mathbf{DU}=\{\lambda\mathbf{e}+\mathbf{u}\otimes\mathbf{t}\,\vdots\;\; \lambda\in\mathbb{R},\ \mathbf{t}\in\mathbf{I}\}.$$

(ii) $\operatorname{Ker}(\mathbf{DU}_{\mathbf{I}}\circ\mathbf{DU})=\operatorname{Ker}\mathbf{DU}$, hence the set $N:=\{x\in M\,\vdots\,\mathbf{DU}_{\mathbf{I}}(\mathbf{U}(x))=\mathbf{0}\}$ of non-accelerated points is a two dimensional affine subspace over $\operatorname{Ker}\mathbf{DU}$.

(iii) $\mathbf{U}$ restricted to N is a constant map having value in $\operatorname{Ker}\mathbf{DU}_{\mathbf{I}}$, i.e. there is an $\mathbf{u}_N\in V(1)$ such that $\mathbf{DU}_{\mathbf{I}}(\mathbf{u}_N)=\mathbf{0}$ and $\mathbf{U}(x)=\mathbf{u}_N$ for all $x\in N$; thus $\operatorname{Ker}\mathbf{DU}$ is spanned by $\mathbf{e}$ and $\mathbf{u}_N$ ⌈find the physical interpretation of $\mathbf{e}$ and $\mathbf{u}_N$⌋.

(iv) If o is an element of N then

$$\mathbf{U}(x)=\mathbf{u}_N+\mathbf{DU}(x-o)\qquad (x\in M)$$

and the acceleration field is characterized by

$$(\mathbf{DU}_{\mathbf{I}}\circ\mathbf{U})(x)=(\mathbf{DU}_{\mathbf{I}}\circ\mathbf{DU})(x-o)\qquad (x\in M);$$

for a given element x of M the element o of N can be chosen in such a way that $\tau(x)=\tau(o)$ and $x-o$ is orthogonal to $\mathbf{e}$.

(v) The $\mathbf{U}$-line passing through $x\in M$ is given by the global world line function

$$t\mapsto o+\mathbf{u}_N\otimes(t-\tau(x))+(\exp(t-\tau(x))\otimes\Omega)(x-o)$$

where o is an element of N such that $\tau(x)=\tau(o)$ and $x-o$ is orthogonal to e.

(vi) The set $\{C_{\mathbf{U}}(x)\,\vdots\,x\in N\}$ is a closed submanifold of $E_{\mathbf{U}}$ and it is an affine space over $\operatorname{Ker}\Omega$ with the addition

$$C_{\mathbf{U}}(x)+\mathbf{c}:=C_{\mathbf{U}}(x+\mathbf{c})\qquad (x\in N,\ \mathbf{c}\in\operatorname{Ker}\Omega)$$

⌈this straight line in the $\mathbf{U}$-space $E_{\mathbf{U}}$ can be interpreted as the axis of rotation⌋;

(vii) $\mathbf{U}$ is not rotation free.

(viii) Take arbitrarily $o\in N$, $\mathbf{d}\in\mathbf{D}^+$, $\mathbf{e}\in\mathbf{I}^+$ and construct the Galilean reference system whose origin, velocity, space unit and time basis are o, $\mathbf{u}_N$, $\mathbf{d}$ and $\mathbf{e}$, respectively, and the space basis is chosen in such a way that $\mathbf{e}_3$ spans $\operatorname{Ker}\Omega$, more closely, $\mathbf{e}_3:=\mathbf{d}\otimes j(\mathbf{e}\otimes\Omega)$ (see V. 3.6.).

Show that in this reference system **U** has the form

$$\mathbb{R}^4 \longrightarrow \mathbb{R}^4, \quad (x^0, x^1, x^2, x^3) \mapsto (1, -\omega x^2, \omega x^1, 0)$$

where ω is the norm of $\mathbf{e} \otimes \Omega$ (see V. 3.3.). The **U**-line passing through $o + \sum_{k=1}^{3} x^k \mathbf{e}_k$ becomes:

$$\{(t, x^1 \cos \omega t - x^2 \sin \omega t, x^1 \sin \omega t + x^2 \cos \omega t, x^3) : \ t \in \mathbb{R}\}.$$

6. Let **U** be a uniformly accelerating observer (see Exercise 4.). Prove the following assertions.

(i) The acceleration field of **U** is a non-zero constant, i.e. there exists an $\mathbf{0} \neq \mathbf{a} \in \frac{\mathbf{E}}{\mathbf{I} \otimes \mathbf{I}}$ such that

$$\mathbf{D}\mathbf{U}_{\mathbf{I}}(\mathbf{U}(x)) = \mathbf{a} \qquad (x \in M).$$

(ii) There is a unique affine map $\mathbf{V} : I \longrightarrow V(1)$ such that $\mathbf{U} = \mathbf{V} \circ \tau$ and $\mathbf{D}\mathbf{V} = \mathbf{a}$ ($\mathbf{D}\mathbf{V}$ is a linear map $\mathbf{I} \longrightarrow \frac{\mathbf{E}}{\mathbf{I}}$, hence it can be viewed as an element of $\frac{\mathbf{E}}{\mathbf{I} \otimes \mathbf{I}}$, see IV. 2.7. (vi) and IV. 2.7. (iii)) (cf. Exercise 3.).

(iii) For all $x, y \in M$

$$\mathbf{U}(y) = \mathbf{U}(x) + \mathbf{a} \otimes \boldsymbol{\tau}(y - x).$$

(iv) The **U**-line passing through $x \in M$ is given by the global world line function

$$t \mapsto x + \mathbf{U}(x) \otimes (t - \tau(x)) + \frac{1}{2} \mathbf{a} \otimes (t - \tau(x)) \otimes (t - \tau(x)).$$

(v) **U** is rotation free.

(vi) Take arbitrarily $o \in M, \mathbf{d} \in \mathbf{D}^+, \mathbf{e} \in \mathbf{I}^+$ and construct the Galilean reference system whose origin, velocity, space unit and time basis are $o, \mathbf{U}(o), \mathbf{d}$ and $\mathbf{e}$, respectively, and the space basis is chosen in such a way that $\mathbf{e}_1 := \frac{1}{a}(\mathbf{a} \otimes \mathbf{e} \otimes \mathbf{e})$ where

$$a := \frac{|\mathbf{a} \otimes \mathbf{e} \otimes \mathbf{e}|}{\mathbf{d}}$$

is the magnitude of **a** measured in the time unit and space unit of the Galilean reference system.

Show that in this reference system **U** has the form

$$\mathbb{R}^4 \longrightarrow \mathbb{R}^4, \quad (x^0, x^1, x^2, x^3) \mapsto (1, a x^0, 0, 0)$$

and the **U**-line passing through the point $o + (x^0(\mathbf{U}(o) \otimes \mathbf{e}) + \sum_{k=1}^{3} x^k \mathbf{e}_k)$ becomes

$$\{(t, x^1 + a x^0 (t - x^0) + \frac{1}{2} a (t - x^0)^2, x^2, x^3) : \ t \in \mathbb{R}\}.$$

In particular, if $x^0=0$, it is

$$\{(t, x^1+\frac{1}{2}at^2, x^2, x^3)\colon\ t\in\mathbb{R}\}.$$

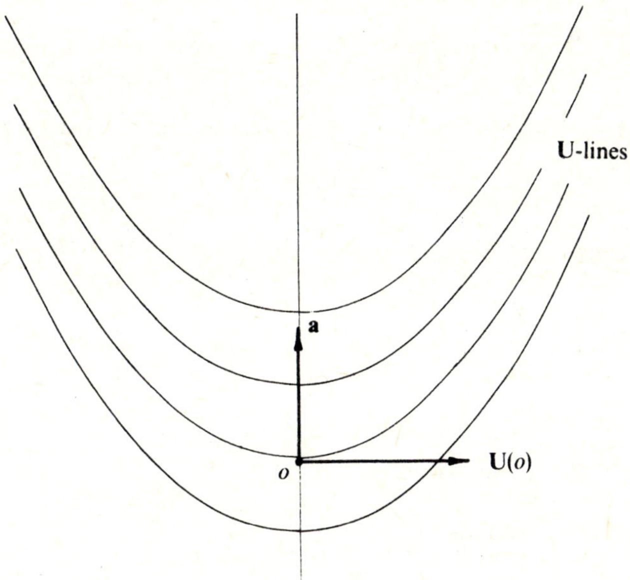

7. Let **U** be an affine observer. Set $D_1\mathbf{U}:=\mathbf{U}$ and

$$D_n\mathbf{U}:=D\mathbf{U}_{(\overset{n-1}{\otimes}\mathbf{I})}\circ D\mathbf{U}_{(\overset{n-2}{\otimes}\mathbf{I})}\circ\ldots\circ D\mathbf{U}_{\mathbf{I}}\circ\mathbf{U}:M\longrightarrow\frac{\mathbf{E}}{\overset{n}{\otimes}\mathbf{I}}.$$

Prove that

(i) for all $x\in M$, the function series

$$\mathbf{t}\mapsto\sum_{n\in\mathbb{N}}\frac{1}{n!}(\overset{n}{\otimes}\mathbf{t})\otimes D_n\mathbf{U}(x)$$

is locally uniformly convergent on **I**;

(ii) the function $I\longrightarrow M$,

$$t\mapsto x+\sum_{n\in\mathbb{N}}\frac{1}{n!}\overset{n}{\otimes}(t-\tau(x))\otimes D_n\mathbf{U}(x)$$

gives the **U**-line passing through $x\in M$; as a consequence, all **U**-lines are global,

(iii)

$$C_{\mathbf{U}}(x+\mathbf{q})=C_{\mathbf{U}}(x)+\mathbf{q}\qquad(x\in M,\ \mathbf{q}\in\mathbf{E})$$

if and only if $\mathbf{q}\in\operatorname{Ker}D\mathbf{U}|_{\mathbf{E}}$.

8. Let $(\mathbf{U}, o)$ be a global rigid observer with origin. Suppose the $\mathbf{U}$-lines are global and use the notations introduced in 4.4. and 4.5. Show that for all $t \in I$

(i) R_t is an element of $\mathrm{SO}(\gamma)$;

(ii) $\Omega(t)$ is a linear mapping antisymmetric with respect to γ, i.e. $\gamma(\mathbf{t}\otimes\Omega(t)\mathbf{q}, \mathbf{q}) = -\gamma(\mathbf{q}, \mathbf{t}\otimes\Omega(t)\mathbf{q})$ for all $\mathbf{t} \in \mathbf{I}$ and $\mathbf{q} \in \mathbf{E}$;

(iii) $\mathbf{U}$ restricted to t is an affine map:

$$\mathbf{U}(x+\mathbf{q}) - \mathbf{U}(x) = \Omega(\tau(x))\mathbf{q} \qquad (x \in M,\ \mathbf{q} \in \mathbf{E}).$$

⌈$\Omega(t)$ is known as the *angular velocity* at the time point t; usually it is identified with an element of $\dfrac{\mathbf{E}}{\mathbf{D}\otimes\mathbf{I}}$ (see V. 3.7.).⌋

9. Let $\mathbf{u}$ and $\mathbf{u}'$ be distinct elements of $V(1)$. Applying the notations of 1.11. and 4.2. put

$$\cos\Theta(\mathbf{w}) := \frac{[\mathbf{v}_{\mathbf{wu}}|\mathbf{v}_{\mathbf{u}'\mathbf{u}}]}{|\mathbf{v}_{\mathbf{wu}}|\otimes|\mathbf{v}_{\mathbf{u}'\mathbf{u}}|},$$

$$(\mathbf{w} \in V(1),\ \mathbf{w} \neq \mathbf{u},\ \mathbf{w} \neq \mathbf{u}')$$

$$-\cos\Theta'(\mathbf{w}) := \frac{[\mathbf{v}_{\mathbf{wu}'}|\mathbf{v}_{\mathbf{uu}'}]}{|\mathbf{v}_{\mathbf{wu}'}|\otimes|\mathbf{v}_{\mathbf{uu}'}|}$$

($\Theta(\mathbf{w})$ is the angle between the relative velocities $\mathbf{w}_{\mathbf{wu}}$ and $\mathbf{v}_{\mathbf{u}'\mathbf{u}}$, $\Theta'(\mathbf{w})$ is the angle between the relative velocities $\mathbf{v}_{\mathbf{wu}'}$ and $-\mathbf{v}_{\mathbf{uu}'} = \mathbf{v}_{\mathbf{u}'\mathbf{u}}$:

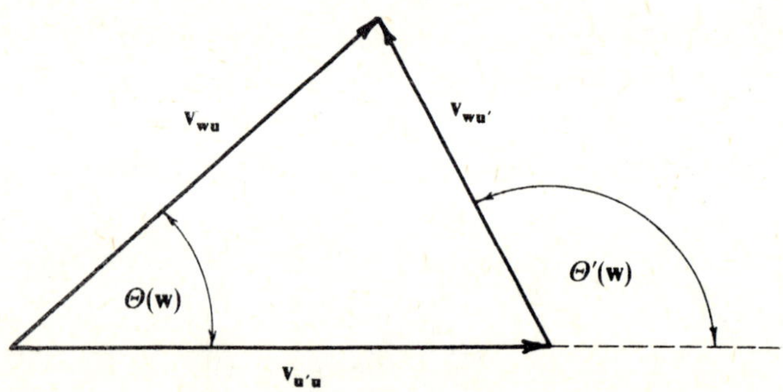

Show that

$$\cos\Theta(\mathbf{w}) = \frac{|\mathbf{v}_{\mathbf{wu}'}|}{|\mathbf{v}_{\mathbf{wu}}|}\cos\Theta'(\mathbf{w}) + \frac{|\mathbf{v}_{\mathbf{u}'\mathbf{u}}|}{|\mathbf{v}_{\mathbf{wu}}|}.$$

As a consequence,

$$\lim_{\mathbf{w}\to\infty}\cos\Theta(\mathbf{w}) = \lim_{\mathbf{w}\to\infty}\cos\Theta'(\mathbf{w}).$$

⌈Two observers with different velocities see a physical object with "infinitely big" relative velocity (a light beam) under the same angle: there is no "aberration" (cf. II. Exercise 15.).⌋

10. Consider the nonrelativistic space-time model $(\mathbb{R}\times\mathbb{R}^3, \mathbb{R}, \mathrm{pr}_{\mathbb{R}}, \mathbb{R}, \langle,\rangle)$ and the following global observer on it:

$$\mathbf{U}(x^0, x^1, x^2, x^3) := (1, -(x^1)^2, 0, 0).$$

Show that the **U**-line passing through (a^0, a^1, a^2, a^3) is

$$\{(t, 0, a^2, a^3) \vdots \quad t\in\mathbb{R}\} \qquad \text{if } a^1=0,$$

$$\left\{\left(t, \frac{1}{t-a^0+\dfrac{1}{a^1}}, a^2, a^3\right) \vdots \quad t>a^0-\frac{1}{a^1}\right\} \text{ if } a^1<0.$$

$$\left\{\left(t, \frac{1}{t-a^0+\dfrac{1}{a^1}}, a^2, a^3\right) \vdots \quad t<a^0-\frac{1}{a^1}\right\} \text{ if } a^1<0.$$

As a consequence, most of the **U**-lines are not global.

11. Derive the splitting formula for 2-tensors and 2-cotensors which are not antisymmetric (see 5.11.) and establish the transformation law of components connected with the variation of elements of $V(1)$.

12. Derive the splitting formula for "mixed 2-tensors", i.e. for elements of $\mathbf{M}\otimes\mathbf{M}^*$ and $\mathbf{M}^*\otimes\mathbf{M}$. Use that $A\in\mathbf{M}\otimes\mathbf{M}^*$ can be considered as a linear map $\mathbf{M}\longrightarrow\mathbf{M}$, thus its splitting corresponding to $\mathbf{u}\in V(1)$ is

$$\mathbf{s}_{\mathbf{u}}\circ A\circ\mathbf{s}_{\mathbf{u}}^{-1} = \begin{pmatrix} (\boldsymbol{\tau}\circ A)_{\mathbf{I}}\mathbf{u} & \boldsymbol{\tau}\circ A|_{\mathbf{E}} \\ A_{\mathbf{I}}\mathbf{u}-((\boldsymbol{\tau}\circ A)_{\mathbf{I}}\mathbf{u})\mathbf{u} & A|_{\mathbf{E}}-\mathbf{u}\otimes(\boldsymbol{\tau}\circ A|_{\mathbf{E}}) \end{pmatrix}.$$

Observe that this formula is very simple if $\boldsymbol{\tau}\circ A=\mathbf{0}$; this case occurs in Theorem 7.21.

Establish the transformation law of components connected with the variation of elements of $V(1)$ and compare the result with that of the previous exercise.

13. Find a necessary and sufficient condition that two affine reference systems be equivalent.

14. Let $(\mathbf{U}, o, K)$ be a natural reference system with origin such that

(i) **U** is a uniformly accelerating observer with acceleration $\mathbf{a}\in\dfrac{\mathbf{E}}{\mathbf{I}\otimes\mathbf{I}}$ (see Exercise 11.);

(ii) there is given an $\mathbf{e}\in\mathbf{I}$,

$$T(\mathbf{t}) := \frac{\mathbf{t}}{\mathbf{e}} \qquad (\mathbf{t}\in\mathbf{I})$$

and $S:\mathbf{E}\longrightarrow\mathbb{R}^3$ is a linear bijection for which

$$S(\mathbf{a}\otimes\mathbf{e}\otimes\mathbf{e}) = (1, 0, 0)$$

holds, and $K=(T\times S)\circ S_o^{\mathbf{U}}\circ S_{\mathbf{U}}$.

Find a necessary and sufficient condition that a reference system with origin $(\mathbf{U}', o', K')$ be equivalent to $(\mathbf{U}, o, K)$.

15. Pose and solve a problem similar to the previous one for reference systems with origin corresponding to uniformly rotating observers.

16. Let $A: \mathbf{M} \longrightarrow \mathbf{M}$ be a linear map such that $\boldsymbol{\tau} \circ A = \mathbf{0}$ (i.e. A can be viewed as a map $\mathbf{M} \longrightarrow \mathbf{E}$). Prove that if $A|_{\mathbf{E}} = 0$ then there is a unique $\mathbf{w} \in \frac{\mathbf{E}}{\mathbf{I}}$ such that $A\mathbf{x} = \mathbf{w} \otimes \boldsymbol{\tau}(\mathbf{x})$ $(\mathbf{x} \in \mathbf{M})$.

17. Let F be a Noether transformation. Show that

$$\text{if ar } F = 1 \quad \text{then} \quad F[T^{\rightarrow}] = T^{\rightarrow}, \quad F[T^{\leftarrow}] = T^{\leftarrow};$$

$$\text{if ar } F = -1 \quad \text{then} \quad F[T^{\rightarrow}] = T^{\leftarrow}, \quad F[T^{\leftarrow}] = T^{\rightarrow}.$$

18. Show that if the Galilean transformation L is in Tn(**E**) (see 7.24. (iv)) then $z_{\mathbf{c},o}(L)$ is the same for all $(\mathbf{c}, o) \in M \times V(1)$ (see 7.20.).

On the contrary, if L is in Tn(**M**) but is not in Tn(**E**), then $z_{\mathbf{c},o}(L)$ depends on **c** and o.

19. Find $m^{-1}_{\mathbf{cc}'oo'} T m_{\mathbf{cc}'oo'}$ and $m^{-1}_{\mathbf{cc}'oo'} S m_{\mathbf{cc}'oo'}$ (see 7.21.) for the split Galilean transformations T and S introduced in 7.17.

20. Find the Galilean transformations that are images by $z^{-1}_{\mathbf{c},o}$ of the split Galilean transformations having the forms listed in 7.23. (use Theorem 7.20.).

21. Prove that the derived Lie algebra of the Galilean group, i.e. $[\mathbf{L}(\mathscr{G}), \mathbf{L}(\mathscr{G})]$ equals the Lie algebra of the instantaneous Galilean group which is $\{X: M \longrightarrow \mathbf{M} \,\vdots\, X \text{ is affine}, \boldsymbol{\tau} \circ X = \mathbf{0}, \mathrm{D}X|_{\mathbf{E}} \in \mathbf{A}(\gamma)\}$.

22. Let F be a Noether transformation. Then F^* is a linear bijection of $\mathbf{M}^*$. Prove that

$$\mathbf{i}^* \circ F^* = (F|_{\mathbf{E}})^* \circ \mathbf{i}^*.$$

23. Using the previous result and Definition 2.3. show that $\mathbf{k} \in \mathbf{M}^*$ is in $\mathbf{I}^*$ (i.e. $\mathbf{i}^*\mathbf{k} = 0$) if and only if $F^*\mathbf{k} = (\text{ar } F)\mathbf{k}$ for all Noether transformations F.

24. Using 7.24. (iv) demonstrate that Noether transformations can be replaced with special vectorial Galilean transformations in the previous assertion.

25. Verify that

$$(\mathbf{t}', \mathbf{q}', \mathbf{v}', R')T(\alpha')(\mathbf{t}, \mathbf{q}, \mathbf{v}, R)T(\alpha) =$$

$$= (\mathbf{t}' + \alpha'\mathbf{t}, \mathbf{q}' + R'\mathbf{q} + \alpha'\mathbf{v}' \otimes \mathbf{t}, \mathbf{v}' + \alpha' R'_{\mathbf{I}}\mathbf{v}, R'R)T(\alpha'\alpha)$$

and

$$[(\mathbf{t}, \mathbf{q}, \mathbf{v}, R)T(\alpha)]^{-1} = (-\alpha\mathbf{t}, -R^{-1}(\mathbf{q} - \mathbf{v} \otimes \mathbf{t}), -\alpha R_{\mathbf{I}}^{-1}\mathbf{v}, R^{-1})T(\alpha)$$

for the elements of the split Galilean group.

26. Let $(\mathbf{t}, \mathbf{q}, \mathbf{v}, B)$ be an element of the Lie algebra of the split Galilean group. Accept the notation $\mathbf{q}_B$, $\mathbf{v}_B$ and $\mathbf{q}_B^{\perp}$, $\mathbf{v}_B^{\perp}$ for the components of $\mathbf{q}$ and $\mathbf{v}$ that are parallel to Ker B and orthogonal to Ker B, respectively. Furthermore let B^{-1} denote the inverse of the restriction of B to the orthocomplement of Ker B. Prove that

$$\exp(\mathbf{t}, \mathbf{q}, \mathbf{v}, B) =$$

$$= \left(\mathbf{t}, \mathbf{q}_B - \mathbf{v}_B^{\perp} \otimes \mathbf{t} + (e^B - \mathrm{id}_{\mathbf{E}})B^{-1}(\mathbf{q}_B^{\perp} + B^{-1}(\mathbf{v}_B^{\perp} \otimes \mathbf{t})),\ \mathbf{v}_B + (e^B - \mathrm{id}_{\mathbf{E}})B^{-1}\mathbf{v}_B^{\perp}, e^B\right).$$

II. SPECIAL RELATIVISTIC SPACE-TIME MODEL

1. Fundamentals

Necessary mathematical tools:

Tensor products and quotients (Chapter IV.)
Affine spaces and affine maps (Section VI. 1.)
Minkowskian vector spaces (Section V. 4.)

Proposition 1.8. involves some notions from the theory of smooth manifolds but it does not touch the essence of this chapter.

1.1. ⌈Nonrelativistic space-time model was constructed on the following basic assumptions corresponding to our simplest experiences:

(i) space-time has an affine structure,

(ii) absolute time exists and has an affine structure,

(iii) there is a Euclidean structure on the set of simultaneous space-time points.

Some of the well known physical insufficiencies of the nonrelativistic space-time model can be well shown in our framework:

(a) it allows "whatever big" relative velocities, more precisely, the relative velocities form a Euclidean vector space $\left(\frac{\mathbf{E}}{\mathbf{I}}\text{ in our notation, see I. 2.4.}\right)$; experiments indicate that the magnitude of relative velocities cannot exceed a given value;

(b) the transformation law of physical quantities modelled by cotensor fields—electromagnetic field, for instance—under change of observers does not correspond to the formulae of I. 5.12. or I. 5.13.

The basic experimental result requiring a new space-time model is that light speed in vacuum is the same for all inertial observers (Michelson–Morley experiment). This fact obliges us to reject the hypothesis of absolute time and suggests the assumptions—we do not detail how—that

(i) space-time has an affine structure,

(ii) there is a Minkowskian structure on space-time.⌋

1.2. Definition. A *special relativistic space-time model* is a four dimensional, oriented and arrow oriented Minkowskian affine space, i.e. a triplet $(M, \mathbf{I}, g)$ where

M is an oriented four dimensional real affine space over the vector space $\mathbf{M}$,

$\mathbf{I}$ is an oriented one dimensional real vector space,
$g: \mathbf{M} \times \mathbf{M} \longrightarrow \mathbf{I} \otimes \mathbf{I}$ is a Minkowskian bilinear map endowed with an arrow orientation (see Section V. 4.).

We assume in the sequel that a special relativistic space-time model $(M, \mathbf{I}, g)$ *is given.*

1.3. M is called *space-time* and the elements of $\mathbf{I}$ are called *space-time lengths.* Note the very important fact that $\mathbf{M}^*$ can be identified with $\dfrac{\mathbf{M}}{\mathbf{I} \otimes \mathbf{I}}$ (see V. 2.4.). Furthermore, $\mathbf{I}^*$ is oriented as the dual of an oriented vector space (see IV. 3.1.). We shall use also the notations

$$S := \{\mathbf{x} \in \mathbf{M} \,\vdots\, g(\mathbf{x}, \mathbf{x}) > \mathbf{0}\},$$

$$T := \{\mathbf{x} \in \mathbf{M} \,\vdots\, g(\mathbf{x}, \mathbf{x}) < \mathbf{0}\},$$

$$L := \{\mathbf{x} \in \mathbf{M} \,\vdots\, g(\mathbf{x}, \mathbf{x}) = \mathbf{0},\ \mathbf{x} \neq \mathbf{0}\}.$$

The distinguished arrow class of T is denoted by $T^{\rightarrow}$ and the other one by $T^{\leftarrow}$.

Since the topological closure of T is $T \cup L \cup \{\mathbf{0}\}$, L can be written in the form $L^{\rightarrow} \cup L^{\leftarrow}$ where $L^{\rightarrow}$ is the boundary of $T^{\rightarrow}$, deprived of zero, and $L^{\leftarrow}$ is the boundary of $T^{\leftarrow}$, deprived of zero.

1.4. Definition. Let $\mathbf{A}$ be a vector space constructed of $\mathbf{I}$ and $\mathbf{I}^*$ by tensor products and tensor quotients. Then the elements of

$\mathbf{A}$	are called	*scalars of type* $\mathbf{A}$,
$\mathbf{A} \otimes \mathbf{M}$	are called	*vectors of type* $\mathbf{A}$.
$\dfrac{\mathbf{M}}{\mathbf{A}}$	are called	*vectors of cotype* $\mathbf{A}$,
$\mathbf{A} \otimes \mathbf{M}^* = \left(\dfrac{\mathbf{M}}{\mathbf{A}}\right)^*$	are called	*covectors of type* $\mathbf{A}$.

Tensors, cotensors, tensor fields and cotensor fields of type $\mathbf{A}$ and of cotype $\mathbf{A}$ are defined as in VI. 3.1.

In the case $\mathbf{A} = \mathbb{R}$ we omit the term "of type $\mathbb{R}$" and "of cotype $\mathbb{R}$".

The elements of S, T and L are called *spacelike*, *timelike* and *lightlike* vectors, respectively. The zero vector is considered spacelike and timelike, too. Vectors in $T^{\rightarrow}$, $L^{\rightarrow}$ and $T^{\leftarrow}$, $L^{\leftarrow}$ are called *positive* and *negative*, respectively.

Remark. Since $\mathbf{M}^*$ can be identified with $\dfrac{\mathbf{M}}{\mathbf{I} \otimes \mathbf{I}}$, a covector of type $\mathbf{A}$ can be considered as a vector of type $\dfrac{\mathbf{A}}{\mathbf{I} \otimes \mathbf{I}}$.

1.5. The bilinear form on $\frac{\mathbf{M}}{\mathbf{I}} \times \frac{\mathbf{M}}{\mathbf{I}}$ arising from g (see Remark V. 2.1. (ii)) is denoted by $(\;|\;)$, and $(\;\|\;) := -(\;|\;)$.

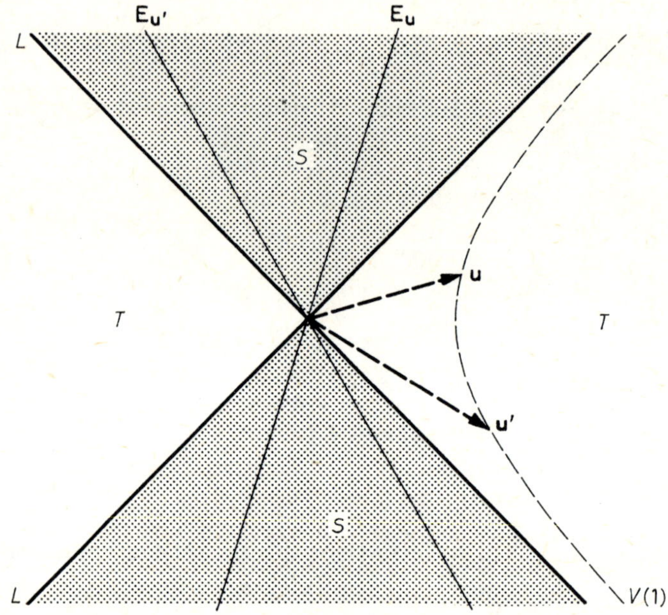

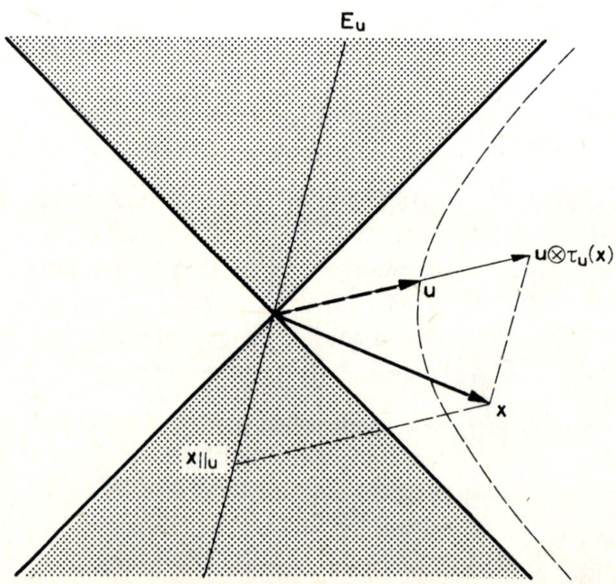

The bilinear map $\frac{\mathbf{M}}{\mathbf{I}} \times \mathbf{M} \longrightarrow \mathbf{I}$ arising from g is denoted by $[\ |\ )$, i.e.

$$m \otimes [\mathbf{u}|\mathbf{x}) := (m \otimes \mathbf{u}|\mathbf{x}) \qquad \left(m \in \mathbf{I}^*,\ \mathbf{u} \in \frac{\mathbf{M}}{\mathbf{I}},\ \mathbf{x} \in \mathbf{M}\right).$$

We introduce the notations

$$V(0) := \left\{\mathbf{w} \in \frac{\mathbf{M}}{\mathbf{I}} : (\mathbf{w} \nmid \mathbf{w}) = 0,\ \mathbf{w} \otimes \mathbf{I}^+ \subset L^{\rightarrow}\right\},$$

$$V(1) := \left\{\mathbf{u} \in \frac{\mathbf{M}}{\mathbf{I}} : (\mathbf{u} \nmid \mathbf{u}) = -1,\ \mathbf{u} \otimes \mathbf{I}^+ \subset T^{\rightarrow}\right\},$$

and for $\mathbf{u} \in V(1)$ we put

$$\mathbf{E}_\mathbf{u} := \{\mathbf{x} \in \mathbf{M} : [\mathbf{u}|\mathbf{x}) = \mathbf{0}\}$$

$$\gamma_\mathbf{u} := g|_{\mathbf{E}_\mathbf{u} \times \mathbf{E}_\mathbf{u}}$$

$$\mathbf{i}_\mathbf{u} : \mathbf{E}_\mathbf{u} \longrightarrow \mathbf{M}, \qquad \mathbf{x} \mapsto \mathbf{x}$$

$$\tau_\mathbf{u} : \mathbf{M} \longrightarrow \mathbf{I}, \qquad \mathbf{x} \mapsto -[\mathbf{u}|\mathbf{x})$$

$$\mathbf{x}\|_\mathbf{u} := \mathbf{x} - \mathbf{u} \otimes \tau_\mathbf{u}(\mathbf{x}) = (\mathrm{id}_\mathbf{M} + \mathbf{u} \otimes \mathbf{u})\mathbf{x} \qquad (\mathbf{x} \in \mathbf{M})$$

where $\mathbf{u} \otimes \mathbf{u} \in \frac{\mathbf{M}}{\mathbf{I}} \otimes \frac{\mathbf{M}}{\mathbf{I}} = \mathbf{M} \otimes \frac{\mathbf{M}}{\mathbf{I} \otimes \mathbf{I}} = \mathbf{M} \otimes \mathbf{M}^*$ is considered as the linear map $\mathbf{M} \longrightarrow \mathbf{M}$, $\mathbf{x} \mapsto \mathbf{u} \otimes [\mathbf{u}|\mathbf{x})$ (see IV. 1.5. (iv)).

1.6. Proposition. Let $\mathbf{u} \in V(1)$. Then $(\mathbf{E}_\mathbf{u}, \mathbf{I}, \gamma_\mathbf{u})$ is a three dimensional Euclidean vector space. Furthermore, $\mathbf{E}_\mathbf{u} = \mathbf{E}_{\mathbf{u}'}$, if and only if $\mathbf{u} = \mathbf{u}'$ ($\mathbf{u}' \in V(1)$). ∎

As a consequence, $\mathbf{E}_\mathbf{u}^*$ can be identified with $\frac{\mathbf{E}_\mathbf{u}}{\mathbf{I} \otimes \mathbf{I}}$.

The Euclidean bilinear form on $\frac{\mathbf{E}_\mathbf{u}}{\mathbf{I}} \times \frac{\mathbf{E}_\mathbf{u}}{\mathbf{I}}$ arising from $\gamma_\mathbf{u}$ is the restriction of $(\ \nmid\ )$. We write

$$|\mathbf{v}| := \sqrt{(\mathbf{v} \nmid \mathbf{v})} \qquad \left(\mathbf{v} \in \frac{\mathbf{E}_\mathbf{u}}{\mathbf{I}}\right)$$

for all $\mathbf{u} \in V(1)$, and

$$B_\mathbf{u} := \left\{\mathbf{v} \in \frac{\mathbf{E}_\mathbf{u}}{\mathbf{I}} : |\mathbf{v}| < 1\right\}.$$

1.7. Definition. Let $\mathbf{u} \in V(1)$. A basis $\mathbf{e}_1, \mathbf{e}_2, \mathbf{e}_3$ in $\mathbf{E}_\mathbf{u}$ is called *positively oriented* if for some (hence for all) non-zero $\mathbf{t} \in \mathbf{I}^+$ the basis $\mathbf{u} \otimes \mathbf{t}, \mathbf{e}_1, \mathbf{e}_2, \mathbf{e}_3$ in $\mathbf{M}$ is positively oriented.

1.8. Proposition. $V(1)$ is a three dimensional submanifold of $\frac{\mathbf{M}}{\mathbf{I}}$. The tangent space of $V(1)$ at $\mathbf{u}$ is $\frac{\mathbf{E_u}}{\mathbf{I}}$. $(V(1), \mathbb{R}, (\ |\ )|_{V(1)})$ is a Riemannian manifold (notation: VII. 4.8. (iii)).

Remark. Pick out arbitrarily a $\mathbf{u} \in V(1)$. Then $V(1) \longrightarrow \frac{\mathbf{E_u}}{\mathbf{I}}$, $\mathbf{u}' \mapsto \mathbf{u}' - (\mathbf{u}'|\mathbf{u})\mathbf{u}$ is a diffeomorphism. Accordingly, $V(1)$ is orientable and it is considered to be oriented in such a way that the above diffeomorphism be orientation preserving. Then the orientation of $V(1)$ determines for all $\mathbf{u} \in V(1)$ the orientation of $\frac{\mathbf{E_u}}{\mathbf{I}}$ (see Remark VII. 2.8. (ii)) that is given by Definitions 1.7. and IV. 3.4.

1.9. The derivative of the above given diffeomorphism $V(1) \longrightarrow \frac{\mathbf{E_u}}{\mathbf{I}}$ establishes a linear bijection $\frac{\mathbf{E_{u'}}}{\mathbf{I}} \longrightarrow \frac{\mathbf{E_u}}{\mathbf{I}}$ and thus a linear bijection $\mathbf{E_{u'}} \longrightarrow \mathbf{E_u}$ for all $\mathbf{u}'$, $\mathbf{u} \in V(1)$. Unfortunately, this linear bijection is not $\gamma_{\mathbf{u}'}-\gamma_{\mathbf{u}}$-orthogonal. Now we give a map that transforms $\mathbf{E_{u'}}$ onto $\mathbf{E_u}$ preserving the Euclidean structure.

Definition. Let $\mathbf{u}, \mathbf{u}' \in V(1)$. Then

$$F(\mathbf{u}, \mathbf{u}') := \mathrm{id}_{\mathbf{M}} + \frac{(\mathbf{u}+\mathbf{u}')\otimes(\mathbf{u}+\mathbf{u}')}{1+(\mathbf{u}|\mathbf{u}')} - 2\mathbf{u}\otimes\mathbf{u}' \ :\ \mathbf{M} \longrightarrow \mathbf{M}$$

is called the *Lorentz boost* from $\mathbf{u}'$ to $\mathbf{u}$ (here $\mathbf{u}\otimes\mathbf{u}' \in \frac{\mathbf{M}}{\mathbf{I}} \otimes \frac{\mathbf{M}}{\mathbf{I}} = \mathbf{M} \otimes \frac{\mathbf{M}}{\mathbf{I}\otimes\mathbf{I}} = \mathbf{M}\otimes\mathbf{M}^*$ is considered as the linear map $\mathbf{M} \longrightarrow \mathbf{M}$, $\mathbf{x} \mapsto \mathbf{u}\otimes[\mathbf{u}'|\mathbf{x})$ (see IV. 1.5. (iv)).

Proposition. (i) $F(\mathbf{u}, \mathbf{u}')$ is a g-orthogonal linear map,

(ii) $F(\mathbf{u}, \mathbf{u}')_{\mathbf{I}}\mathbf{u}' = \mathbf{u}$, $\quad F(\mathbf{u}, \mathbf{u}')[\mathbf{E_{u'}}] = \mathbf{E_u}$,

(iii) $F(\mathbf{u}, \mathbf{u}')$ and $F(\mathbf{u}, \mathbf{u}')|_{\mathbf{E_{u'}}}$ are orientation preserving,

(iv) if $\mathbf{u} \neq \mathbf{u}'$ then the plane spanned by $\mathbf{u}\otimes\mathbf{I}$ and $\mathbf{u}'\otimes\mathbf{I}$ is invariant under $F(\mathbf{u}, \mathbf{u}')$ and the restriction of $F(\mathbf{u}, \mathbf{u}')$ to the two dimensional subspace $\mathbf{E_u} \cap \mathbf{E_{u'}}$ is the identity,

(v) $F(\mathbf{u}, \mathbf{u}) = \mathrm{id}_{\mathbf{M}}$, $\qquad F(\mathbf{u}, \mathbf{u}')^{-1} = F(\mathbf{u}', \mathbf{u})$.

Remarks. (i) The listed properties allow us to say that $F(\mathbf{u}, \mathbf{u}')$ maps $\mathbf{E_{u'}}$ onto $\mathbf{E_u}$ "without rotation and reflection" (see also Exercise 2).

(ii) It is useful to know that

$$F(\mathbf{u}, \mathbf{u}')|_{\mathbf{E_{u'}}} = \left(\mathrm{id}_{\mathbf{M}} + \frac{(\mathbf{u}+\mathbf{u}')\otimes\mathbf{u}}{1+(\mathbf{u}|\mathbf{u}')}\right)\Bigg|_{\mathbf{E_{u'}}}.$$

(iii) A Lorentz boost is a proper Lorentz transformation (see 7.4.).

1.10. Proposition. Let $\mathbf{c} \in V(1)$. Let us orient $\mathbf{E_c}$ according to Definition 1.7. and attach to $\mathbf{I} \times \mathbf{E_c}$ the orientation determined by the orientations of $\mathbf{I}$ and $\mathbf{E_c}$. Then the map

$$\mathbf{s_c} : \mathbf{M} \longrightarrow \mathbf{I} \times \mathbf{E_c}, \qquad \mathbf{x} \mapsto (\boldsymbol{\tau}_\mathbf{c}(\mathbf{x}), \mathbf{x}\|_\mathbf{c})$$

is an orientation preserving linear bijection whose inverse is

$$(\mathbf{t}, \mathbf{q}) \mapsto \mathbf{c} \otimes \mathbf{t} + \mathbf{q} \qquad ((\mathbf{t}, \mathbf{q}) \in \mathbf{I} \times \mathbf{E_c}).$$

Furthermore, if $o \in M$, the map

$$s_{\mathbf{c},o} := \mathbf{s_c} \circ O_o : M \longrightarrow \mathbf{I} \times \mathbf{E_c}, \qquad x \mapsto (\boldsymbol{\tau}_\mathbf{c}(x-o), (x-o)\|_\mathbf{c})$$

is an orientation preserving affine bijection whose inverse is

$$(\mathbf{t}, \mathbf{q}) \mapsto o + \mathbf{c} \otimes \mathbf{t} + \mathbf{q} \qquad ((\mathbf{t}, \mathbf{q}) \in \mathbf{I} \times \mathbf{E_c})$$

and the derivative of $s_{\mathbf{c},o}$ equals $\mathbf{s_c}$.

Remark. Note the similarity of Proposition I. 1.13. and II. 1.10. However, there is an important difference between them: in the nonrelativistic space-time model the range of $\mathbf{s_c}$ is the same for all $\mathbf{c}$ whereas here it depends on $\mathbf{c}$.

1.11. The question posed for nonrelativistic space-time models in I. 1.10. arises for special relativistic space-time models as well. That is why we need the following definition.

Definition. Two special relativistic space-time models $(M, \mathbf{I}, g)$ and $(M', \mathbf{I}', g')$ are *isomorphic* if there are

(i) an orientation and arrow preserving affine bijection $F : M \longrightarrow M'$

(ii) an orientation preserving linear bijection $Z : \mathbf{I} \longrightarrow \mathbf{I}'$

in such a way that

$$g' \circ (DF \times DF) = (Z \otimes Z) \circ g$$

(i.e. DF is a $g-g'$-orthogonal map).

The pair (F, Z) is an *isomorphism* between the two special relativistic space-time models. If the two models coincide, an isomorphism is called an *automorphism*. An automorphism of the form $(F, \mathrm{id}_\mathbf{I})$ is called *strict*.

1.12. Proposition. Let $(M, \mathbf{I}, g)$ be a special relativistic space-time model. Then for all $\mathbf{u} \in V(1)$ the triplet $(\mathbf{I} \times \mathbf{E_u}, \mathbf{I}, g_\mathbf{u})$ is a special relativistic space-time model where

$$g_\mathbf{u}((\mathbf{t}, \mathbf{q}), (\mathbf{t}', \mathbf{q}')) := -\mathbf{t} \otimes \mathbf{t}' + \gamma_\mathbf{u}(\mathbf{q}, \mathbf{q}')$$

$$(\mathbf{t}, \mathbf{t}' \in \mathbf{I}, \mathbf{q}, \mathbf{q}' \in \mathbf{E_u})$$

and the arrow orientation is chosen in such a way that a timelike vector $(\mathbf{t}, \mathbf{q})$ is positive if and only if $\mathbf{t} \in \mathbf{I}^+$. Furthermore, if $o \in M$ then $(s_{\mathbf{u},o}, \mathrm{id}_\mathbf{I})$ is an isomorphism between the two special relativistic space-time models.

Proof. The assertions are obvious except that $s_{\mathbf{u},o}$ is arrow preserving.

Since $\mathbf{u}\otimes\mathbf{t}$ is in $T^{\rightarrow}$ if and only if $\mathbf{t}$ is in $\mathbf{I}^+$, $\mathbf{s_u}(\mathbf{u}\otimes\mathbf{t})=(\mathbf{t}, \mathbf{0})$ has positive arrow if and only if $\mathbf{u}\otimes\mathbf{t}\in T^{\rightarrow}$. If $\mathbf{x}\in T^{\rightarrow}$ then $\mathbf{x}$ has the same arrow as $\mathbf{u}\otimes\mathbf{t}$ for $\mathbf{t}\in\mathbf{I}^+$, i.e. we have

$$\mathbf{0} > g(\mathbf{x}, \mathbf{u}\otimes\mathbf{t}) = g_{\mathbf{u}}(\mathbf{s_u}(\mathbf{x}), \mathbf{s_u}(\mathbf{u}\otimes\mathbf{t})),$$

thus $\mathbf{s_u}(\mathbf{x})$ has the same arrow as $\mathbf{s_u}(\mathbf{u}\otimes\mathbf{t})$ which proves that $s_{\mathbf{u},o}$ is arrow preserving.

1.13. Proposition. Let $(\ |\ )$ denote the usual Minkowskian form on $\mathbb{R}^4$, i.e.

$$((x^0, x^1, x^2, x^3) | (y^0, y^1, y^2, y^3)) = -x^0y^0 + \sum_{k=1}^{3} x^k y^k.$$

Then $(\mathbb{R}^4, \mathbb{R}, (\ |\ ))$ is a special relativistic space-time model (where the orientations are the natural ones and the arrow orientation is given in such a way that a timelike vector is positive if and only if its zeroth component is positive) and an arbitrary special relativistic space-time model is isomorphic to it.

Proof. Let $(M, \mathbf{I}, g)$ be a special relativistic space-time model. Take an element o of M, a positive element $\mathbf{e}$ of $\mathbf{I}$, a positively oriented g-orthogonal basis $\mathbf{e}_0, \mathbf{e}_1, \mathbf{e}_2. \mathbf{e}_3$ of $\mathbf{M}$, normed to $\mathbf{e}$, such that $\mathbf{e}_0$ has positive arrow. Then the maps

$$M \longrightarrow \mathbb{R}^4, \qquad x \longmapsto \left(\frac{g(\mathbf{e}_k, x-o)}{g(\mathbf{e}_k, \mathbf{e}_k)}\right)_{k=0,1,2,3}$$

$$\mathbf{I} \longrightarrow \mathbb{R}, \qquad \mathbf{t} \longmapsto \frac{\mathbf{t}}{\mathbf{e}}$$

constitute an isomorphism between the two special relativistic space-time models.

Remark. Because of the form of the Minkowskian form it is convenient to consider $\mathbb{R}^4$ as $\mathbb{R}\times\mathbb{R}^3$. Doing so we can prove that $(\mathbf{I}\times\mathbf{E_u}, \mathbf{I}, g_{\mathbf{u}})$ for every $\mathbf{u}\in V(1)$ is isomorphic to $(\mathbb{R}\times\mathbb{R}^3, \mathbb{R}, (\ |\ ))$, similarly as we did in I. 1.15.

1.14. The previous assertion has the following very important corollary.

Theorem. Any two special relativistic space-time models are isomorphic.

1.15. Now we see that special relativistic space-time model as a mathematical structure is unique. For investigations and applications of the special relativistic space-time model we can use an arbitrarily given one. However, an actual model such as $(\mathbf{I}\times\mathbf{E_u}, \mathbf{I}, g_{\mathbf{u}})$ or $(\mathbb{R}\times\mathbb{R}^3, \mathbb{R}, (\ |\ ))$ has a number of additional structures, too. First of all, space-time is a vector space and has the form of a Cartesian product. These models tempt us to attribute an origin to space-time and consider space-time as a Cartesian product of time and space though we well know that neither time nor space exist. In the latter model space-time lengths have a canonical unit, namely the number one in $\mathbb{R}$.

To avoid such confusions we should keep away from similar specially constructed models; we can and will use only the structures listed in Definition 1.2. for the investigation and applications of the special relativistic space-time model.

2. World lines

Necessary mathematical tools:

Tensor products and quotients (Chapter IV.)
Differentiation in affine spaces (Sections VI. 2., VI. 3.)
Elements of the theory of smooth manifolds (Section VII. 1.)

The elements of the theory of Riemannian manifolds involved in Proposition 2.2. and Definition 2.5. can be evaded by Remark 2.5. (i).

2.1. ⌈We are to define the notions connected with a masspoint, similarly as we did it in Section I. 2.

Our task is now more difficult because the physical meaning of a special relativistic space-time model is not so clear. As we mentioned in 1.1., the Minkowskian structure on space-time reflects somehow that light spreads isotropically and with the same speed for all inertial observers.

We accept without further clarification that the history of a light signal is modelled by a lightlike straight line in space-time, i.e. by a straight line directed by a lightlike vector, and the history of a masspoint is modelled by a line in space-time whose all tangent vectors are timelike.⌋

2.2. Definition. (i) A *light signal* is a segment of a straight line directed by a lightlike vector.

(ii) A connected one dimensional submanifold of M is called a *world line* if all its tangent vectors are timelike.

A world line which is a straight line segment is called *inertial.*

Proposition. If C is a world line then $(C, \mathbf{I}, -g|_C)$ (notation: VII. 4.8. (iii)) is a Riemannian manifold.

2.3. Proposition. Let C be a world line. Choose arbitrarily a point x of C. Then $y-x$ is timelike for all $y \in C$, in other words, $C \setminus \{x\} \subset x+T$.

Proof. There are a neighbourhood N of x in C, an interval Z of $\mathbb{R}$, $0 \in Z$, a diffeomorphism $z: Z \longrightarrow N$, $z(0)=x$. Since $\dot{z}(0)$ is a tangent vector of C at x, we easily find that there exists a neighbourhood $N_0 \subset N$ of x such that $y-x \in T$ for all $y \in N_0 \setminus \{x\}$.

If $N_0 \neq C$, let x_1 be a boundary point of N_0. Then x_1 has a neighbourhood N_1 such that $y-x_1 \in T$ for all $y \in N_1 \setminus \{x_1\}$. Suppose now $y \in N_0 \cap N_1$, $y \neq x$, $y \neq x_1$. Then $y \in x+T$ and $x_1 \in y+T$; thus $x_1 \in x+T$ and we easily derive that $y-x \in T$ for all $y \in N_0 \cup N_1$, $y \neq x$. We can continue to step forward to get the desired result.

2.4. Definition 1. Let C be a world line. If $x, y \in C$ we say that x *precedes* y, in notations $x \underset{C}{\longrightarrow} y$, if $y \in x + T^{\rightarrow}$.

If $x \underset{C}{\longrightarrow} y$ then

$$C]x, y[:=\{z \in C : x \underset{C}{\longrightarrow} z \underset{C}{\longrightarrow} y\}$$

is called the *interval* of C between x and y.

Proposition. (i) $\underset{C}{\longrightarrow}$ is a total ordering on C.

(ii) C is orientable; an atlas of C is oriented if and only if all corresponding coordinatizations are monoton increasing or decreasing maps with respect to the orderings of C and $\mathbb{R}$.

Definition 2. C is considered to be oriented by monoton increasing coordinatizations.

2.5. Definition. Let $x, y \in M$, $y-x \in T$, and suppose C is a world line passing through x and y (i.e. $x, y \in C$). Then

$$\mathbf{t}_C(x, y) := \begin{cases} \int\limits_{C]x,y[} \sqrt[\otimes]{|g|_C|} & \text{if } x \underset{C}{\longrightarrow} y \\ 0 & \text{if } x = y \\ -\int\limits_{C]y,x[} \sqrt[\otimes]{|g|_C|} & \text{if } y \underset{C}{\longrightarrow} x \end{cases}$$

is called the *time passed between x and y along C*. The time passed between x and y along a straight line is called the *inertial time between x and y* and is denoted by $\mathbf{t}(x, y)$.

Remarks. (i) Recall that the above integral can be computed with the aid of a parametrization $p : \mathbf{I} \rightarrowtail C \subset M$ as follows:

$$\int\limits_{C]x,y[} \sqrt[\otimes]{|g|_C|} = \int\limits_{p^{-1}(x)}^{p^{-1}(y)} \sqrt{-(\dot{p}(\mathbf{t}) \cdot \dot{p}(\mathbf{t}))}\, d\mathbf{t}$$

where $\dot{p}$ is the derivative of p in the sense of VI. 2.6. $\left(\text{thus } \dot{p}(\mathbf{t}) \text{ is in } \frac{\mathbf{M}}{\mathbf{I}}\right)$, $(\cdot)$ is the bilinear form introduced in 1.5., and we integrate by the $\mathbf{I}$-valued Lebesgue meausure on $\mathbf{I}$ (i.e. the measure of an interval $[\mathbf{t}_1, \mathbf{t}_2] \subset \mathbf{I}$ is $\mathbf{t}_2 - \mathbf{t}_1$).

(ii) The inertial time between x and y is $\sqrt[\otimes]{-g(y-x, y-x)}$.

2.6. Proposition. Let $x \in M$, $y \in x + T^{\rightarrow}$, and suppose C is a world line passing through x and y. Then

$$\mathbf{t}_C(x, y) \leq \mathbf{t}(x, y)$$

and equality holds if and only if $C]x, y[$ is a subset of the straight line connecting x and y.

Proof. Let z be an element of $x+T^{\rightarrow}$ such that $y \in z+T^{\rightarrow}$. Then

$$\mathbf{t}(x, z)+\mathbf{t}(z, y) \leq \mathbf{t}(x, y)$$

and equality holds if and only if z is on the straight line passing through x and y (see V. 4.7.). As a consequence, the time passed between x and y along a broken line (defined as the sum of times passed along the corresponding straight line pieces) is smaller then the inertial time between x and y.

The definition of $\mathbf{t}_C(x, y)$ as an integral involves that $\mathbf{t}_C(x, y)$ can be obtained as an infimum of times passed between x and y along broken lines; hence our assertion is true.

2.7. Proposition. Let C be a world line. Then there is a smooth map $r:\mathbf{I} \rightarrowtail M$ defined on a connected subset such that C is the range of r and $\dot{r}(\mathbf{t}) \in V(1)$ for all $\mathbf{t} \in \operatorname{Dom} r$.

Furthermore, if r_1 and r_2 are maps with these properties then there exists an element $\mathbf{t}_0$ of $\mathbf{I}$ such that $\operatorname{Dom} r_2 = \mathbf{t}_0 + \operatorname{Dom} r_1$ and $r_2(\mathbf{t}) = r_1(\mathbf{t}-\mathbf{t}_0)$ $(\mathbf{t} \in \operatorname{Dom} r_2)$.

Proof. Pick out an element x_0 of C. The map $C \longrightarrow \mathbf{I}$, $x \mapsto \mathbf{t}_C(x_0, x)$ is clearly smooth and injective, its derivative is nowhere zero. Let us denote its inverse by $r:\mathbf{I} \rightarrowtail C$. Then r is a parametrization of C. The ordering on C involves that the timelike vector of cotype $\mathbf{I}$ $\dot{r}(\mathbf{t})$ is positive (i.e. $\dot{r}(\mathbf{t}) \otimes \mathbf{s}$ is positive for positive $\mathbf{s} \in \mathbf{I}$) and we have

$$\int_{]0,\mathbf{t}[} \sqrt[\otimes]{-(\dot{r}(\mathbf{t}') \mid \dot{r}(\mathbf{t}'))}\, d\mathbf{t}' = \mathbf{t}_C(x_0, r(\mathbf{t})) = \mathbf{t} \qquad (\mathbf{t} \in \operatorname{Dom} r).$$

As a consequence, r satisfies all the desired requirements.

Suppose now that r_1 and r_2 are maps with the listed properties. Choose an element x_0 of C and put $\mathbf{t}_0 := r_2^{-1}(x_0) - r_1^{-1}(x_0)$. The equalities

$$r_1^{-1}(x) - r_1^{-1}(x_0) = \mathbf{t}_C(x_0, x) = r_2^{-1}(x) - r_2^{-1}(x_0)$$

involve

$$r_1^{-1}(x) = r_2^{-1}(x) - \mathbf{t}_0 \quad (x \in C).$$

Substituting $x := r_2(\mathbf{t})$ $(\mathbf{t} \in \operatorname{Dom} r_2)$ we get the desired result relative to r_1 and r_2.

2.8. Definition. A smooth function $r:\mathbf{I} \rightarrowtail M$ having the properties

(i) $\operatorname{Dom} r$ is connected,

(ii) $\dot{r}(\mathbf{t}) \in V(1)$ $\qquad (\mathbf{t} \in \operatorname{Dom} r)$

is called a *world line function.*

Remark. According to Proposition 2.7. a world line determines a world line function uniquely up to a translation of its domain.

Proposition. Let r be a world line function. Then $\ddot{r}(\mathbf{t})$ is a spacelike vector of cotype $\mathbf{I} \otimes \mathbf{I}$, more closely, it is in $\dfrac{\mathbf{E}_{\dot{r}(\mathbf{t})}}{\mathbf{I} \otimes \mathbf{I}}$ for all $\mathbf{t} \in \operatorname{Dom} r$.

Proof. Differentiate the function $\mathbf{t} \mapsto (\dot{r}(\mathbf{t}) \mid \dot{r}(\mathbf{t})) = -1$ to get the desired result.

2.9. ⌈It is evident by Definitions 2.5. and 2.8. that if r is a world line function then "its variable is the time passing along r" or as it is often called, the *proper time* of the world line. Hence $\dot{r}(\mathbf{t})$ and $\ddot{r}(\mathbf{t})$ can be interpreted as the value of the (absolute) velocity and (absolute) acceleration at $\mathbf{t} \in \mathrm{Dom}\, r$, respectively, of a masspoint having the world line function r as its history.⌋

That is why we call the elements of $V(1)$ *velocities* and the spacelike elements of $\frac{\mathbf{M}}{\mathbf{I}\otimes\mathbf{I}}$ *accelerations.*

Notice that

— the set of velocities is a hypersurface—but not a hyperplane—in $\frac{\mathbf{M}}{\mathbf{I}}$, all velocities are related to the Minkowskian structure in the same way, hence all velocities play equal role, there is no distinguished velocity; a velocity cannot be big or small compared with a given one, only relative velocities can be compared in such a way (see 4.2.),

— the Minkowskian structure distinguishes accelerations by their magnitudes; the magnitude of an acceleration

$$\mathbf{a} \in \frac{\mathbf{M}}{\mathbf{I}\otimes\mathbf{I}} = \mathbf{M}^* \quad \text{is} \quad \sqrt{g^{(*)}(\mathbf{a},\mathbf{a})} \in \mathbf{I}^* = \frac{\mathbb{R}}{\mathbf{I}}.$$

There is a distinguished acceleration: the zero one.

2.10. ⌈Before saying a few words about special relativistic units we recall how tensorial products and quotients enter in measuring magnitudes of physical quantities (see Introduction 2. and I. 2.5.).

In our special relativistic space-time model there is a single set for space lengths and time lengths. This is clarified physically by the fact that light speed is a universal constant allowing us to identify time lengths and space lengths physically: chosen a time unit, space unit will be the distance convered by a light signal during a unit time.

$\mathbf{I}$ is the vector space of space-time lengths. As a consequence, the magnitudes of relative velocities are in $\frac{\mathbf{I}}{\mathbf{I}} = \mathbb{R}$, the magnitudes of accelerations are in $\frac{\mathbf{I}}{\mathbf{I}\otimes\mathbf{I}} = \frac{\mathbb{R}}{\mathbf{I}} = \mathbf{I}^*$.

The mass of a masspoint will be considered as a positive element of $\mathbf{I}^* = \frac{\mathbb{R}}{\mathbf{I}}$; this corresponds to the choice of units of measurements in which the Planck constant is a real number (dimensionless). A justification for this choice will appear in volume Models in Mechanics. The vector space of force magnitudes is $\mathbf{I}^*\otimes\mathbf{I}^* = \frac{\mathbb{R}}{\mathbf{I}\otimes\mathbf{I}}$.⌋

2.11. A world line C and a corresponding world line function r should be called rotation free if the acceleration do not change direction, do not rotate (cf. I. 2.6.). A convenient definition would be that $\frac{\ddot{r}}{|\ddot{r}|}$ is a Fermi–Walker parametrized vector field along C (see VII. 10.9.). However, some difficulties may arise for $\mathbf{t} \in \mathrm{Dom}\, r$, $\ddot{r}(\mathbf{t})=0$. Hence we write the Fermi–Walker differential equation for $\frac{\ddot{r}}{|\ddot{r}|}$, we transform it to obtain a differential equation which can be defined on the whole Dom r.

Definition. A world line function r and the corresponding world line is called *rotation free* if

$$|\mathbf{a}|^{②} \otimes \dot{\mathbf{a}} - (\mathbf{a} \cdot \dot{\mathbf{a}}) \otimes \mathbf{a} = |\mathbf{a}|^{④} \otimes \mathbf{u}$$

where $\mathbf{u} := \dot{r}$, $\mathbf{a} := \ddot{r}$, $|\mathbf{a}|^{②} := |\mathbf{a}| \otimes |\mathbf{a}|$ etc. and $\mathbf{a} \cdot \dot{\mathbf{a}}$ is the product of $\mathbf{a}$ and $\dot{\mathbf{a}}$ deriving from the bilinear map g.

2.12. Definition. (i) A smooth covector field $\mathbf{K} : M \rightarrowtail \mathbf{M}^*$ is called a *potential.*

The potential $\mathbf{K}$ is *static* if there are a $\mathbf{c} \in V(1)$, an $o \in M$ and a map $\mathbf{Z} : \mathbf{E_c} \to \mathbf{M}^*$ such that

$$\mathbf{K}(x) = \mathbf{Z}((x-o)\|_{\mathbf{c}}) \qquad (x \in \mathrm{Dom}\, \mathbf{K}).$$

(ii) A smooth map

$$\mathbf{f} : M \times V(1) \rightarrowtail \frac{\mathbf{M}^*}{\mathbf{I}} = \frac{\mathbf{M}}{\mathbf{I} \otimes \mathbf{I} \otimes \mathbf{I}}$$

for which

$$\mathbf{f}(x, \mathbf{u}) \in \frac{\mathbf{E_u}}{\mathbf{I} \otimes \mathbf{I} \otimes \mathbf{I}} \qquad ((x, \mathbf{u}) \in \mathrm{Dom}\, \mathbf{f})$$

holds is called a *force field.* The force field $\mathbf{f}$ is said to have a potential if there exists a potential $\mathbf{K}$ such that

$$\mathbf{f}(x, \mathbf{u}) = \mathrm{d}\mathbf{K}[x]_{\mathbf{I}}\mathbf{u} \qquad ((x, \mathbf{u}) \in \mathrm{Dom}\, \mathbf{f})$$

where $\mathrm{d}\mathbf{K}$ is the exterior derivative of $\mathbf{K}$ (see V. 5.2.).

(iii) A positive element of $\mathbf{I}^*$ is called a *mass.*

2.13. In a nonrelativistic space-time model a force field can be independent of either of its variables, in particular, it can be a constant map. In the present case, on the contrary, a non-zero force field cannot be independent of its second variable, in particular, it cannot be a constant map.

⌈A nonrelativistic constant force field models a physical force which is the same "ever and everywhere". A special relativistic force field $\mathbf{f}$ could be called "physically constant" if the Lorentz boosts map its corresponding values into each other, i.e. the canonical identification of $\mathbf{E}_{\mathbf{u}'}$ with $\mathbf{E}_{\mathbf{u}}$ sends the value of **f** at **u**′ into the value of **f** at **u** (see 1.9.):

$$F(\mathbf{u},\mathbf{u}')_{\mathbf{I}\otimes\mathbf{I}\otimes\mathbf{I}}\mathbf{f}(x,\mathbf{u}')=\mathbf{f}(x,\mathbf{u}) \qquad ((x,\mathbf{u}'),\ (x,\mathbf{u})\in \mathrm{Dom}\,\mathbf{f})$$

However, such a non-zero force field cannot exist (see Exercise 3.): *there is no non-zero special relativistic constant force field*! ⌋

2.14. Definition. A world line function r is called *Einsteinian* if there are

(i) a force field **f**,

(ii) a mass m

such that r satisfies the second order differential equation (see VI. 3.8. and VI. 3.9.)

$$m\otimes\ddot{x}=\mathbf{f}\circ(x,\dot{x})$$

or

$$\dot{x}=\mathbf{u}\,, \qquad \dot{\mathbf{u}}=\frac{\mathbf{f}\circ(x,\mathbf{u})}{m}.$$

2.15. Let r be a world line function. It is very important to know, both from theoretical and practical point of view, whether the acceleration $\ddot{r}$ is the non-trivial part of $\dddot{r}$ or $\nabla_{\dot{r}}\dot{r}$ (see VII. 9.9.).

Since a special relativistic space-time model is a particular general relativistic space-time model, and there $\nabla_{\dot{r}}\dot{r}$ is accepted as acceleration, we can state that $\ddot{r}$ represents $\nabla_{\dot{r}}\dot{r}$.

2.16. Later we need a notion which is "somehow complementary" to the notion of world lines.

Definition. A connected three dimensional submanifold of M is called a *world surface* if all its tangent vectors are spacelike.

Proposition. If D is a world surface then $(D,\mathbf{I},g|_D)$ (notation: VII. 4.8. (iii)) is a Riemannian manifold.

2.17. Proposition. Let D be a world surface. Choose arbitrarily a point x of D. Then $y-x$ is spacelike for all $y\in D$, in other words, $D\setminus\{x\}\subset x+S$.

Proof. We can argue as we did in the proof of Proposition 2.3.

2.18. Proposition. Let C and D be a world line and a world surface, respectively. Then $C\cap D$ is either void or a singleton.

Proof. Let x be an element of $C\cap D$. Since the intersection of $x+T$ and $x+S$ is evidently void, Propositions 2.3. and 2.17. imply that there is no other element in $C\cap D$.

3. A. Regular observers

Necessary mathematical tools:

Tensor products and quotients (Chapter IV.)
Affine spaces, differentiation (Sections VI. 1., VI. 2.)
Elements of the theory of manifolds (Section VII. 1.)
Pseudo-Riemannian manifolds (Section VII. 10.)

The essence of Propositions 3.4. and 3.5. can be grasped without the involved knowledge on smooth manifolds (foliations).

Though some results of the theory of pseudo-Riemannian manifolds is indispensable to understand regular observers, dealing only with inertial observers we can remain in the theory of affine spaces. The reader interested only in inertial observers is asked to replace definitions 3.6. (ii), 3.7. (i) – (ii) and 3.13. with remarks 3.7. (ii), 3.7. (i) and 3.13., respectively.

3.1. Definition. An *observer* is a smooth map $\mathbf{U}: M \rightarrowtail V(1) \subset \frac{\mathbf{M}}{\mathbf{I}}$ (i.e. a vector field of cotype $\mathbf{I}$); depending on whether $\mathrm{Dom}\,\mathbf{U} \neq M$ or $\mathrm{Dom}\,\mathbf{U} = M$, the observer is called *local* or *global*, respectively.

The maximal integral curves of $\mathbf{U}$ are called $\mathbf{U}$-*lines*.

The $\mathbf{U}$-line passing through $x \in \mathrm{Dom}\,\mathbf{U}$ is denoted by $C_{\mathbf{U}}(x)$ and $E_{\mathbf{U}}$ stands for the set of $\mathbf{U}$-lines.

Remark. The integral curves of an observer are evidently world lines.

⌈On the base of 2.7. we can imagine that an observer is a *velocity field*. Furthermore, $\mathrm{D}\mathbf{U}[\cdot]_{\mathbf{I}} \circ \mathbf{U}(\cdot): M \rightarrowtail \frac{\mathbf{M}}{\mathbf{I} \otimes \mathbf{I}}$ can be considered as the *acceleration field* of the observer $\mathbf{U}$.⌋

3.2. Definition. Let $\mathbf{U}$ be an observer. A three dimensional submanifold D in $\mathrm{Dom}\,\mathbf{U}$ is called a $\mathbf{U}$-*surface* if the tangent space of D at x is $\mathbf{E}_{\mathbf{U}(x)}$ (notation: 1.5.) for all $x \in D$ and if it is maximal, i.e. it is connected and is not contained properly in a submanifold with the same property.

$\mathbf{U}$ is *regular* if for all $x \in \mathrm{Dom}\,\mathbf{U}$ there is a unique $\mathbf{U}$-surface containing x.

If $\mathbf{U}$ is regular, the $\mathbf{U}$-surface passing through $x \in \mathrm{Dom}\,\mathbf{U}$ is denoted by $\tau_{\mathbf{U}}(x)$ and $I_{\mathbf{U}}$ stands for the set of $\mathbf{U}$-surfaces.

Remark. By Proposition VII. 7.4. we have that an observer $\mathbf{U}$ is regular if and only if the following is satisfied:

if $\mathbf{X}$ and $\mathbf{Y}$ are vector fields in $\mathrm{Dom}\,\mathbf{U}$ such that $[\mathbf{U}(x)|\mathbf{X}(x)) = [\mathbf{U}(x)|\mathbf{Y}(x)) = \mathbf{0}$ then $[\mathbf{U}(x)|[\mathbf{X}, \mathbf{Y}](x)) = \mathbf{0}$ for all $x \in \mathrm{Dom}\,\mathbf{X} \cap \mathrm{Dom}\,\mathbf{Y}$.

3.3. Proposition. Suppose $\mathbf{U}$ is an observer and t is a $\mathbf{U}$-surface. Then t is a world surface, hence $(t, \mathbf{I}, g|_t)$ is a Riemannian manifold (notation: VII. 4.8. (iii)).∎

In the sequel we shall use the notation $\gamma_t := g|_t$.

3.4. Proposition. Let **U** be a regular observer. Then $E_{\mathbf{U}}$ and $I_{\mathbf{U}}$ are manifolds with the smooth structure described in VII. 7.7.

Proof. **U**-lines and **U**-surfaces are transversal to each other by definition. Furthermore, (see 2.18.) the intersection of a **U**-line and a **U**-surface is either void or a singleton.

3.5. Theorem. Let **U** be a regular observer. Then the map

$$S_{\mathbf{U}}: \mathrm{Dom}\, \mathbf{U} \longrightarrow I_{\mathbf{U}} \times E_{\mathbf{U}}, \quad x \mapsto (\tau_{\mathbf{U}}(x), C_{\mathbf{U}}(x))$$

is a little diffeomorphism whose inverse has the form

$$(t, q) \mapsto t \cap q \qquad ((t, q) \in \mathrm{Ran}\, S_{\mathbf{U}} \subset I_{\mathbf{U}} \times E_{\mathbf{U}})$$

(we denote the unique element of the singleton $t \cap q$ by $t \cap q$ as well).

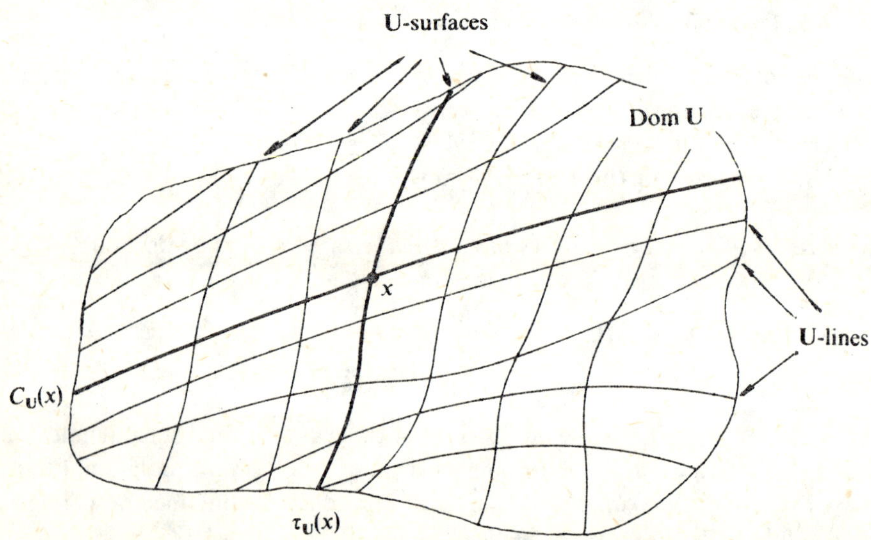

Furthermore,

$$\mathrm{D}S_{\mathbf{U}}[x] = \mathbf{s}_{\mathbf{U}(x)} = (\boldsymbol{\tau}_{\mathbf{U}(x)}, (\cdot)\|_{\mathbf{U}(x)}) \qquad (x \in \mathrm{Dom}\, \mathbf{U}).$$

Proof. The first part of the theorem is a reformulation of VII. 7.8. for the present special case.

Then tangent space of $I_{\mathbf{U}} \times E_{\mathbf{U}}$ at $(\tau_{\mathbf{U}}(x), C_{\mathbf{U}}(x))$ can be identified with $(\mathbf{U}(x) \otimes \mathbf{I}) \times E_{\mathbf{U}(x)}$ (see Remark VII. 7.10. (v)). The linear subspace $\mathbf{U}(x) \otimes \mathbf{I}$ of $\mathbf{M}$ can be identified with $\mathbf{I}$ trivially. Apply Remark VII. 7.10. (i) to get the formula for the derivative of $S_{\mathbf{U}}$.

3.6. ⌈The material points of a physical observer constitute the space of the observer; hence the space of the physical observer modelled by **U** must be modelled by the set of **U**-lines (cf. I. 3.5.).

The time of an observer can be defined by a notion of simultaneity; the best we can do to give a time point of the observer is to point out the set of corresponding simultaneous space-time points. Simultaneity is determined physically by light signals; we shall see later a straightforward method for inertial observers and this method suggests us the general notion of simultaneity.⌋

Definition. (i) Let **U** be a regular observer. $I_{\mathbf{U}}$ and $E_{\mathbf{U}}$ are called **U**-*time* and **U**-*space* or *time* and *space relative* to **U**, respectively, and $S_{\mathbf{U}}$ is the *splitting* of (a part of) space-time into **U**-time and **U**-space.

If q is a **U**-space point and t is a **U**-time point then $t \cap q$ is the *instantaneous value* of q at t; t itself is called sometimes the *instantaneous* **U**-*space* at the **U**-time point t. The elements of a **U**-time point are called *U-simultaneous space-time points.*

(ii) The *distance* of the **U**-space points q' and q at the **U**-time point t makes sense if neither of $t \cap q'$ and $t \cap q$ is void and they are sufficiently near to each other (see VII. 10.6.) and then it is the geodesic distance of $t \cap q'$ and $t \cap q$ according to the Riemannian metric γ_t (see 3.3. and VII. 10.9.).

Remark. It is a very interesting fact that though the elements of a **U**-time point (**U**-surface) are simultaneous with respect to **U**, the times passed between two **U**-time points (**U**-surfaces) along different **U**-space points (**U**-lines) are different in general (see Exercise 7.)

$$\mathbf{t}_{C_1}(t_1 \cap C_1, t_2 \cap C_1) \neq \mathbf{t}_{C_2}(t_1 \cap C_2, t_2 \cap C_2).$$

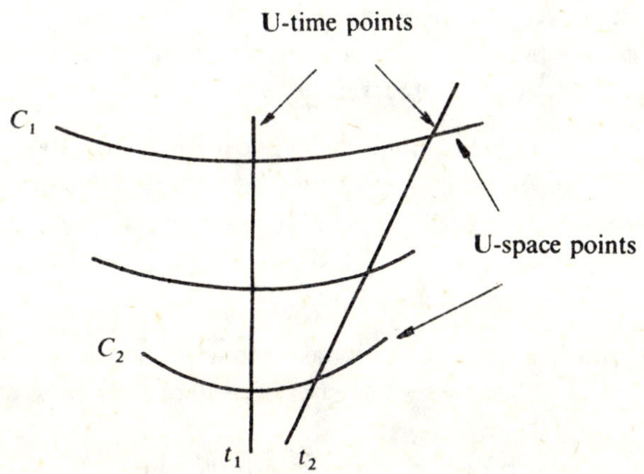

3.7. Definition. A regular observer **U** is called

(i) *rigid* if the distances of **U**-space points at given **U**-time points are independent of the **U**-time points;

(ii) *quasi rotation free* if for all **U**-space points q_1 and q_2 and for all **U**-time points t, s such that $t \cap q_1$ and $t \cap q_2$, $s \cap q_1$ and $s \cap q_2$ can be connected by unique geodesics c_t and c_s corresponding to γ_t and γ_s (see VII. 10.60), the tangent vector of c_s at $s \cap q_i$ is the Fermi-Walker translation of the tangent vector of c_t at $t \cap q_i$ along q_i, $i=1, 2$;
rotation free if it is quasi rotation free and all **U**-lines are rotation free;

(iii) *inertial* if **U** is a constant map or, equivalently, the **U**-lines are parallel straight line segments.

Remarks. (i) Inertial observers are rigid and rotation free.

(ii) The simultaneous **U**-spaces or **U**-time points of an inertial observer **U** are subsets of affine hyperplanes. Then the distance of the **U**-space points q_1 and q_2 at the **U**-time point t is $\sqrt{g(t \cap q_1 - t \cap q_2, t \cap q_1 - t \cap q_2)}$.

3.8. We introduced notions similar to those in nonrelativistic space-time model (see I. 3.5., I. 3.7.). Some of the objects entering in the definitions were complicated in the nonrelativistic case and most of the objects in the definitions here are much more complicated.

A very important difference showing the hardness of the special relativistic space-time model is that the instantaneous **U**-spaces of a global observer **U** are Euclidean affine spaces in the nonrelativistic case (see Proposition I. 3.8.) whereas generally they are curved Riemmanian spaces in the special relativistic case:

Proposition. Let **U** be a global regular observer. Then all the instantaneous **U**-spaces are affine subspaces of M if and only if **U** is inertial.∎

Furthermore, in the special relativistic space-time model the properties "rigid" and "rotation free" are very unpleasant; practically one can handle solely the inertial observers. ⌈That is why one treats in usual special relativistic physics exclusively inertial observers; it is not true, however, what is often stated, that special relativistic physics concerns only inertial observers.⌋

3.9. The vectorization of time and **U**-space is a nice tool to simplify the structures involved in splitting of space-time in the nonrelativistic case (see I. 3.9.). Such a trick cannot be generally applied in special relativistic space-time model: there is no distinguished subspace **E** of **M** and **U**-time is not related to **I**. The only exceptions are again the inertial observers.

3.10. Proposition. Let **U** be a global inertial observer and assume $\mathbf{c} \in V(1)$ is the constant value of **U**. Then $E_{\mathbf{U}}$ is the set of affine lines directed by $\mathbf{c}\otimes\mathbf{I}$, $I_{\mathbf{U}}$ is the set of affine hyperplanes directed by $\mathbf{E_c}$.

$E_{\mathbf{U}}$ is an affine space over $\mathbf{E_c}$ with the subtraction

$$q'-q := (x'-x)\|_{\mathbf{c}} \qquad (q', q \in E_{\mathbf{U}},\ x' \in q',\ x \in q).$$

$I_{\mathbf{U}}$ is an affine space over **I**, with the subtraction

$$t'-t := \boldsymbol{\tau}_{\mathbf{c}}(x'-x) \qquad (t', t \in I_{\mathbf{U}},\ x' \in t',\ x \in t).$$

S_U is an affine bijection,

$$S_U(x) = (x + \mathbf{E_c}, x + \mathbf{c} \otimes \mathbf{I}) \qquad (x \in M)$$

and $DS_U = \mathbf{s_c}$.

3.11. Definition. A *global inertial observer with origin* is a pair $(\mathbf{U}, o)$ where $\mathbf{U}$ is a global inertial observer and o is an element of M.

Proposition. Let $(\mathbf{U}, o)$ be a global inertial observer with origin and assume $\mathbf{c} \in V(1)$ is the constant value of $\mathbf{U}$. Then

$$S_o^{\mathbf{U}} : I_{\mathbf{U}} \times E_{\mathbf{U}} \longrightarrow \mathbf{I} \times \mathbf{E_c}, \quad (t, q) \mapsto (t - \tau_{\mathbf{U}}(o),\ q - C_{\mathbf{U}}(o))$$

is an affine bijection, $DS_o^{\mathbf{U}} = \mathrm{id}_{\mathbf{I} \times \mathbf{E_c}}$ and $S_o^{\mathbf{U}} \circ S_{\mathbf{U}} = s_{\mathbf{c},o}$ (see 1.10.).

3.12. Definition. Let $\mathbf{U}$ be a global inertial observer with the constant velocity $\mathbf{c} \in V(1)$. Then the *time observed by* $\mathbf{U}$ *between* $x, y \in M$ is $\tau_{\mathbf{U}}(y) - \tau_{\mathbf{U}}(x) = \tau_{\mathbf{c}}(y - x)$.∎

It is a trivial fact that if $x \in M, y \in x + T^{\rightarrow}$ then

$$\tau_{\mathbf{c}}(y - x) = \mathbf{t}(x, y)\,(\mathbf{c}' \,\mathbb{H}\, \mathbf{c})$$

(notation: Definition 2.5.) where $\mathbf{c}' := \dfrac{y - x}{\mathbf{t}(x, y)}$ is an element of $V(1)$. Hence the inertial time between x and y is shorter then the time observed by $\mathbf{U}$ between x and y (see V. 4.4. and Remark 4.2. (ii)).

⌈If $\mathbf{U}'$ is a global inertial observer whose velocity is $\mathbf{c}'$ then the time observed by $\mathbf{U}'$ between x and y coincides with the inertial time between x and y. That is why one usually speaks about *time dilation*: it seems to me that time passes more slowly for someone moving relative to me, if we both are inertial. However, the other one sees me to move, hence it is my time that seems to him to pass more slowly. The famous twin paradoxon was constructed on this illusory time dilation.

We can be convinced that time passes equally for all inertial observers. However, two inertial twins can meet either for ever or only once, hence they are either in the same situation or they cannot be compared after their meeting which of them is younger. To meet again, at least one of them must stop being inertial. The inertial time between two space-time points is longer then the non-inertial one (see 2.6.) but this is an absolute fact and has nothing to do with the illusory time dilation.⌋

3.13. Definition. Let C_1 and C_2 be two world lines. Suppose $\mathbf{U}$ is a regular observer whose domain contains C_1 and C_2. Then the *distance between* C_1 *and* C_2 at the $\mathbf{U}$-time point t is the geodesic distance (if exists, see VII. 10.6. and VII. 10.8.) of $t \cap C_1$ and $t \cap C_2$ according to the Riemannian metric γ_t.

Remark. If $\mathbf{U}$ is an inertial observer then a $\mathbf{U}$-time point t is a hyperplane and the distance between C_1 and C_2 observed by $\mathbf{U}$ is

$$\sqrt{g(t \cap C_1 - t \cap C_2,\, t \cap C_1 - t \cap C_2)}\,.$$

3.14. Proposition. Let C_1 and C_2 be parallel straight lines directed by $\mathbf{c}'\otimes\mathbf{I}$, $\mathbf{c}'\in V(1)$. Suppose $\mathbf{U}$ is a global inertial observer with constant velocity $\mathbf{c}\in V(1)$. Then the distance of C_1 and C_2 observed by $\mathbf{U}$ at every $\mathbf{U}$-time point is the same and equals

$$\sqrt[\otimes]{g(\mathbf{d},\mathbf{d})-\frac{[\mathbf{c}|\mathbf{d})\otimes[\mathbf{c}|\mathbf{d})}{(\mathbf{c}'\dashv\vdash\mathbf{c})^2}}$$

where $\mathbf{d}$ is the element of $\mathbf{E}_{\mathbf{c}'}$ that $C_2=C_1+\mathbf{d}$ holds for.

Proof. Let t be a $\mathbf{U}$-time point, i.e. an affine subspace of M over $\mathbf{E}_{\mathbf{c}}$. If x is the single element of the intersection of t and C_1 (see 2.18) then $x+\mathbf{d}+\mathbf{c}'\otimes\frac{[\mathbf{c}|\mathbf{d})}{(\mathbf{c}'\dashv\vdash\mathbf{c})}$ is the single element of the intersection of t and $C_1+\mathbf{d}$. The distance between C_1 and $C_1+\mathbf{d}$ at t equals

$$\sqrt[\otimes]{g\left(x+\mathbf{d}+\mathbf{c}'\otimes\frac{[\mathbf{c}|\mathbf{d})}{(\mathbf{c}'\dashv\vdash\mathbf{c})}-x,\ x+\mathbf{d}+\mathbf{c}'\otimes\frac{[\mathbf{c}|\mathbf{d})}{(\mathbf{c}'\dashv\vdash\mathbf{c})}-x\right)}$$

which reduces to the desired formula.

⌈If $\mathbf{U}'$ is a global inertial observer with velocity $\mathbf{c}'$ then the distance between C_1 and $C_1+\mathbf{d}$ is $\sqrt[\otimes]{g(\mathbf{d},\mathbf{d})}$. In that case C_1 and $C_1+\mathbf{d}$ are $\mathbf{U}'$-lines, i.e. the masspoints represented by them rest with respect to the observer. If $\mathbf{c}\neq\mathbf{c}'$, the masspoints move with respect to $\mathbf{U}$. The result of our Proposition says that the motion length is smaller then the rest length: this is the so called *Lorentz contraction.*⌋

The usually cited formula for Lorentz contraction concerns the case when $\mathbf{d}$ is in the plane spanned by $\mathbf{c}\otimes\mathbf{I}$ and $\mathbf{c}'\otimes\mathbf{I}$, i.e. $\mathbf{d}=(\mathbf{c}-\mathbf{c}'(\mathbf{c}'\dashv\vdash\mathbf{c}))\otimes\mathbf{t}$ for some $\mathbf{t}\in\mathbf{I}$. Then

$$g(\mathbf{d},\mathbf{d})=((\mathbf{c}'\dashv\vdash\mathbf{c})^2-1)\mathbf{t}\otimes\mathbf{t},$$

$[\mathbf{c}|\mathbf{d})=((\mathbf{c}'\dashv\vdash\mathbf{c})^2-1)\mathbf{t}$, hence

$$\sqrt[\otimes]{g(\mathbf{d},\mathbf{d})-\frac{[\mathbf{c}|\mathbf{d})\otimes[\mathbf{c}|\mathbf{d})}{(\mathbf{c}'\dashv\vdash\mathbf{c})^2}}=\mathbf{t}\sqrt{\frac{(\mathbf{c}'\dashv\vdash\mathbf{c})^2-1}{(\mathbf{c}'\dashv\vdash\mathbf{c})^2}}=\frac{\sqrt[\otimes]{g(\mathbf{d},\mathbf{d})}}{(\mathbf{c}'\dashv\vdash\mathbf{c})}.$$

(see also Remark 4.2. (ii)).

3.15. ⌈Lastly we show how simultaneity is defined by an inertial observer. Suppose $\mathbf{c}$ is the velocity of the observer. The space-time points x and y are considered simultaneous if for certain non-zero $\mathbf{t}\in\mathbf{I}$ a light signal starting from $x-\mathbf{c}\otimes\mathbf{t}$ and reflecting at y reaches $x+\mathbf{c}\otimes\mathbf{t}$.

Put $\mathbf{z}:=y-x$. Then $\mathbf{z}-\mathbf{c}\otimes\mathbf{t}$ and $\mathbf{z}+\mathbf{c}\otimes\mathbf{t}$ are lightlike vectors, thus

$$g(\mathbf{z},\mathbf{z})-2[\mathbf{c}|\mathbf{z})\otimes\mathbf{t}+\mathbf{t}\otimes\mathbf{t}=g(\mathbf{z},\mathbf{z})+2[\mathbf{c}|\mathbf{z})\otimes\mathbf{t}+\mathbf{t}\otimes\mathbf{t}=\mathbf{0}$$

As a consequence, $[\mathbf{c}|\mathbf{z})=\mathbf{0}$, in other words $(y-x)\in\mathbf{E}_{\mathbf{c}}$.

We obtained that simultaneous space-time points are in a hyperplane directed by $\mathbf{E}_{\mathbf{c}}$.

Let now **U** be an arbitrary observer, $x \in \text{Dom}\,\mathbf{U}$. Then a space-time point y in a neighbourhood of x is considered approximately simultaneous with x if $y-x$ is in $\mathbf{E}_{\mathbf{U}(x)}$. This straightforward consideration suggests the general definition given in 3.6. ⌋

3.B. General observers

Necessary mathematical tools:

Tensor products and quotients (Chapter IV.)
Foliations (Section VII. 7.)
Theory of smooth manifolds (Sections VII. 1.–VII. 4.)
Pseudo-Riemannian manifolds (Sections VII. 10.)

This section needs the more advanced mathematical tools. The further sections can be understood without the results of this one, hence the reader who is not familiar with the involved mathematical tools can skip this section.

3.16. So far we have considered only regular observers. It may happen that an observer **U** has no **U**-surfaces. Such observers are not so strange, one meets them in the simplest applications (see Exercises 12. and 13.). If **U**-surfaces do not exist, we cannot define either simultaneity or **U**-time; however, $E_{\mathbf{U}}$ can and should be considered as **U**-space which can be turned into a three dimensional manifold even in this case.

3.17. Definition. A three dimensional foliation F in M (see VII. 7.3.) such that $F[x]$ is a spacelike linear subspace of $\mathbf{M}$ for all $x \in \text{Dom}\,F$ is called a *time foliation*. The set of maximal integral manifolds of a time foliation is called an *artificial time* on Dom F.

3.18. An observer **U** and a time foliation on Dom **U** are transversal to each other. Arguing as in 3.4. and 3.5. we can prove:

Proposition. Let **U** be an observer and let J be an artificial time on Dom **U**. Then $E_{\mathbf{U}}$ and J are manifolds with the smooth structure described in VII. 7.7. If $\tau_J(x)$ denotes the world surface in J passing through $x \in \text{Dom}\,\mathbf{U}$ then the map

$$S_{J,\mathbf{U}}: \text{Dom}\,\mathbf{U} \longrightarrow J \times E_{\mathbf{U}} \qquad x \mapsto (\tau_J(x), C_{\mathbf{U}}(x))$$

is a little diffeomorphism.

Remark. According to VII. 7.9. the smooth structure on $E_{\mathbf{U}}$ is independent of the artificial time, and if **U** is regular it is the same given by the aid of **U**-time.

3.19. Let **U** be an observer and suppose J is an artificial time on Dom **U**. Then $G_{J,\mathbf{U}} := (S_{J,\mathbf{U}}^{-1})^* g$ is an Einsteinian metric on Ran $S_{J,\mathbf{U}} \subset J \times E_{\mathbf{U}}$. The corresponding bilinear map on the tangent space $T_{(t,q)}(J \times E_{\mathbf{U}})$ ($(t,q) \in \operatorname{Ran} S_{J,\mathbf{U}}$) can be written in the form (see V. 1.2.)

$$G_{J,\mathbf{U}}[t,q] = \begin{pmatrix} g_{JJ}[t,q] & g_{J\mathbf{U}}[t,q] \\ g_{\mathbf{U}J}[t,q] & g^J_{\mathbf{UU}}[t,q] \end{pmatrix}$$

where

$$\begin{aligned} g_{JJ}[t,q] &: T_t(J) \times T_t(J) \longrightarrow \mathbf{I} \otimes \mathbf{I}\,, \\ g_{J\mathbf{U}}[t,q] &: T_t(J) \times T_q(E_{\mathbf{U}}) \longrightarrow \mathbf{I} \otimes \mathbf{I}\,, \\ g_{\mathbf{U}J}[t,q] &: T_q(E_{\mathbf{U}}) \times T_t(J) \longrightarrow \mathbf{I} \otimes \mathbf{I}\,, \\ g^J_{\mathbf{UU}}[t,q] &: T_q(E_{\mathbf{U}}) \times T_q(E_{\mathbf{U}}) \longrightarrow \mathbf{I} \otimes \mathbf{I} \end{aligned}$$

are bilinear maps, $g_{\mathbf{U}J}[t,q] = g_{J\mathbf{U}}[t,q]^*$, (notation: V. 1.1.), $g_{JJ}[t,q]$ is negative definite, $g^J_{\mathbf{UU}}[t,q]$ is positive definite.

Proposition. $\mathbf{U}$ is regular and J is the $\mathbf{U}$-time if and only if $g_{J\mathbf{U}}[t,q] = g_{\mathbf{U}J}[q,t] = 0$ for all $(t,q) \in \operatorname{Ran} S_{J,\mathbf{U}}$.

3.20. For all $(t,q) \in \operatorname{Ran} S_{J,\mathbf{U}}$ we introduce

$$-\frac{g_{\mathbf{U}J}[q,t] \otimes g_{J\mathbf{U}}[t,q]}{g_{JJ}[t,q]} : T_q(E_{\mathbf{U}}) \times T_q(E_{\mathbf{U}}) \longrightarrow \mathbf{I} \otimes \mathbf{I}\,,$$

$$(\mathbf{q}, \mathbf{q}') \mapsto -\frac{g_{\mathbf{U}J}[q,t]\,(\mathbf{q}, \mathbf{t}) \otimes g_{J\mathbf{U}}[t,q]\,(\mathbf{t}, \mathbf{q}')}{g_{JJ}[t,q]\,(\mathbf{t}, \mathbf{t})}$$

where $\mathbf{t}$ is any non-zero element of $T_t(J)$.

We easily find that it is a symmetric positive definite bilinear map, hence we have:

Proposition. Let $t \in J$. Then

$$\gamma^J_t[\,\cdot\,] := g^J_{\mathbf{UU}}[t, \,\cdot\,] - \frac{g_{\mathbf{U}J}[\,\cdot\,, t] \otimes g_{J\mathbf{U}}[t, \,\cdot\,]}{g_{JJ}[t, \,\cdot\,]}$$

is a Riemannian metric on the open submanifold of $E_{\mathbf{U}}$:

$$\{q \in E_{\mathbf{U}} \,\vdots\, (t,q) \in \operatorname{Ran} S_{J,\mathbf{U}}\}\,.$$

If $\mathbf{U}$ is regular and J is the $\mathbf{U}$-time then $\gamma^J_t = g^J_{\mathbf{UU}}[t, \,\cdot\,]$, and if the above set is identified with the $\mathbf{U}$-surface t then $\gamma^J_t[q] = \gamma_t[t \cap q]$ (see 3.3.).

3.21. Proposition. Let $\mathbf{U}$ be an observer. Let J and I be artificial times on $\operatorname{Dom} \mathbf{U}$. Then

$$\gamma^J_t[q] = \gamma^I_{\tau_I(t \cap q)}[q] \qquad ((t,q) \in \operatorname{Ran} S_{J,\mathbf{U}})\,.$$

Proof.

$$F := S_{I,\mathbf{U}} \circ S^{-1}_{J,\mathbf{U}} : J \times E_{\mathbf{U}} \longrightarrow I \times E_{\mathbf{U}}\,, \qquad (t,q) \mapsto (\tau_I(t \cap q), q)$$

is a diffeomorphism between Ran $S_{J,\mathbf{U}}$ and Ran $S_{I,\mathbf{U}}$. Its derivative at the point (t, q) can be written in the form

$$DF[t, q] = \begin{pmatrix} A_q[t] & B_t[q] \\ 0 & \mathrm{id}_{T_q(E_\mathbf{U})} \end{pmatrix}$$

where

$$A_q[t] : T_t(J) \longrightarrow T_{\tau_I(t \cap q)}(I)$$

is the derivative at t of the map $J \longrightarrow I$, $t \mapsto \tau_I(t \cap q)$,

$$B_t[q] : T_q(E_\mathbf{U}) \longrightarrow T_{\tau_I(t \cap q)}(I)$$

is the derivative at q of the map $E_\mathbf{U} \longrightarrow I$, $q \mapsto \tau_I(t \cap q)$.

Since $G_{J,\mathbf{U}} = (S_{I,\mathbf{U}} \circ S_{J,\mathbf{U}}^{-1})^* G_{I,\mathbf{U}}$, using the matrix representations we have

$$G_{J,\mathbf{U}}[t, q] = DF[t, q]^* G_{I,\mathbf{U}}[\tau_I(t \cap q), q]\, DF[t, q]$$

from which we easily deduce the desired result.

3.22. Remarks. (i) Detailing the previous equality we can see that the Riemannian metrics $g^J_{\mathbf{UU}}[t, \cdot\,]$ $(t \in J)$ on $E_\mathbf{U}$ badly depend on J, in general, whereas γ^J_t $(t \in J)$ do not: the set of Riemannian metrics $\{\gamma^J_t : t \in J\}$ is essentially independent of J.

In particular, if for some artificial time J, γ^J_t is the same for all $t \in J$ on $\bigcap_{t \in J} \{q \in E_\mathbf{U} : (t, q) \in \text{Ran } S_{J,\mathbf{U}}\}$, i.e. roughly speaking, $t \mapsto \gamma^J_t$ is a constant map, then the same is true for all other artificial times, too.

(ii) If $t \in J$, we can identify $\{q \in E_\mathbf{U} : (t, q) \in \text{Ran } S_{J,\mathbf{U}}\}$ with t. Doing so we can imagine that γ^J_t is a Riemannian metric given on t, i.e. we have equipped the world surfaces of J with Riemannian metrics.

3.23. Definition. Let $\mathbf{U}$ be an observer. Then $E_\mathbf{U}$ is called **U**-*space* or *space relative* to **U**. If J is an artificial time on Dom **U** then $S_{J,\mathbf{U}}$ is the *splitting* of (a part of) space-time into artificial time and **U**-space.

The *distance* of the **U**-space points q and q' at the artificial time point $t \in J$ makes sense if neither of $t \cap q$ and $t \cap q'$ is void and they are sufficiently near to each other (see VII. 10.6.) and then it is the geodesic distance of q and q' according to the Riemannian metric γ^J_t (or, if we make the identification mentioned in the previous remark, the geodesic distance of $t \cap q$ and $t \cap q'$).

3.24. Definition. An observer **U** is called

(i) *rigid* if the distances of **U**-space points at given artificial time points are independent of the artificial time points, or, equivalently, if the Riemennian metrics on **U**-space given in Proposition 3.20. are independent of the artificial time points,

(ii) *inertial* if **U** is a constant map or, equivalently, the **U**-lines are parallel straight line segments.

Remarks. (i) In 3.7. we defined when a regular observer is called inertial, now we introduced the notion of inertial observers in general and we see that an inertial observer is regular.

(ii) The notion of rigidity given here coincides for regular observer with that given in 3.7.

4. Relative velocities

Necessary mathematical tools:

Tensor products and quotients (Chapter IV.)
Differentiation in affine spaces (Section VI. 2.)

4.1. ⌈Let **U** be a global inertial observer with velocity $\mathbf{u}$ and suppose C is an inertial world line. Then the position of the masspoint with world line C at the **U**-time point t in **U**-space is $r_{\mathbf{U}}(t) := t \cap C + \mathbf{u} \otimes \mathbf{I}$. The velocity of the masspoint relative to the observer is the derivative of the map $I_{\mathbf{U}} \longrightarrow E_{\mathbf{U}}$, $t \mapsto r_{\mathbf{U}}(t)$.

Suppose C is directed by $\mathbf{u}' \in V(1)$. Then for some $x \in M$, $C = \{x + \mathbf{u}' \otimes \mathbf{t}' \,\vdots\, \mathbf{t}' \in \mathbf{I}\}$. Since the difference of $t \cap C$ and $t \cap C_{\mathbf{U}}(x) = x + u \otimes (t - \tau_{\mathbf{U}}(x))$ is in $\mathbf{E}_{\mathbf{u}}$, we easily derive that $t \cap C = x + \mathbf{u}' \otimes \dfrac{t - \tau_{\mathbf{U}}(x)}{(\mathbf{u}' \cdot \mathbf{u})}$. As a consequence, the subtraction defined in $E_{\mathbf{U}}$ (see 3.10.) involves that

$$\lim_{\mathbf{h} \to 0} \frac{r_{\mathbf{U}}(t + \mathbf{h}) - r_{\mathbf{U}}(t)}{\mathbf{h}} = \frac{\mathbf{u}'}{(\mathbf{u}' \cdot \mathbf{u})} - \mathbf{u} \qquad (t \in I_{\mathbf{U}}).$$

This suggests to introduce the following notion.⌋

4.2. Definition. Let $\mathbf{u}$, $\mathbf{u}'$ be elements of $V(1)$. Then we call

$$\mathbf{v}_{\mathbf{u}'\mathbf{u}} := \frac{\mathbf{u}'}{(\mathbf{u}' \cdot \mathbf{u})} - \mathbf{u}$$

the *relative velocity of* $\mathbf{u}'$ *with respect to* $\mathbf{u}$.

Proposition. Suppose $\mathbf{u}$ and $\mathbf{u}'$ are elements of $V(1)$. Then

(i) $\mathbf{v}_{\mathbf{u}'\mathbf{u}}$ is an element of $\dfrac{\mathbf{E}_{\mathbf{u}}}{\mathbf{I}}$,

(ii) $\mathbf{v}_{\mathbf{u}'\mathbf{u}} = -\mathbf{v}_{\mathbf{u}\mathbf{u}'}$ if and only if $\mathbf{u} = \mathbf{u}'$,

(iii) $|\mathbf{v}_{\mathbf{u}'\mathbf{u}}|^2 = |\mathbf{v}_{\mathbf{u}\mathbf{u}'}|^2 = 1 - \dfrac{1}{(\mathbf{u}' \cdot \mathbf{u})^2} < 1$,

where $|\ |$ denotes the Euclidean norm deriving from $\gamma_{\mathbf{u}}$ and $\gamma_{\mathbf{u}'}$ (see 1.6.).

Remarks. (i) Note the very important fact that non-zero relative velocities $\mathbf{v}_{\mathbf{u}'\mathbf{u}}$ and $\mathbf{v}_{\mathbf{u}\mathbf{u}'}$ are not additive inverses of each other in $\frac{\mathbf{M}}{\mathbf{I}}$, in contradistinction to the nonrelativistic case (see Proposition I. 4.2. (ii)); the magnitude of the two relative velocities are equal and less than 1.

(ii) We have

$$(\mathbf{u}' \cdot \mathbf{u}) = \frac{1}{\sqrt{1-|\mathbf{v}_{\mathbf{u}'\mathbf{u}}|^2}} \qquad (\mathbf{u}', \mathbf{u} \in V(1)).$$

The time dilation and the length contraction formulae given in 3.12. and 3.14. obtain their usual form if we put $\mathbf{v}^2 := |\mathbf{v}_{\mathbf{u}'\mathbf{u}}|^2$.

4.3. In nonrelativistic space-time model the relative velocities form a vector space, namely $\frac{\mathbf{E}}{\mathbf{I}}$. The vector addition of relative velocities (see Proposition I. 4.2. (iii)) is a nice feature of nonrelativistic space-time model.

In special relativistic space-time model the relative velocities do not form a vector space. Given a $\mathbf{u} \in V(1)$, the relative velocities with respect to $\mathbf{u}$ constitute the open unit ball of the Euclidean vector space $\frac{\mathbf{E}_{\mathbf{u}}}{\mathbf{I}}$ which is the tangent space of $V(1)$ at $\mathbf{u}$. For different $\mathbf{u}$ and $\mathbf{u}'$ the tangent spaces are different but they can be considered as subspaces of $\frac{\mathbf{M}}{\mathbf{I}}$. The set of all relative velocities is the union of the open unit balls of the tangent spaces of $V(1)$; it is a subset of the spacelike vectors of $\frac{\mathbf{M}}{\mathbf{I}}$.

As a consequence of the structure of the set of relative velocities, the "addition of relative velocities" is not a vector addition, i.e. if $\mathbf{u}$, $\mathbf{u}'$ and $\mathbf{u}''$ are different elements of $V(1)$ then

$$\mathbf{v}_{\mathbf{u}''\mathbf{u}} \neq \mathbf{v}_{\mathbf{u}''\mathbf{u}'} + \mathbf{v}_{\mathbf{u}'\mathbf{u}}.$$

A rather unpleasant but straightforward calculation yields the following result.

Proposition. Let $\mathbf{u}$, $\mathbf{u}'$ and $\mathbf{u}''$ be elements of $V(1)$. Then

$$\mathbf{v}_{\mathbf{u}''\mathbf{u}} = \frac{\left(\mathbf{v}_{\mathbf{u}''\mathbf{u}'} - \frac{(\mathbf{v}_{\mathbf{u}''\mathbf{u}'} \cdot \mathbf{v}_{\mathbf{u}\mathbf{u}'})}{|\mathbf{v}_{\mathbf{u}\mathbf{u}'}|^2}\mathbf{v}_{\mathbf{u}\mathbf{u}'}\right)\sqrt{1-|\mathbf{v}_{\mathbf{u}\mathbf{u}'}|^2} + \left(1 - \frac{(\mathbf{v}_{\mathbf{u}''\mathbf{u}'} \cdot \mathbf{v}_{\mathbf{u}\mathbf{u}'})}{|\mathbf{v}_{\mathbf{u}\mathbf{u}'}|^2}\right)\mathbf{v}_{\mathbf{u}'\mathbf{u}}}{1+(\mathbf{v}_{\mathbf{u}''\mathbf{u}'} \cdot \mathbf{v}_{\mathbf{u}\mathbf{u}'})}$$

Remarks. (i) It is a very interesting fact that $\mathbf{v}_{\mathbf{u}''\mathbf{u}}$ cannot be expressed as a function of $\mathbf{v}_{\mathbf{u}''\mathbf{u}'}$ and $\mathbf{v}_{\mathbf{u}'\mathbf{u}}$; we need $\mathbf{v}_{\mathbf{u}\mathbf{u}'}$ as well.

(ii) Putting $\mathbf{v}'' := \mathbf{v}_{\mathbf{u}''\mathbf{u}}$, $\mathbf{v}' := \mathbf{v}_{\mathbf{u}''\mathbf{u}'}$, $\mathbf{v} := \mathbf{v}_{\mathbf{u}'\mathbf{u}}$, $\mathbf{v} := -\mathbf{v}_{\mathbf{u}\mathbf{u}'}$ (but this is impossible!) we obtain the usual formula for addition of relative velocities ([5], formulae (2.55.) and (2.56.)).

4.4. ⌈In usual literature one does not make difference between $\mathbf{v}_{\mathbf{uu}'}$ and $-\mathbf{v}_{\mathbf{u}'\mathbf{u}}$; one speaks about a relative velocity $\mathbf{v}$ and takes it for granted that if a masspoint moves with velocity $\mathbf{v}$ relative to an other masspoint then that second one moves with velocity $-\mathbf{v}$ relative to the first one. It is a consequence of a tricky manipulation that there is no contradiction in doing so: vectors are given by components with respect to convenient bases and then the components of $\mathbf{v}_{\mathbf{uu}'}$ and $\mathbf{v}_{\mathbf{u}'\mathbf{u}}$ become opposite to each other.

How to choose bases in $\mathbf{E}_{\mathbf{u}}$ and $\mathbf{E}_{\mathbf{u}'}$ to get this relation between components of relative velocities? We need a method for transferring an orthogonal basis from $\mathbf{E}_{\mathbf{u}'}$ into $\mathbf{E}_{\mathbf{u}}$ in some even way. This is done by the Lorentz boost $F(\mathbf{u}, \mathbf{u}')$ (see 1.9.).⌋

4.5. $F(\mathbf{u}, \mathbf{u}')_{\mathbf{I}}$ establishes a canonical identification of the tangent spaces of $V(1)$ at $\mathbf{u}'$ and $\mathbf{u}$. Under this canonical identification the relative velocity $\mathbf{v}_{\mathbf{u}'\mathbf{u}}$ becomes the additive inverse of $\mathbf{v}_{\mathbf{u}'\mathbf{u}}$:

Proposition.

$$F(\mathbf{u}, \mathbf{u}')_{\mathbf{I}}\mathbf{v}_{\mathbf{uu}'} = -\mathbf{v}_{\mathbf{u}'\mathbf{u}} \,. \blacksquare$$

It is worth mentioning that the addition of relative velocities is not a vector addition either after such a canonical identification, i.e.

$$\mathbf{v}_{\mathbf{u}''\mathbf{u}} \neq F(\mathbf{u}, \mathbf{u}')_{\mathbf{I}}\mathbf{v}_{\mathbf{u}''\mathbf{u}'} + \mathbf{v}_{\mathbf{u}'\mathbf{u}} \,.$$

4.6. Let $\mathbf{U}$ be a global inertial observer and suppose C is a light signal (see 2.1.). Then we can repeat the arguments of 4.1. to arrive at the definition of the relative velocity of the light signal C with respect to the observer $\mathbf{U}$.

Definition. Let $\mathbf{u} \in V(1)$, $\mathbf{w} \in V(0)$. Then we call

$$\mathbf{v}_{\mathbf{wu}} := \frac{\mathbf{w}}{(\mathbf{w} \mid\!\mid \mathbf{u})} - \mathbf{u}$$

the *relative velocity of* $\mathbf{w}$ *with respect to* $\mathbf{u}$.

Proposition. Let $\mathbf{u} \in V(1)$, $\mathbf{w} \in V(0)$. Then $\mathbf{v}_{\mathbf{wu}}$ is an element of $\frac{\mathbf{E}_{\mathbf{u}}}{\mathbf{I}}$ and $|\mathbf{v}_{\mathbf{wu}}| = 1$.

4.7. Proposition. Proposition 4.3. (the addition formula of relative velocities) is valid even if $\mathbf{u}, \mathbf{u}' \in V(1)$ and $\mathbf{u}'' \in V(0)$.

4.8. Let $\mathbf{U}$ be a regular observer and let r be a world line function. Then $\tau_{\mathbf{U}} \circ r$ is smooth and, according to 3.5., $\mathrm{D}(\tau_{\mathbf{U}} \circ r) = (\mathbf{U} \circ r \mid\!\mid \dot{r})$ is nowhere zero. Hence $\tau_{\mathbf{U}} \circ r$ is invertible and we put

$$z_{\mathbf{U}} := (\tau_{\mathbf{U}} \circ r)^{-1} : I_{\mathbf{U}} \rightarrowtail \mathbf{I} \,.$$

$z_{\mathbf{U}}$ gives the dependence of the proper time of r on $\mathbf{U}$-time.

Definition.

$$r_{\mathbf{U}}: I_{\mathbf{U}} \rightarrowtail E_{\mathbf{U}}, \qquad t \mapsto C_{\mathbf{U}}(r(z_{\mathbf{U}}(t)))$$

$$\mathbf{v}_{\mathbf{U}}: I_{\mathbf{U}} \rightarrowtail \frac{\mathbf{E}_{\mathbf{U}(\cdot)}}{\mathbf{I}}, \qquad t \mapsto \mathbf{v}_{\dot{r}(z_{\mathbf{U}}(t))\mathbf{U}(r(z_{\mathbf{U}}(t)))}$$

are called the **U**-projections of r and $\dot{r}$, respectively (the last symbol denotes the relative velocity of $\dot{r}(z_{\mathbf{U}}(t))$ with respect to $\mathbf{U}(r(z_{\mathbf{U}}(t)))$.

Proposition. (i) $\dot{z}_{\mathbf{U}} = \sqrt{1 - |\mathbf{v}_{\mathbf{U}}|^2}$,

(ii) $\dot{r}_{\mathbf{U}} = \mathbf{v}_{\mathbf{U}}$

Proof. Apply the formula $\dot{z}_{\mathbf{U}} = \dfrac{1}{\mathrm{D}(\tau_{\mathbf{U}} \circ r) \circ z_{\mathbf{U}}}$ and Remark 4.2. (ii).

5. Splitting of tensor fields

Necessary mathematical tools:

Tensor products and quotients (Chapter IV.)

The formulae of splittings given in 5.7. and 5.8. can be understood without the involved knowledge on smooth manifolds.

5.1. An observer splits not only the corresponding space-time region but **M** and **M*** and their various tensor products, too. This splitting, however, depends on the points of the domain of the observer, or better to say, on "the instantaneous velocities" of the observer, i.e. on the values of the vector field that the observer is.

After some preparation we give the splitting due to an observer in 5.7.

5.2. We recall that for $\mathbf{u} \in V(1)$ the map

$$\mathbf{s}_{\mathbf{u}}: \mathbf{M} \longrightarrow \mathbf{I} \times \mathbf{E}_{\mathbf{u}}, \quad \mathbf{x} \mapsto (\tau_{\mathbf{u}}(\mathbf{x}), \mathbf{x}\|_{\mathbf{u}}) = (-[\mathbf{u}|\mathbf{x}), \mathbf{x} + \mathbf{u} \otimes [\mathbf{u}|\mathbf{x}))$$

is a linear bijection whose inverse is

$$(\mathbf{t}, \mathbf{q}) \mapsto \mathbf{u} \otimes \mathbf{t} + \mathbf{q} \qquad ((\mathbf{t}, \mathbf{q}) \in \mathbf{I} \times \mathbf{E}_{\mathbf{u}})$$

(see 1.10.). Observe that $\mathbf{u} \otimes \tau_{\mathbf{u}}(\mathbf{x})$ and $\mathbf{x}\|_{\mathbf{u}}$ are projections of $\mathbf{x}$ parallel to the complementary subspaces $\mathbf{E}_{\mathbf{u}}$ and $\mathbf{u} \otimes \mathbf{I}$.

Definition. $\tau_{\mathbf{u}}(\mathbf{x})$ and $\mathbf{x}\|_{\mathbf{u}}$ are called the **u**-*time component* and the **u**-*space component* of $\mathbf{x} \in \mathbf{M}$, respectively. $\mathbf{s}_{\mathbf{u}}$ is the *splitting* of vectors into time and space components corresponding to **u**.

Remark. If A is a space of scalars then $\frac{\mathbf{M}}{\mathbf{A}}$ is split into $\frac{\mathbf{I}}{\mathbf{A}} \times \frac{\mathbf{E_u}}{\mathbf{A}}$ by $(\mathbf{s_u})_{\mathbf{A}} := = \frac{\mathbf{s_u}}{\mathrm{id}_{\mathbf{A}}}$ (see IV. 2.10. (ii).)

In particular, $(\mathbf{s_u})_{\mathbf{I}}$ maps $\frac{\mathbf{M}}{\mathbf{I}}$ into $\mathbb{R} \times \frac{\mathbf{E_u}}{\mathbf{I}}$ and

$$(\mathbf{s_u})_{\mathbf{I}}(\mathbf{u}') = ((\mathbf{u}' \cdot \mathbf{u}), \mathbf{u}' - (\mathbf{u}' \cdot \mathbf{u})\mathbf{u}) = \frac{1}{\sqrt{1 - |\mathbf{v_{u'u}}|^2}}(1, \mathbf{v_{u'u}})$$

for $\mathbf{u}, \mathbf{u}' \in V(1)$. Compare this formula with that of Remark I. 5.2.

5.3. In contradistinction to the nonrelativistic case, here both **u**-time components and **u**-space components vary with **u**, without exception. The so called transformation law that shows how the **u**-time components and **u**-space components vary with **u** can be well seen from the following formulae.

5.4. Proposition. Let $\mathbf{u}, \mathbf{u}' \in V(1)$. Then

$$(\mathbf{s_{u'}} \circ \mathbf{s_u}^{-1})(\mathbf{t}, \mathbf{q}) = ((\mathbf{u}' \cdot \mathbf{u})\mathbf{t} - [\mathbf{u}' | \mathbf{q}), (\mathbf{u} - (\mathbf{u}' \cdot \mathbf{u})\mathbf{u}') \otimes \mathbf{t} + \mathbf{q} + \mathbf{u}' \otimes [\mathbf{u}' | \mathbf{q})) =$$

$$= \Bigg(\frac{1}{\sqrt{1 - |\mathbf{v_{u'u}}|^2}}(\mathbf{t} - [\mathbf{v_{u'u}} | \mathbf{q})),$$

$$\frac{1}{\sqrt{1 - |\mathbf{v_{u'u}}|^2}} \left(\mathbf{v_{uu'}} \otimes \mathbf{t} - \frac{\mathbf{v_{uu'}} + \mathbf{v_{u'u}}\sqrt{1 - |\mathbf{v_{u'u}}|^2}}{|\mathbf{v_{u'u}}|^2} \otimes [\mathbf{v_{u'u}} | \mathbf{q}) \right) + \mathbf{q} \Bigg),$$

$$((\mathbf{t}, \mathbf{q}) \in \mathbf{I} \times \mathbf{E_u}).$$

Proof. Since $[\mathbf{u} | \mathbf{q}) = \mathbf{0}$, we can substitute $(\mathbf{u}' \cdot \mathbf{u}) \left[\frac{\mathbf{u}'}{(\mathbf{u}' \cdot \mathbf{u})} - \mathbf{u} \,\middle|\, \mathbf{q} \right)$ for $[\mathbf{u}' | \mathbf{q})$. Furthermore,

$$(\mathbf{u}' \cdot \mathbf{u}) = \frac{1}{\sqrt{1 - |\mathbf{v_{u'u}}|^2}}$$

and

$$\mathbf{u}' \left(\frac{1}{(\mathbf{u}' \cdot \mathbf{u})} - (\mathbf{u}' \cdot \mathbf{u}) \right) = \mathbf{v_{u'u}} + \frac{\mathbf{v_{uu'}}}{\sqrt{1 - |\mathbf{v_{u'u}}|^2}},$$

whence an easy computation yields the final result.

5.5. The above transformation formula is nearly the same as the usually cited one ([5], formula (2.25.)). The essential difference is that here both $\mathbf{v_{u'u}}$ and $\mathbf{v_{uu'}}$ appear and neither of them can be eliminated. This is the consequence of the fact that $\mathbf{v_{u'u}}$ and $\mathbf{v_{uu'}}$ are not additive inverses of each other and that $\mathbf{s_{u'}} \circ \mathbf{s_u}^{-1}$ maps $\mathbf{I} \times \mathbf{E_u}$ into $\mathbf{I} \times \mathbf{E_{u'}}$, hence $(\mathbf{t}, \mathbf{q})$ and $(\mathbf{s_{u'}} \circ \mathbf{s_u}^{-1})(\mathbf{t}, \mathbf{q})$ are not in the same vector space. This important and non-negligible fact does not appear in usual literature because there the vectors are identified with their components corresponding to convenient bases, in other words both $\mathbf{E_u}$ and $\mathbf{E_{u'}}$ are identified with $\mathbb{R}^3$. To be able to compare the **u**-space component and the **u**′-space component of a vector we must apply the Lorentz boost from **u**′ to **u**, i.e. identify $\mathbf{E_{u'}}$ with $\mathbf{E_u}$.

A simple calculation shows the Lorentz boost $F(\mathbf{u}, \mathbf{u}')$ applied to the second component of $(\mathbf{s_{u'}} \circ \mathbf{s_u}^{-1})(\mathbf{t}, \mathbf{q})$ converts $\mathbf{v_{uu'}}$ into $-\mathbf{v_{u'u}}$ and makes no other alteration. Hence we get a formula in which only $\mathbf{v_{u'u}}$ occurs.

Putting

$$\mathbf{v} := -\mathbf{v_{u'u}},$$

we obtain the usual transformation law that we can extract, e.g., from Definition 7.14. substituting $\mathrm{id}_{\mathbf{E_u}}$ for R.

5.6. The elements of $\mathbf{M}^*$ can be split by $\mathbf{r_u} := (\mathbf{s_u}^{-1})^*$ into $(\mathbf{I} \times \mathbf{E_u})^* = \mathbf{I}^* \times \mathbf{E_u^*}$ and we get

$$\mathbf{r_u}(\mathbf{k}) = (\mathbf{k_I}(\mathbf{u}), \mathbf{k}|_{\mathbf{E_u}}) = (\mathbf{k_I}(\mathbf{u}), \mathbf{i_u^*}(\mathbf{k})) \qquad (\mathbf{k} \in \mathbf{M}^*).$$

Since $\mathbf{M}^*$ can be identified with $\dfrac{\mathbf{M}}{\mathbf{I}\otimes\mathbf{I}}$, $\mathbf{M}^*$ can be split also by $(\mathbf{s_u})_{\mathbf{I}\otimes\mathbf{I}}$ into $\dfrac{\mathbf{I}}{\mathbf{I}\otimes\mathbf{I}} \times \dfrac{\mathbf{E_u}}{\mathbf{I}\otimes\mathbf{I}} = \mathbf{I}^* \times \mathbf{E_u^*}$. In other words, both $\mathbf{r_u}$ and $(\mathbf{s_u})_{\mathbf{I}\otimes\mathbf{I}}$ map $\mathbf{M}^*$ onto $\mathbf{I}^* \times \mathbf{E_u^*}$. The relation between the two splittings is simple:

Proposition. Put

$$e_{\mathbf{u}} : \mathbf{I}^* \times \mathbf{E_u^*} \longrightarrow \mathbf{I}^* \times \mathbf{E_u^*}, \qquad (\mathbf{a}, \mathbf{p}) \mapsto (-\mathbf{a}, \mathbf{p}).$$

Then

$$\mathbf{r_u} = e_{\mathbf{u}} \circ (\mathbf{s_u})_{\mathbf{I}\otimes\mathbf{I}}.$$

As a consequence, the transformation law showing how the splitting $\mathbf{r_u}$ of $\mathbf{M}^*$ depends on **u** can be easily obtained from Proposition 5.4.

Remark. In usual literature (in a somewhat different setting) the split forms of $\mathbf{k} \in \mathbf{M}^*$ by $\mathbf{r_u}$ and $(\mathbf{s_u})_{\mathbf{I}\otimes\mathbf{I}}$ are called the *covariant* and *contravariant* components of **k**, respectively.

5.7. Let **U** be a regular observer. The splitting of (Dom **U**) × **M** and (Dom **U**) × $\mathbf{M}^*$ corresponding to **U** can be obtained by the maps appearing in the following proposition whose proof is evident from the smoothness of **U** and $S_{\mathbf{U}}$.

Proposition. The maps

$$(\mathrm{Dom}\,\mathbf{U}) \times \mathbf{M} \longrightarrow (I_{\mathbf{U}} \times E_{\mathbf{U}}) \times (\mathbf{I} \times \mathbf{E}_{\mathbf{U}(.)}),$$

$$(x, \mathbf{x}) \mapsto (S_{\mathbf{U}}(x), \mathbf{s}_{\mathbf{U}(x)}(\mathbf{x})) = (\tau_{\mathbf{U}}(x), C_{\mathbf{U}}(x), \boldsymbol{\tau}_{\mathbf{U}(x)}(\mathbf{x}), \mathbf{x}\|_{\mathbf{U}(x)});$$

$$(\mathrm{Dom}\,\mathbf{U}) \times \mathbf{M}^* \longrightarrow (I_{\mathbf{U}} \times E_{\mathbf{U}}) \times (\mathbf{I}^* \times \mathbf{E}^*_{\mathbf{U}(\cdot)}),$$

$$(x, \mathbf{k}) \mapsto (S_{\mathbf{U}}(x), \mathbf{r}_{\mathbf{U}(x)}(\mathbf{k})) = (\tau_{\mathbf{U}}(x), C_{\mathbf{U}}(x), \mathbf{k_I}(\mathbf{U}(x)), \mathbf{k}|_{\mathbf{E}_{\mathbf{U}(x)}})$$

are smooth injections. (The notation $E_{\mathbf{U}(\cdot)}$ in the above maps is not totally correct. The maps can be better interpreted by the following consideration.)

Remarks. (i) $(I_\mathbf{U} \times E_\mathbf{U}) \times (\mathbf{I} \times E_{\mathbf{U}(\cdot)})$ in the above formulae stands for the tangent bundle of $I_\mathbf{U} \times E_\mathbf{U}$ and the splitting of (Dom **U**) × **M** (the tangent bundle of Dom **U**) is $DS_\mathbf{U}$, the derivative of $S_\mathbf{U}$ (see Proposition 3.5.).

Similarly, the splitting of (Dom **U**) × **M*** (the contangent bundle of Dom **U**) is $DS_\mathbf{U}^{*-1}$ (see VII. 2.20.).

(ii) Compare these maps with the corresponding maps in nonrelativistic space-time model (see I. 5.7.) and notice that the present ones are more complicated if **U** is not inertial.

(iii) Similar maps can be written for a general (not necessarily regular) observer **U**; then if J is an artificial time on Dom **U**, we should replace $S_\mathbf{U}$ by $S_{J,\mathbf{U}}$ and $\tau_\mathbf{U}$ by τ_J.

5.8. Vector fields and covector fields are split by a regular observer **U** as follows.

Let $\mathbf{X} : \text{Dom}\,\mathbf{U} \rightarrowtail \mathbf{M}$ be a vector field and $\mathbf{K} : \text{Dom}\,\mathbf{U} \rightarrowtail \mathbf{M}^*$ a covector field. They are transferred by the observer **U** into the maps

$$I_\mathbf{U} \times E_\mathbf{U} \rightarrowtail \mathbf{I} \times \mathbf{E}_{\mathbf{U}(\cdot)}, \qquad (t, q) \mapsto (\tau_{\mathbf{U}(t\cap q)}(\mathbf{X}(t\cap q)), \mathbf{X}(t\cap q)\|_{\mathbf{U}(t\cap q)}),$$

$$I_\mathbf{U} \times E_\mathbf{U} \rightarrowtail \mathbf{I}^* \times \mathbf{E}^*_{\mathbf{U}(\cdot)}, \qquad (t, q) \mapsto (\mathbf{K}(t\cap q)_\mathbf{I}(\mathbf{U}(t\cap q)), \mathbf{K}(t\cap q)|_{\mathbf{E}_{\mathbf{U}(t\cap q)}}),$$

respectively. The following diagram helps us to see what is done:

$$I_\mathbf{U} \times E_\mathbf{U} \xrightarrow{\;S_\mathbf{U}^{-1}\;} M \xrightarrow{\;\mathbf{X}\;} \mathbf{M} \xrightarrow{\;\mathbf{s}_{\mathbf{U}(\cdot)}\;} \mathbf{I} \times \mathbf{E}_{\mathbf{U}(\cdot)}$$

$$I_\mathbf{U} \times E_\mathbf{U} \xrightarrow{\;S_\mathbf{U}^{-1}\;} M \xrightarrow{\;\mathbf{K}\;} \mathbf{M}^* \xrightarrow{\;\mathbf{r}_{\mathbf{U}(\cdot)}\;} \mathbf{I}^* \times \mathbf{E}^*_{\mathbf{U}(\cdot)}.$$

In terms of the theory of smooth manifolds we can say that the split form of **X** and **K** by **U** are $(S_\mathbf{U})_*\mathbf{X}$ and $(S_\mathbf{U}^{-1})^*\mathbf{K}$, respectively (see VII. 3.9. and VII. 4.8.).

5.9. Observers manipulate tensor fields and cotensor fields, too. We shall detail only the splitting of 2-tensors and 2-cotensors, the general formula can be derived from it.

Let $\mathbf{u} \in V(1)$. Then

$$\mathbf{s}_\mathbf{u} \otimes \mathbf{s}_\mathbf{u} : \mathbf{M} \otimes \mathbf{M} \longrightarrow (\mathbf{I} \times \mathbf{E}_\mathbf{u}) \otimes (\mathbf{I} \times \mathbf{E}_\mathbf{u}) =$$
$$= (\mathbf{I} \otimes \mathbf{I}) \times (\mathbf{I} \otimes \mathbf{E}_\mathbf{u}) \times (\mathbf{E}_\mathbf{u} \otimes \mathbf{I}) \times (\mathbf{E}_\mathbf{u} \otimes \mathbf{E}_\mathbf{u}),$$

$$\mathbf{r}_\mathbf{u} \otimes \mathbf{r}_\mathbf{u} : \mathbf{M}^* \otimes \mathbf{M}^* \longrightarrow (\mathbf{I}^* \times \mathbf{E}^*_\mathbf{u}) \otimes (\mathbf{I}^* \times \mathbf{E}^*_\mathbf{u}) =$$
$$= (\mathbf{I}^* \otimes \mathbf{I}^*) \times (\mathbf{I}^* \otimes \mathbf{E}^*_\mathbf{u}) \times (\mathbf{E}^*_\mathbf{u} \otimes \mathbf{I}^*) \times (\mathbf{E}^*_\mathbf{u} \otimes \mathbf{E}^*_\mathbf{u})$$

are the splittings of 2-tensors and 2-cotensors according to **u**.

We shall use similar notations to those in I. 5.10. Thus antisymmetric 2-tensors split by **u** take the form

$$(\mathbf{a}, \mathbf{A}) := \begin{pmatrix} \mathbf{0} & -\mathbf{a}^* \\ \mathbf{a} & \mathbf{A} \end{pmatrix} \qquad (\mathbf{a} \in \mathbf{E_u} \otimes \mathbf{I},\ \mathbf{A} \in \mathbf{E_u} \wedge \mathbf{E_u}),$$

and antisymmetric 2-cotensors split by **u** take the form

$$(\mathbf{z}, \mathbf{Z}) := \begin{pmatrix} \mathbf{0} & -\mathbf{z}^* \\ \mathbf{z} & \mathbf{Z} \end{pmatrix} \qquad (\mathbf{z} \in \mathbf{E_u^*} \otimes \mathbf{I}^*,\ \mathbf{Z} \in \mathbf{E_u^*} \wedge \mathbf{E_u^*}).$$

5.10. Using similar conventions and considerations as in I. 5.11., we can state:

Proposition. Let T and F be an antisymmetric 2-tensor and 2-cotensor, respectively. Then for $\mathbf{u} \in V(1)$ we have

$$(\mathbf{s_u} \otimes \mathbf{s_u})(T) = ((\mathrm{id}_\mathbf{M} \otimes \tau_\mathbf{u})(T),\ T - (\mathrm{id}_\mathbf{M} \otimes \tau_\mathbf{u})(T) \wedge \mathbf{u})$$

(cf. IV. 2.8. (iii)) and

$$(\mathbf{r_u} \otimes \mathbf{r_u})(F) = ((\mathbf{i_u^*} \circ F)_\mathbf{I}(\mathbf{u}),\ (\mathbf{i_u^*} \circ F)|_{\mathbf{E_u}}) =$$
$$= (F|_{\mathbf{E_u} \times \mathbf{u} \otimes \mathbf{I}},\ F|_{\mathbf{E_u} \times \mathbf{E_u}}).$$

Definition. We shall employ the following names

$(\mathrm{id}_\mathbf{M} \otimes \tau_\mathbf{u})(T)$	is the **u**-space-time component of T,
$T - (\mathrm{id}_\mathbf{M} \otimes \tau_\mathbf{u})(T) \wedge \mathbf{u}$	is the **u**-space-space component of T,
$(\mathbf{i_u^*} \circ F)_\mathbf{I}(\mathbf{u})$ or $F\|_{\mathbf{E_u} \times \mathbf{u} \otimes \mathbf{I}}$	is the **u**-space-time component of F,
$(\mathbf{i_u^*} \circ F)\|_{\mathbf{E_u}}$ or $F\|_{\mathbf{E_u} \times \mathbf{E_u}}$	is the **u**-space-space component of F.

5.11. In contradistinction to the nonrelativistic case, here both **u**-space-time and **u**-space-space components vary with **u**. The transformation law showing how the split components of antisymmetric two-tensors vary with the elements of $V(1)$ can be clearly seen from the following formulae. The transformation formulae for antisymmetric two-cotensors is similar; it can be obtained, for instance, by the identification of $\mathbf{M}^*$ with $\dfrac{\mathbf{M}}{\mathbf{I} \otimes \mathbf{I}}$ (cf. 5.6.).

Proposition. Let $\mathbf{u}, \mathbf{u}' \in V(1)$. Then, for $\mathbf{a} \in \mathbf{E}_\mathbf{u} \otimes \mathbf{I}$, $\mathbf{A} \in \mathbf{E}_\mathbf{u} \wedge \mathbf{E}_\mathbf{u}$, we have

$$(\mathbf{s}_{\mathbf{u}'} \otimes \mathbf{s}_{\mathbf{u}'}) \circ (\mathbf{s}_\mathbf{u} \otimes \mathbf{s}_\mathbf{u})^{-1}(\mathbf{a}, \mathbf{A}) =$$

$$= ((\mathbf{u}' \,\mathsf{H}\, \mathbf{u})\mathbf{a} + \mathbf{u}' \otimes g(\mathbf{u}', \mathbf{a}) + \mathbf{A}^\mathbf{I}(\mathbf{u}'),$$

$$\mathbf{a} \wedge (\mathbf{u} - (\mathbf{u}' \,\mathsf{H}\, \mathbf{u})\mathbf{u}') - (\mathbf{u} \wedge \mathbf{u}') \otimes g(\mathbf{u}', \mathbf{a}) + \mathbf{A} - \mathbf{A}^\mathbf{I}(\mathbf{u}') \wedge \mathbf{u}')) =$$

$$= \left(\frac{1}{\sqrt{1 - |\mathbf{v}_{\mathbf{u}'\mathbf{u}}|^2}} \left(\mathbf{a} - \frac{\mathbf{v}_{\mathbf{u}\mathbf{u}'} + \mathbf{v}_{\mathbf{u}'\mathbf{u}} \sqrt{1 - |\mathbf{v}_{\mathbf{u}'\mathbf{u}}|^2}}{|\mathbf{v}_{\mathbf{u}'\mathbf{u}}|^2} \otimes g(\mathbf{v}_{\mathbf{u}'\mathbf{u}}, \mathbf{a}) + \mathbf{A}^\mathbf{I}(\mathbf{v}_{\mathbf{u}'\mathbf{u}}) \right),\right.$$

$$\frac{1}{\sqrt{1 - |\mathbf{v}_{\mathbf{u}'\mathbf{u}}|^2}} \left(\mathbf{a} \wedge \mathbf{v}_{\mathbf{u}\mathbf{u}'} + \mathbf{A}^\mathbf{I}(\mathbf{v}_{\mathbf{u}'\mathbf{u}}) \wedge \frac{\mathbf{v}_{\mathbf{u}\mathbf{u}'} + \mathbf{v}_{\mathbf{u}'\mathbf{u}} \sqrt{1 - |\mathbf{v}_{\mathbf{u}'\mathbf{u}}|^2}}{|\mathbf{v}_{\mathbf{u}'\mathbf{u}}|^2} \right) + \mathbf{A} -$$

$$\left. - \frac{\mathbf{v}_{\mathbf{u}'\mathbf{u}} \wedge \mathbf{v}_{\mathbf{u}\mathbf{u}'}}{|\mathbf{v}_{\mathbf{u}'\mathbf{u}}|^2 \sqrt{1 - |\mathbf{v}_{\mathbf{u}'\mathbf{u}}|^2}} \otimes g(\mathbf{v}_{\mathbf{u}'\mathbf{u}}, \mathbf{a}) \right)$$

where $\mathbf{A} \in \mathbf{E}_\mathbf{u} \wedge \mathbf{E}_\mathbf{u} \subset \mathbf{M} \wedge \mathbf{M}$ is considered as a linear map $\mathbf{M}^* \longrightarrow \mathbf{M}$ (see IV. 1.5. (iv)) and $\mathbf{M}^*$ is identified with $\frac{\mathbf{M}}{\mathbf{I} \otimes \mathbf{I}}$, hence $\mathbf{A}^\mathbf{I} := \mathbf{A} \otimes \mathrm{id}_\mathbf{I}$ maps $\frac{\mathbf{M}}{\mathbf{I}}$ into $\mathbf{M} \otimes \mathbf{I}$; moreover we used the notation $g(\mathbf{w}, \mathbf{a}) := g(\mathbf{w} \otimes \mathbf{t}, \frac{\mathbf{a}}{\mathbf{t}})$ for $\mathbf{w} \in \frac{\mathbf{M}}{\mathbf{I}}$, $\mathbf{a} \in \mathbf{M} \otimes \mathbf{I}$ and for non-zero $\mathbf{t} \in \mathbf{I}$.

Proof. We easily deduce that

$$(\mathbf{s}_\mathbf{u} \otimes \mathbf{s}_\mathbf{u})^{-1}(\mathbf{a}, \mathbf{A}) = \mathbf{a} \wedge \mathbf{u} + \mathbf{A}$$

which implies our first equality. Since $g(\mathbf{u}, \mathbf{a}) = \mathbf{0}$, $\mathbf{A}^\mathbf{I}(\mathbf{u}) = \mathbf{0}$, we can substitute $(\mathbf{u}' \,\mathsf{H}\, \mathbf{u}) g\left(\frac{\mathbf{u}'}{(\mathbf{u}' \,\mathsf{H}\, \mathbf{u})} - \mathbf{u}, \mathbf{a} \right)$ for $g(\mathbf{u}', \mathbf{a})$ and $(\mathbf{u}' \,\mathsf{H}\, \mathbf{u}) \mathbf{A}^\mathbf{I}\left(\frac{\mathbf{u}'}{(\mathbf{u}' \,\mathsf{H}\, \mathbf{u})} - \mathbf{u} \right)$ for $\mathbf{A}^\mathbf{I}(\mathbf{u}')$. Furthermore, we can express $(\mathbf{u}' \,\mathsf{H}\, \mathbf{u})$, $\mathbf{u}$ and $\mathbf{u}'$ by $\mathbf{v}_{\mathbf{u}\mathbf{u}'}$ and $\mathbf{v}_{\mathbf{u}'\mathbf{u}}$ as in the proof of Proposition 5.4. to obtain the final result.

5.12. The above transformation formula resembles the usually cited one ([5], formula (4.84′)). The essential difference is that here both $\mathbf{v}_{\mathbf{u}'\mathbf{u}}$ and $\mathbf{v}_{\mathbf{u}\mathbf{u}'}$ appear and neither of them can be eliminated. This is the consequence of the fact mentioned in 5.5. The term containing $\mathbf{v}_{\mathbf{u}'\mathbf{u}} \wedge \mathbf{v}_{\mathbf{u}\mathbf{u}'}$ is very remarkable: such a term is missing from the usual transformation formula; evidently, if $\mathbf{v}_{\mathbf{u}\mathbf{u}'}$ were equal to $-\mathbf{v}_{\mathbf{u}'\mathbf{u}}$ then this term would be zero.

To get the usual transformation formula—where both $\mathbf{E}_\mathbf{u}$ and $\mathbf{E}_{\mathbf{u}'}$ are identified with $\mathbb{R}^3$—we must identify $\mathbf{E}_{\mathbf{u}'}$ with $\mathbf{E}_\mathbf{u}$ by the Lorentz boost $F(\mathbf{u}, \mathbf{u}')$. A simple calculation shows that $F(\mathbf{u}, \mathbf{u}') \otimes \mathrm{id}_\mathbf{I}$ and $F(\mathbf{u}, \mathbf{u}') \otimes F(\mathbf{u}, \mathbf{u}')$ applied to the first and second component of $(\mathbf{s}_{\mathbf{u}'} \otimes \mathbf{s}_{\mathbf{u}'}) \circ (\mathbf{s}_\mathbf{u} \otimes \mathbf{s}_\mathbf{u})^{-1}(\mathbf{a}, \mathbf{A})$, respectively, converts $\mathbf{v}_{\mathbf{u}\mathbf{u}'}$ into $-\mathbf{v}_{\mathbf{u}'\mathbf{u}}$ and makes no other alteration. Hence we get a formula in which only $\mathbf{v}_{\mathbf{u}'\mathbf{u}}$ occurs and the curious term drops out.

5.13. We make the following comments on splitting.

⌈All the fields used in physics: force fields, potentials, electromagnetic fields are split by different observers in different ways. There is no potential which can be considered as a scalar field (cf. I. 5.13.).

We emphasize as in I. 5.13. that the so called scalar potential and vector potential are relative notions, they depend on observers, they are split components of a potential which is an absolute object (with respect to its space-time properties). Similarly, electromagnetic field is absolute, electric field and magnetic field appear only to observers. We would endeavour to avoid these relative notions.⌋

6. Reference systems

Necessary mathematical tools:

Tensor products and quotients (Chapter IV.)
Affine spaces, differentiation (Sections VI. 1., VI. 2.)

The notion of Poincaré transformations appears here; the Poincaré group is defined and investigated in the next section.

6.1. ⌈To describe physical phenomena, an observer coordinatizes his own time and space. The coordinatization is realized physically by measuring time with a clock (having a dial) and by introducing numbered reference lines in space. As a result, a time point will be represented by a number and a space point will be represented by a triplet of numbers.⌋

6.2. Definition. A *regular reference system* is a pair $(\mathbf{U}, K)$ where $\mathbf{U}$ is a regular observer and $K : \mathrm{Dom}\,\mathbf{U} \rightarrowtail \mathbb{R}^4$ is a *coordinatization*, i.e. $\mathrm{Dom}\,K$ and $\mathrm{Ran}\,K$ are open subsets and K is a diffeomorphism between $\mathrm{Dom}\,K$ and $\mathrm{Ran}\,K$.

The regular reference system $(\mathbf{U}, K)$ is called

(i) *natural*, if there are maps $T: I_{\mathbf{U}} \rightarrowtail \mathbb{R}$ and $S: E_{\mathbf{U}} \rightarrowtail \mathbb{R}^3$ such that

$$K = (T \times S) \circ S_{\mathbf{U}},$$

i.e. $\mathbf{U}$-time and $\mathbf{U}$-space are coordinatized separately; in this case we say that the coordinatization K is *adapted* to the observer $\mathbf{U}$;

(ii) *global* if $\mathbf{U}$ is global and $\mathrm{Dom}\,K = M$;

(iii) *rigid*, *rotation free* and *inertial* if $\mathbf{U}$ is rigid, rotation free and inertial, respectively.

Remark. We can define *reference systems* with observers that are not regular in a similar way.

6.3. As we have mentioned, non-inertial observers are rather ill, hence one uses them practically never. Reference systems connected with global inertial observers, too, may have bad properties if their coordinatization is unpleasant. The adapted coordinatizations which are affine are of the simplest kind. They can be defined and treated like in nonrelativistic space-time model (see I. 6.4.). We shall deal here only with the simplest ones.

Definition. A *Lorentz reference system* is a global, inertial, natural reference system such that the coordinatization is an affine map whose derivative is a $g-(\nmid)-$ orthogonal map $\mathbf{M} \longrightarrow \mathbb{R}^4$ (cf. 1.13.).

6.4. Proposition. Let (U, K) be a Lorentz reference system and suppose $\mathbf{c} \in V(1)$ is the constant value of U. Then there are an $o \in M$, a non-zero $\mathbf{e} \in \mathbf{I}$, a g-orthogonal basis $\mathbf{e}_0, \mathbf{e}_1, \mathbf{e}_2, \mathbf{e}_3$ in $\mathbf{M}$, normed to $\mathbf{e}$; all these elements are uniquely determined and

$$\mathbf{e}_0 = \mathbf{c} \otimes \mathbf{e},$$

$$K(x) = \left(\frac{g(\mathbf{e}_k, x - o)}{g(\mathbf{e}_k, \mathbf{e}_k)} \right)_{k=0,1,2,3}.$$

Proof. Let o be the unique element of M that $K(o)=0$ holds for. Then $K(x) = \mathrm{D}K(x-o)$ $(x \in M)$. Let $\mathbf{e}_k$ $(k=0, 1, 2, 3)$ be the elements of $\mathbf{M}$ whose images by $\mathrm{D}K$ are the standard basis elements of $\mathbb{R}^4$. Then the $g-(\nmid)$-orthogonality of $\mathrm{D}K$ implies that $\{\mathbf{e}_k : k=0, 1, 2, 3\}$ form a g-orthogonal basis, normed to some $\delta \in \mathbf{I}^+$. Since K is adapted to U, $\mathbf{e}_0$ is an element of $\mathbf{c} \otimes \mathbf{I}$. According to whether the timelike vector $\mathbf{e}_0$ is positive or negative, we put $\mathbf{e} := \delta$ or $\mathbf{e} := -\delta$ respectively, and the proof is finished.

Definition. Using the above notations, we call $\mathbf{c} \in V(1)$, $o \in M$, $|\mathbf{e}| \in \mathbf{I}^+$, and $\{\mathbf{e}_k : k=0, 1, 2, 3\}$ the *velocity*, the *origin*, the *space-time unit* and the *space-time basis* of the Lorentz system, respectively.

6.5. Proposition. If (U, K) is a Lorentz reference system and L is a Poincaré transformation then $(\mathrm{ar}(L)\mathrm{D}L_{\mathbf{I}} \circ \mathrm{U}, K \circ L^{-1})$ is a Lorentz reference system, too. The space-time units of the two systems are equal.

Proof. It is trivial that $\mathrm{ar}(L)\mathrm{D}L_{\mathbf{I}}(\mathbf{u}) \in V(1)$ for $\mathbf{u} \in V(1)$, hence $\mathrm{ar}(L)\mathrm{D}L_{\mathbf{I}} \circ \mathrm{U}$ is a global inertial observer. Furthermore $K \circ L^{-1}$ is evidently an affine map whose derivative is g-$(\nmid)$-orthogonal. Lastly, it is an easy task to show that $K \circ L^{-1}$ is adapted to $\mathrm{ar}(L)\mathrm{D}L_{\mathbf{I}} \circ \mathrm{U}$, thus they form a Lorentz reference system.

If o and $\{\mathbf{e}_k : k=0, 1, 2, 3\}$ are the origin and the space-time basis of (U, K), then we deduce at once that $L(o)$ and $\{\mathrm{D}L(\mathbf{e}_k) : k=0, 1, 2, 3\}$ are the origin and space-time basis of $(\mathrm{ar}(L)\mathrm{D}L_{\mathbf{I}} \circ \mathrm{U}, K \circ L^{-1})$. As a consequence, the space-time units of the two systems are equal.

6.6. Proposition. If (U, K) and (U', K') are Lorentz reference systems with equal space-time units then there exists a unique Poincaré transformation L such that $\mathrm{U}' = \mathrm{ar}(L)\mathrm{D}L_{\mathbf{I}} \circ \mathrm{U}$, $K' = K \circ L^{-1}$.

Proof. Let o, o' and $\{\mathbf{e}_k : k=0,1,2,3\}$, $\{\mathbf{e}'_k : k=0,1,2,3\}$ be the correspon-ding origins and space-time bases. Define the Lorentz transformation F by $F(\mathbf{e}_k) := \mathbf{e}'_k$ $(k=0,1,2,3)$. Then

$$L(x) := o' + F(x-o) \qquad (x \in M)$$

is a Poincaré transformation having the desired properties.

If N is another Poincaré transformation such that $K' = K \circ N^{-1} = K \circ L^{-1}$ then by composing the equality from the left by K^{-1} we get that $N = L$.

6.7. Now we are to define a notion of some comparison between reference systems. We refer to I. 6.9. and I. 6.10. for the motivations and the necessary mathematical tools.

Definition. Let $(\mathbf{U}, K)$ and $(\mathbf{U}', K')$ be two reference systems. A *morphism* from $(\mathbf{U}, K)$ into $(\mathbf{U}', K')$ is a pair (F, λ) where

(i) $F : \mathrm{Dom}\,\mathbf{U} \longrightarrow \mathrm{Dom}\,\mathbf{U}'$ is a little diffeomorphism,

(ii) $\lambda : F[\mathrm{Dom}\,\mathbf{U}] \longrightarrow \mathbb{R} \setminus \{0\}$ is a smooth map such that

(a) $F_* \mathbf{U} \subset \lambda \mathbf{U}'$,

(b) $K \circ F^{-1} \subset K'$.

If F is surjective (i.e. F is a diffeomorphism between $\mathrm{Dom}\,\mathbf{U}$ and $\mathrm{Dom}\,\mathbf{U}'$) and equality holds in (b) then (F, λ) is called an *isomorphism* and the two reference systems are *isomorphic* by (F, λ). ∎

Note that (b) implies F maps $\mathrm{Dom}\,K$ into $\mathrm{Dom}\,K'$.

6.8. Proposition. If L is a Poincaré transformation and (L, λ) is a morphism of reference systems then λ is a constant map, $\lambda = \mathrm{ar}\,L$.

Proof. Suppose $\mathbf{U}$ and $\mathbf{U}'$ are the observers corresponding to the reference systems. Then

$$DL_{\mathbf{I}}\mathbf{U}(L^{-1}(x)) = \lambda(x)\mathbf{U}'(x) \quad (x \in L[\mathrm{Dom}\,\mathbf{U}] \subset \mathrm{Dom}\,\mathbf{U}').$$

$\mathbf{U}(L^{-1}(x))$ and $\mathbf{U}'(x)$ are elements of $V(1)$. If $\mathrm{ar}\,L = 1$ then the left side, too, is in $V(1)$, hence $\lambda(x) = 1$. If $\mathrm{ar}\,L = -1$, then the negative of the left side is in $V(1)$, hence $\lambda(x) = -1$. ∎

As a consequence, in such a case it is convenient to consider the Poincaré transformation L as a morphism instead of $(L, \mathrm{ar}\,L)$.

6.9. Definition. Two reference systems isomorphic by a proper Poincaré transformation are called *equivalent*.

7. The Lorentz and Poincaré groups

Necessary mathematical tools:

Tensor products and quotients (Chapter IV.)
Elements of group theory
Elements of Lie group theory (Section VIII. 1.)
Some knowledge on certain Lie groups (Section VIII. 4.)

7.1. ⌈Lorentz and Poincaré transformations play a fundamental role in special relativistic physics; they are considered as "symmetries" of space-time. Unfortunately, one usually takes $\mathbb{R}\times\mathbb{R}^3$ as space-time, thus Lorentz and Poincaré transformations are given as transformations of $\mathbb{R}\times\mathbb{R}^3$. Consequently, the usually defined transformation groups of $\mathbb{R}\times\mathbb{R}^3$ ([12], p. 172)—let us call them *arithmetic Lorentz group* and *arithmetic Poincaré group*—have a number of additional structures that the Lorentz and Poincaré groups defined by means of a special relativistic space-time model have not.

We emphasize that using a special relativistic space-time model we must define the Lorentz and Poincaré groups only in terms of the structure given in Definition 1.2.⌋

7.2. In the sequel we consider a special relativistic space-time model $(M, \mathbf{I}, g)$.

The g-orthogonal group of $\mathbf{M}$ (see VIII. 4.11.) is called the *Lorentz group* and is denoted by $\mathscr{L}$. $\mathscr{L}$ is a six dimensional Lie group having $\mathbf{A}(g)$ as its Lie algebra.

The sets S, T and L (see 1.3.) are invariant under Lorentz transformations. Furthermore, it is evident that if F is a Lorentz transformation and the elements $\mathbf{x}$ and $\mathbf{y}$ of T have the same arrow then $F\mathbf{x}$ and $F\mathbf{y}$ have the same arrow. As a consequence, Lorentz transformations either preserve or reverse the arrow.

7.3. Proposition. Let F be a Lorentz transformation. Then $F[\mathbf{E_u}]$ equals $\mathbf{E}_{F\cdot\mathbf{u}}$ or $\mathbf{E}_{-F\cdot\mathbf{u}}$ according to whether F preserves or reverses the arrow, and $F|_{\mathbf{E_u}}$ is either orientation preserving or orientation reversing (with respect to the orientation described in 1.7.) for all $\mathbf{u}\in V(1)$ simultaneously.

Proof. Suppose F is arrow preserving and orientation preserving. Then $F|_{\mathbf{E_u}}$ is evidently orientation preserving for all $\mathbf{u}\in V(1)$. Continue to take into account all the possibilities of preserving and reversing the arrow and the orientation, to obtain the desired result.

7.4. Definition. For $F\in\mathscr{L}$ we put

$$\operatorname{ar} F := \begin{cases} +1 & \text{if } F \text{ is arrow preserving,} \\ -1 & \text{if } F \text{ is not arrow preserving,} \end{cases}$$

$$\operatorname{sign} F := \begin{cases} +1 & \text{if } F|_{\mathbf{E_u}} \text{ is orientation preserving,} \\ -1 & \text{if } F|_{\mathbf{E_u}} \text{ is not orientation preserving} \end{cases}$$

and we introduce

$$\mathscr{L}^{+\rightarrow} := \{F \in \mathscr{L} \,\vdots\, \operatorname{sign} F = \operatorname{ar} F = 1\},$$

$$\mathscr{L}^{+\leftarrow} := \{F \in \mathscr{L} \,\vdots\, \operatorname{sign} F = -\operatorname{ar} F = 1\},$$

$$\mathscr{L}^{-\rightarrow} := \{F \in \mathscr{L} \,\vdots\, \operatorname{sign} F = -\operatorname{ar} F = -1\},$$

$$\mathscr{L}^{-\leftarrow} := \{F \in \mathscr{L} \,\vdots\, \operatorname{sign} F = \operatorname{ar} F = -1\}.$$

$\mathscr{L}^{+\rightarrow}$ is called the *proper Lorentz group*.
$\mathscr{L}^{\rightarrow} := \mathscr{L}^{+\rightarrow} \cup \mathscr{L}^{-\rightarrow}$ is the *orthocronous Lorentz group*.

7.5. Proposition. The Lorentz group has four connected components which are the above listed subsets. The identity component is the proper Lorentz group. The elements of $\mathscr{L}^{+\rightarrow}$ and $\mathscr{L}^{-\leftarrow}$ preserve the orientation of $\mathbf{M}$ whereas the elements of $\mathscr{L}^{+\leftarrow}$ and $\mathscr{L}^{-\rightarrow}$ do not.

Proof. Only the assertion concerning the components of $\mathscr{L}$ is not trivial. It will be proved in 7.19. by establishing a Lie group isomorphism from $\mathscr{L}$ onto an other Lie group for which the result is immediate.

7.6. Definition. The *Poincaré group* $\mathscr{P}$ is the subgroup of the bijective affine transformations of M defined by

$$\mathscr{P} := \{L \in \mathrm{GA}(M) \,\vdots\, \mathrm{D}L \in \mathscr{L}\}.$$

7.7. Definition. For $L \in \mathscr{P}$ we put

$$\operatorname{ar} L := \operatorname{ar} \mathrm{D}L, \qquad \operatorname{sign} L := \operatorname{sign} \mathrm{D}L$$

and we introduce $\mathscr{P}^{+\rightarrow}, \mathscr{P}^{+\leftarrow}, \mathscr{P}^{-\rightarrow}, \mathscr{P}^{-\leftarrow}$ as the subsets of $\mathscr{P}$ consisting of the elements whose derivative is in $\mathscr{L}^{+\rightarrow}, \mathscr{L}^{+\leftarrow}, \mathscr{L}^{-\rightarrow}, \mathscr{L}^{-\leftarrow}$, respectively.

$\mathscr{P}^{+\rightarrow}$ is called the *proper Poincaré group*.

$\mathscr{P}^{\rightarrow} := \mathscr{P}^{+\rightarrow} \cup \mathscr{P}^{-\rightarrow}$ is the *orthocronous Poincaré group*.

7.8. Theorem. The Poincaré group is a ten dimensional Lie group having four connected components which are the above listed subsets. The identity component is the proper Poincaré group. The elements of $\mathscr{P}^{+\rightarrow}$ and $\mathscr{P}^{-\leftarrow}$ preserve the orientation of M whereas the elements of $\mathscr{P}^{+\leftarrow}$ and $\mathscr{P}^{-\rightarrow}$ do not.

Proof. Combine Proposition 7.5. with VIII. 4.10. (ii)–(iii).

7.9. Theorem. The Lie algebra of the Poincaré group is

$$\mathsf{L}(\mathscr{P}) = \{X : M \longrightarrow \mathbf{M} \,\vdots\, X \text{ is affine, } \mathrm{D}X \in \mathsf{A}(g)\}$$

equipped with the commutator

$$[X, Y] = (\mathrm{D}X) \circ Y - (\mathrm{D}Y) \circ X.$$

Proof. See VIII. 4.8. and VIII. 4.10. (iii).

7.10. Proposition. The proper Poincaré group can be considered as the subgroup of the automorphism group of $(M, \mathbf{I}, g)$ consisting of the strict automorphisms.

Proof. If $L \in \mathscr{P}^{+\rightarrow}$ then $(L, \mathrm{id}_{\mathbf{I}})$ is a strict automorphism.

Conversely, if $(F, \mathrm{id}_{\mathbf{I}})$ is a strict automorphism then F must be a proper Poincaré transformation.

7.11. Theorem 7.8. is based on the fact that the Poincaré group—a subgroup of the affine group of M—can be transferred into a subgroup of the affine group of $\mathbf{M}$ (see (VIII. 4.8.). This possibility is very useful in applications and reveals the nature of the relation between the Poincaré group and the arithmetic Poincaré group, hence we examine it more closely.

7.12. Definition. The *vectorial Poincaré group* $\mathscr{P}_V$ is $\mathbf{M} \times \mathscr{L}$ equipped with the multiplication law

$$(\mathbf{x}', F')(\mathbf{x}, F) := (\mathbf{x}' + F'\mathbf{x}, F'F).$$

Remarks. (i) The vectorial Poincaré group is the semi-direct product of $\mathrm{Tn}(\mathbf{M})$ (the additive group of $\mathbf{M}$) and the Lorentz group $\mathscr{L}$ (see VIII. 4.5.).

(ii) The elements of the vectorial Poincaré group can be considered in a natural way as affine transformations of $\mathbf{M}$:

$$\hat{\mathbf{x}} \mapsto (\mathbf{x}, F)(\hat{\mathbf{x}}) := F\hat{\mathbf{x}} + \mathbf{x}.$$

Then the group multiplication coincides with the composition.

(iii) The Lie algebra of $\mathscr{P}_V$ is $\mathbf{M} \times \mathbf{A}(g)$ equipped with the commutator

$$[(\mathbf{x}', A'), (\mathbf{x}, A)] = (A'\mathbf{x} - A\mathbf{x}', A'A - AA').$$

(see VIII. 4.4. and VIII. 4.5.); the elements of $\mathbf{L}(\mathscr{P}_V)$ can be considered as affine transformations of $\mathbf{M}$.

(iv) We can introduce the subsets $\mathscr{P}_V^{+\rightarrow}, \mathscr{P}_V^{+\leftarrow}, \mathscr{P}_V^{-\rightarrow}, \mathscr{P}_V^{-\leftarrow}$ by an evident definition based on Definition 7.4.

7.13. We know that given $o \in M$, we can establish isomorphisms z_o: $\mathscr{P} \longrightarrow \mathscr{P}_V$ and $w_o: \mathbf{L}(\mathscr{P}) \longrightarrow \mathbf{L}(\mathscr{P}_V)$; for different o and o' the isomorphisms are different as it is shown in VIII. 4.7. and VIII. 4.8. (i).

The topological properties of $\mathscr{L}$ (implying the topological properties of $\mathscr{P}$) can be seen by giving a further Lie group isomorphism which maps $\mathscr{P}_V$ onto a group whose underlying set exhibits clearly its topological properties. Moreover, this isomorphism plays an important role in applications, too.

7.14. Definition. Let $\mathbf{c} \in V(1)$. The *proper* **c**-*split Lorentz group* $\mathscr{L}_0(\mathbf{c})$ is $B_{\mathbf{c}} \times \mathrm{SO}(\gamma_{\mathbf{c}})$ considered as a linear transformation group of $\mathbf{I} \times \mathbf{E}_{\mathbf{c}}$ given by

$$(\hat{\mathbf{t}}, \hat{\mathbf{q}}) \mapsto (\mathbf{v}, R)(\hat{\mathbf{t}}, \hat{\mathbf{q}}) := \left(\kappa(\mathbf{v})(\hat{\mathbf{t}} + [\mathbf{v} | R\hat{\mathbf{q}}]),\ \kappa(\mathbf{v})\mathbf{v} \otimes \hat{\mathbf{t}} + R\hat{\mathbf{q}} + \frac{\kappa(\mathbf{v}) - 1}{|\mathbf{v}|^2} \mathbf{v} \otimes [\mathbf{v} | R\hat{\mathbf{q}}]\right)$$

where we used notations introduced in 1.5. and 1.6., and

$$\kappa(\mathbf{v}) := \frac{1}{\sqrt{1-|\mathbf{v}|^2}}.$$

The *proper* **c**-*split Poincaré group* $\mathscr{P}_0(\mathbf{c})$ is $\mathbf{I} \times \mathbf{E_c} \times B_\mathbf{c} \times \mathrm{SO}(\gamma_\mathbf{c})$ considered as an affine transformation group of $\mathbf{I} \times \mathbf{E_c}$ given by

$$(\hat{\mathbf{t}}, \hat{\mathbf{q}}) \mapsto (\mathbf{t}, \mathbf{q}, \mathbf{v}, R)(\hat{\mathbf{t}}, \hat{\mathbf{q}}) := (\mathbf{v}, R)(\hat{\mathbf{t}}, \hat{\mathbf{q}}) + (\mathbf{t}, \mathbf{q}).$$

7.15. Definition. The **c**-*split Lorentz group* $\mathscr{L}_s(\mathbf{c})$ and the **c**-*split Poincaré group* $\mathscr{P}_s(\mathbf{c})$ are the transformation groups of $\mathbf{I} \times \mathbf{E_c}$ generated by $\mathscr{L}_0(\mathbf{c})$ and $\mathscr{P}_0(\mathbf{c})$, respectively, and by the two inversions

$$T(\mathbf{c})(\hat{\mathbf{t}}, \hat{\mathbf{q}}) := (-\hat{\mathbf{t}}, \hat{\mathbf{q}}), \qquad ((\hat{\mathbf{t}}, \hat{\mathbf{q}}) \in \mathbf{I} \times \mathbf{E_c})$$
$$S(\mathbf{c})(\hat{\mathbf{t}}, \hat{\mathbf{q}}) := (\hat{\mathbf{t}}, -\hat{\mathbf{q}}).$$

Remark. $S(\mathbf{c})$ can be incorporated in the given form of proper **c**-split Poincaré transformations, namely $S(\mathbf{c}) := (\mathbf{0}, \mathbf{0}, \mathbf{0}, -\mathrm{id}_{\mathbf{E_c}})$; in other words, $\mathscr{P}_0(\mathbf{c})$ extended by $S(\mathbf{c})$ is $\mathbf{I} \times \mathbf{E_c} \times B_\mathbf{c} \times \mathrm{O}(\gamma_\mathbf{c})$.

On the other hand, $T(\mathbf{c})$ cannot be handled in a similar way.

Every element of $\mathscr{P}_s(\mathbf{c})$ can be written in the form

$$(\mathbf{t}, \mathbf{q}, \mathbf{v}, R)\, T(\alpha)$$

where $(\mathbf{t}, \mathbf{q}, \mathbf{v}, R) \in \mathbf{I} \times \mathbf{E_c} \times B_\mathbf{c} \times \mathrm{O}(\gamma_\mathbf{c})$, $\alpha = \pm 1$ and

$$T(\alpha) := \begin{cases} \mathrm{id}_{\mathbf{I} \times \mathbf{E_c}} & \text{if } \alpha = 1, \\ T(\mathbf{c}) & \text{if } \alpha = -1. \end{cases}$$

7.16. Remark. Notice the similarity between the split Galilean group and the **c**-split Poincaré group, and notice their following differences, too:

(ii) the underlying set of the **c**-split Poincaré group depends on $\mathbf{c} \in V(1)$ whereas the underlying set of the split Galilean group is unique;

(ii) the action of a split Galilean transformation is very simple, hence the multiplication law of the group is simple as well; on the other hand, the action of a **c**-split Poincaré transformation is complicated, hence the multiplication law of the group (which is omitted because we do not need it) is complicated as well;

(iii) $\{\mathbf{0}\} \times \{\mathbf{0}\} \times \frac{\mathbf{E}}{\mathbf{I}} \times \{\mathrm{id}_\mathbf{E}\}$ is a commutative subgroup of $\mathscr{G}_s$ whereas the elements of $\{\mathbf{0}\} \times \{\mathbf{0}\} \times B_\mathbf{c} \times \{\mathrm{id}_{\mathbf{E_c}}\}$ do not commute in general and they do not constitute a subgroup of $\mathscr{P}_s(\mathbf{c})$.

7.17. Proposition. The **c**-split Poincaré group is a ten dimensional Lie group having four connected components, namely

$$\mathscr{P}_0(\mathbf{c}), \quad T(\mathbf{c})\mathscr{P}_0(\mathbf{c}), \quad S(\mathbf{c})\mathscr{P}_0(\mathbf{c}), \quad T(\mathbf{c})S(\mathbf{c})\mathscr{P}_0(\mathbf{c}).$$

Proof. $\mathscr{P}_0(\mathbf{c})$ is evidently ten dimensional and connected; the other three subsets, as images of $\mathscr{P}_0(\mathbf{c})$ by continuous maps, are connected as well. We end the proof by noting the obvious fact that the subsets in question are disjoint and their union is the whole group.

7.18. Proposition. The Lie algebra $\mathsf{L}(\mathscr{P}_s(\mathbf{c}))$ of $\mathscr{P}_s(\mathbf{c})$ is $\mathbf{I}\times\mathbf{E}_\mathbf{c}\times\frac{\mathbf{E}_\mathbf{c}}{\mathbf{I}}\times\mathbf{A}(\gamma_\mathbf{c})$ equipped with the commutator

$$[(\mathbf{t}',\mathbf{q}',\mathbf{v}',B'),(\mathbf{t},\mathbf{q},\mathbf{v},B)]=$$
$$=([\mathbf{v}'|\mathbf{q})-[\mathbf{v}|\mathbf{q}'),\mathbf{v}'\otimes\mathbf{t}-\mathbf{v}\otimes\mathbf{t}'+B'\mathbf{q}-B\mathbf{q}',B_1'\mathbf{v}-B_1\mathbf{v}',B'B-BB'+\mathbf{v}'\wedge\mathbf{v}).$$

Proof. The underlying set of $\mathsf{L}(\mathscr{P}_s(\mathbf{c}))$ is $\mathbf{I}\times\mathbf{E}_\mathbf{c}\times\frac{\mathbf{E}_\mathbf{c}}{\mathbf{I}}\times\mathbf{A}(\gamma_\mathbf{c})$ because the tangent space at a point of a Cartesian product of manifolds is the Cartesian product of the corresponding tangent spaces.

The elements of $\mathsf{L}(\mathscr{P}_s(\mathbf{c}))$ can be considered as affine transformation of $\mathbf{I}\times\mathbf{E}_\mathbf{c}$, as follows:

$$(\hat{\mathbf{t}},\hat{\mathbf{q}})\mapsto(\mathbf{t},\mathbf{q},\mathbf{v},B)(\hat{\mathbf{t}},\hat{\mathbf{q}}):=([\mathbf{v}|\hat{\mathbf{q}})+\mathbf{t},\mathbf{v}\otimes\hat{\mathbf{t}}+B\hat{\mathbf{q}}+\mathbf{q}).$$

Then the commutator has the form

$$[(\mathbf{t}',\mathbf{q}',\mathbf{v}',B'),(\mathbf{t},\mathbf{q},\mathbf{v},B)]=$$
$$=\mathrm{D}(\mathbf{t}',\mathbf{q}',\mathbf{v}',B')\circ(\mathbf{t},\mathbf{q},\mathbf{v},B)-\mathrm{D}(\mathbf{t},\mathbf{q},\mathbf{v},B)\circ(\mathbf{t}',\mathbf{q}',\mathbf{v}',B')$$

where $\mathrm{D}(\mathbf{t},\mathbf{q},\mathbf{v},B)$ is the derivative of the affine map $(\mathbf{t},\mathbf{q},\mathbf{v},B)$ and equals $(\mathbf{0},\mathbf{0},\mathbf{v},B)$.

7.19. A straightforward but laborious calculation leads us to the following important results.

Theorem. Let $\mathbf{c}\in V(1)$. Then

$$\mathbf{z}_\mathbf{c}:\mathscr{P}_V\longrightarrow\mathscr{P}_s(\mathbf{c}),\qquad(\mathbf{x},F)\mapsto\mathbf{s}_\mathbf{c}\circ(\mathbf{x},F)\circ\mathbf{s}_\mathbf{c}^{-1}$$

(see 1.10.) is a Lie group isomorphism, in detail

$$\mathbf{z}_\mathbf{c}(\mathbf{x},F)=\left(\boldsymbol{\tau}_\mathbf{c}(\mathbf{x}),\mathbf{x}\|_\mathbf{c},\frac{F_\mathbf{I}\mathbf{c}}{(\mathbf{c}\cdot F_\mathbf{I}\mathbf{c})}-\mathbf{c},R(\mathbf{c},F)\right)T(\mathrm{ar}\,F)\qquad\circledast$$

for $(\mathbf{x},F)\in\mathscr{P}_V$, where

$$R(\mathbf{c},F):=\left[\mathrm{id}_\mathbf{M}+\mathbf{c}\otimes\mathbf{c}-\frac{(\mathbf{c}\cdot F_\mathbf{I}\mathbf{c})}{(\mathbf{c}\cdot F_\mathbf{I}\mathbf{c})+1}\left(\frac{F_\mathbf{I}\mathbf{c}}{(\mathbf{c}\cdot F_\mathbf{I}\mathbf{c})}-\mathbf{c}\right)\otimes\left(\frac{F_\mathbf{I}\mathbf{c}}{(\mathbf{c}\cdot F_\mathbf{I}\mathbf{c})}-\mathbf{c}\right)\right]F|_{\mathbf{E}_\mathbf{c}}$$

(notations: 1.5.).

Furthermore we have

$$\mathbf{z}_\mathbf{c}[\mathscr{P}_V^{+\rightarrow}]=\mathscr{P}_0(\mathbf{c})\qquad\mathbf{z}_\mathbf{c}[\mathscr{P}_V^{+\leftarrow}]=T(\mathbf{c})\mathscr{P}_0(\mathbf{c}),$$
$$\mathbf{z}_\mathbf{c}[\mathscr{P}_V^{-\rightarrow}]=S(\mathbf{c})\mathscr{P}_0(\mathbf{c}),\qquad\mathbf{z}_\mathbf{c}[\mathscr{P}_V^{-\leftarrow}]=T(\mathbf{c})S(\mathbf{c})\mathscr{P}_0(\mathbf{c}).$$

As a consequence, for $o \in M$,

$$z_{\mathbf{c},o} := \mathbf{z}_{\mathbf{c}} \circ z_o : \mathscr{P} \longrightarrow \mathscr{P}_s(\mathbf{c})$$

(see 7.13.) is a Lie group isomorphism as well. We can obtain $z_{\mathbf{c},o}(L)$ for $L \in \mathscr{P}$ from the right hand side of ⊛ by substituting $L(o) - o$ for $\mathbf{x}$ and $\mathrm{D}L$ for F.

Furthermore we have

$$z_{\mathbf{c},o}[\mathscr{P}^{+\rightarrow}] = \mathscr{P}_0(\mathbf{c}), \qquad z_{\mathbf{c},o}[\mathscr{P}^{+\leftarrow}] = T(\mathbf{c})\mathscr{P}_0(\mathbf{c}),$$

$$z_{\mathbf{c},o}[\mathscr{P}^{-\rightarrow}] = S(\mathbf{c})\mathscr{P}_0(\mathbf{c}), \qquad z_{\mathbf{c},o}[\mathscr{P}^{-\leftarrow}] = T(\mathbf{c})S(\mathbf{c})\mathscr{P}_0(\mathbf{c}).$$

Remarks. (i) Our theorem furnishes a proof of Theorem 7.8.

(ii) Since the Lorentz group, in the form $\{(\mathbf{0}, F) : F \in \mathscr{L}\}$, is a subgroup of the vectorial Poincaré group, a simple consideration shows how the Lorentz group is mapped by $\mathbf{z}_{\mathbf{c}}$ onto the group generated by $\{0\} \times \{\mathbf{0}\} \times B_{\mathbf{c}} \times \mathrm{SO}(\gamma_{\mathbf{c}})$ and $S(\mathbf{c})$ and $T(\mathbf{c})$, providing a proof of Proposition 7.5.

7.20. Consider the elements of $\mathsf{L}(\mathscr{P}_V)$ and $\mathsf{L}(\mathscr{P}_s(\mathbf{c}))$ as affine transformations of $\mathbf{M}$ and $\mathbf{I} \times \mathbf{E}_{\mathbf{c}}$, respectively, (see VIII. 4.4. and 7.18.) to derive, by a calculation similar to the previous one, the following important results.

Theorem. Let $\mathbf{c} \in V(1)$. Then

$$\mathbf{w}_{\mathbf{c}} : \mathsf{L}(\mathscr{P}_V) \longrightarrow \mathsf{L}(\mathscr{P}_s(\mathbf{c})), \quad (\mathbf{x}, A) \mapsto \mathbf{s}_{\mathbf{c}} \circ (\mathbf{x}, A) \circ \mathbf{s}_{\mathbf{c}}^{-1}$$

(see 1.10.) is a Lie algebra isomorphism, in detail

$$\mathbf{w}_{\mathbf{c}}(\mathbf{x}, A) = (\tau_{\mathbf{c}}(\mathbf{x}), \mathbf{x}\|_{\mathbf{c}}, A_{\mathbf{I}}\mathbf{c}, (\mathrm{id}_{\mathbf{M}} + \mathbf{c} \otimes \mathbf{c})A|_{\mathbf{E}_{\mathbf{c}}}) \qquad ⊛⊛$$

for $(\mathbf{x}, A) \in \mathsf{L}(\mathscr{P}_V)$.

As a consequence, for $o \in M$,

$$w_{\mathbf{c},o} := \mathbf{w}_{\mathbf{c}} \circ w_o : \mathsf{L}(\mathscr{P}) \longrightarrow \mathsf{L}(\mathscr{P}_s(\mathbf{c}))$$

(see 7.13.) is a Lie algebra isomorphism as well. We can obtain $w_{\mathbf{c},o}(X)$ for $X \in \mathsf{L}(\mathscr{P})$ from the right hand side of ⊛ ⊛ by substituting $X(o)$ for $\mathbf{x}$ and $\mathrm{D}X$ for A.

7.21. The isomorphisms constructed in Theorem 7.19. and 7.20. depend heavily on the elements of M and $V(1)$. If $o, o' \in M$ and $\mathbf{c}, \mathbf{c}' \in V(1)$ then

$$z_{\mathbf{c}',o'}(L) = m_{\mathbf{cc}'oo'}^{-1} z_{\mathbf{c},o}(L)\, m_{\mathbf{cc}'oo'} \qquad (L \in \mathscr{P}),$$

$$w_{\mathbf{c}',o'}(X) = m_{\mathbf{cc}'}^{-1}\, w_{\mathbf{c},o}(X)\, m_{\mathbf{cc}'oo'} \qquad (X \in \mathsf{L}(\mathscr{P})),$$

where

$$m_{\mathbf{cc}'} := \mathbf{s}_{\mathbf{c}} \circ \mathbf{s}_{\mathbf{c}'}^{-1}, \qquad m_{\mathbf{cc}'oo'} := s_{\mathbf{c},o} \circ s_{\mathbf{c}',o'}^{-1}.$$

In contradistinction to the nonrelativistic case (see I. 7.22.) $m_{\mathbf{cc}'}$ and $m_{\mathbf{cc}'oo'}$ are neither $\mathbf{c}$-split nor $\mathbf{c}'$-split Poincaré transformations: they map $\mathbf{I} \times \mathbf{E}_{\mathbf{c}'}$ onto $\mathbf{I} \times \mathbf{E}_{\mathbf{c}}$. The formula of transferring a $\mathbf{c}'$-split Poincaré transformation into a $\mathbf{c}$-split one is rather complicated.

7.22. The Poincaré group is isomorphic as a Lie group to the **c**-split Poincaré group for all $\mathbf{c} \in V(1)$; evidently, there is *no canonical isomorphism*. Given $(\mathbf{c}, o) \in V(1) \times M$, we can establish an isomorphism but for different $(\mathbf{c}, o)$ and $(\mathbf{c}', o')$ the isomorphisms are different—even their ranges differ—and all elements of $V(1) \times M$ have the "same right" to generate an isomorphism.

One can prove that the **c**-split Poincaré group is the "Poincaré group of the special relativistic space-time model $(\mathbf{I} \times \mathbf{E_c}, \mathbf{I}, g_\mathbf{c})$". Due to the additional structures of this special model (see 1.15.) the **c**-split Poincaré group has a number of additional structures that the Poincaré group has not.

The form of **c**-split Poincaré transformations, in accordance with the usual terminology for arithmetic Poincaré transformations, tempts us to consider elements

$(\mathbf{t}, \mathbf{0}, \mathbf{0}, \mathrm{id}_{\mathbf{E_c}})$	as a time translation,
$(\mathbf{0}, \mathbf{q}, \mathbf{0}, \mathrm{id}_{\mathbf{E_c}})$	as a space translation,
$(\mathbf{0}, \mathbf{0}, \mathbf{v}, \mathrm{id}_{\mathbf{E_c}})$	as a special Lorentz transformation,
$(\mathbf{0}, \mathbf{0}, \mathbf{0}, R)$	as a space rotation,
$T(\mathbf{c})$	as time inversion,
$S(\mathbf{c})$	as space inversion.

However, the **c**-split Poincaré transformations are not transformations of our space-time; the above notions arise from the additional structure of $(\mathbf{I} \times \mathbf{E_c}, \mathbf{I}, g_\mathbf{c})$. Evidently, the Poincaré group does not contain time translations and time inversion (there is no time), space translations, space rotations and space inversion (there is no space), and does not contain even special Lorentz transformations and space-time inversion. ⌈In customary special relativistic physics one uses such elements of the arithmetic Poincaré group and one attaches to them a physical meaning. We shall see in volume Models in Mechanics the physical significance of the non-existence of such elements of the Poincaré group.⌋

Only space-time translations are in the Poincaré group; they constitute the subgroup $\{L \in \mathscr{P} : \mathrm{D}L = \mathrm{id}_\mathbf{M}\}$.

The Lorentz group (but not the Poincaré group) contains an element, namely $-\mathrm{id}_\mathbf{M}$, which can be considered as "vectorial space-time inversion" but there are no "vectorial time inversion" and "vectorial space inversion" either in the Lorentz group.

On the base of 7.19. we can symbolically say that the Poincaré transformations

in $\mathscr{P}^{+\rightarrow}$ do not invert either time or space,
in $\mathscr{P}^{+\leftarrow}$ invert time and do not invert space,
in $\mathscr{P}^{-\rightarrow}$ do not invert time and invert space,
in $\mathscr{P}^{-\leftarrow}$ invert both time and space,

but there are no elements which one could call time inversion and space inversion.

8. Discussion

8.1. One often asserts in usual physics that nonrelativistic physics is the limit of special relativistic physics as the light speed tends to infinity, or that nonrelativistic physics is the limit of special relativistic physics, valid for relative velocities much smaller then the light speed.

Can we give such an exact statement concerning our space-time models? The answer is no.

We have two different mathematical structures. There is no reasonable way of introducing a topology or convergence notion even on a class of mathematical structures of the same kind (e.g., on the class of groups, and to say that a sequence of groups converges to a given one) and it is quite unnatural to introduce a convergence notion on a class consisting of structures of different kind (e.g., to say that a sequence of groups tends to an algebra).

There is no natural limit procedure in which a class of special relativistic space-time models converge to a nonrelativistic one.

8.2. The real meaning of the usually cited "limit" can be seen from the following consideration.

Let us fix a special relativistic global inertial observer having the velocity $\mathbf{u} \in V(1)$. Suppose he chooses an element o of M and splits space-time into $\mathbf{I} \times \mathbf{E}_{\mathbf{u}}$ according to Proposition 3.11.

Now the observer renames $\mathbf{I}$ to $\mathbf{D}$ and introduces a new one dimensional oriented real vector space $\mathbf{I}$, and chooses a positive element c of $\frac{\mathbf{D}}{\mathbf{I}}$. He identifies $\mathbf{I}$ with $\mathbf{D}$ by the map $\mathbf{t} \mapsto c \otimes \mathbf{t}$ and considers $\mathbf{I}$ as his time. As a consequence, if $\mathbf{u}' \in V(1)$, then the observer calls $\mathbf{v} := c \otimes \mathbf{v}_{\mathbf{u}'\mathbf{u}}$ the relative velocity of $\mathbf{u}'$ with respect to himself. Since the supremum of magnitudes of relative velocities is now c, c is the "light speed in the new units".

Substituting $\frac{\mathbf{v}}{c}$ for $\mathbf{v}_{\mathbf{u}'\mathbf{u}}$ or for $\mathbf{v}$ and $c \otimes \mathbf{t}$ for $\mathbf{t}$ in the formulae of Definition 7.14. and Proposition 5.4. etc. and then letting c to tend to infinity, we get the corresponding nonrelativistic formulae.

Here the limit procedure has an exact meaning, and we see that not a space-time model converges to another but some formulae (i.e. elements of real vector spaces) of a space-time model converge to those of another one.

We can say that the consequences of special relativistic space-time model concerning an inertial observer who observes only relative velocities small compared with the light speed are (nearly) the same as the consequences of nonrelativistic space-time model.

Exercises

1. Let $\mathbf{u}$, $\mathbf{u}'$ and $\mathbf{u}''$ be different elements of $V(1)$. Verify that, in general, the composition of the Lorentz boost from $\mathbf{u}$ to $\mathbf{u}'$ and the Lorentz boost from $\mathbf{u}'$ to $\mathbf{u}''$ is not the Lorentz boost from $\mathbf{u}$ to $\mathbf{u}''$:

$$F(\mathbf{u}'', \mathbf{u}')F(\mathbf{u}', \mathbf{u}) \neq F(\mathbf{u}'', \mathbf{u}) \quad \text{in general.}$$

Prove that a necessary condition for the equality is that $\mathbf{u}''$ be in the plane generated by $\mathbf{u}$ and $\mathbf{u}'$, and look for a sufficient condition.

2. Show that $\mathbf{z}_{\mathbf{u}}(0, F(\mathbf{u}', \mathbf{u})) = (\mathbf{0}, \mathbf{0}, \mathbf{v}_{\mathbf{u}'\mathbf{u}}, \mathrm{id}_{\mathbf{E}_{\mathbf{u}}})$ (see 7.19. and 4.2.) for all $\mathbf{u}, \mathbf{u}' \in V(1)$.

3. Let $\mathbf{c} \in V(1)$ and $\mathbf{z} \in \mathbf{E}_{\mathbf{c}}$. Put

$$\mathbf{f}(\mathbf{u}) := F(\mathbf{u}, \mathbf{c})\mathbf{z} \qquad (\mathbf{u} \in V(1)).$$

Find elements $\mathbf{u}$ and $\mathbf{u}'$ of $V(1)$ such that $\mathbf{f}(\mathbf{u}') \neq F(\mathbf{u}', \mathbf{u})\mathbf{f}(\mathbf{u})$.

4. Let r be a world line function. Show that if $\mathbf{X}$ is a Fermi-Walker parametrized vector field along the world line corresponding to r then

$$\mathbf{X}(\mathbf{t}+\mathbf{h}) = F(\dot{r}(\mathbf{t}+\mathbf{h}), \dot{r}(\mathbf{t}))\mathbf{X}(\mathbf{t}) + \mathrm{ordo}\,(\mathbf{h})$$

$$(\mathbf{t}, \mathbf{t}+\mathbf{h} \in \mathrm{Dom}\, r)$$

(see 1.9. and VII. 10.9.).

5. Let $o \in M$, $\mathbf{c} \in V(1)$. Verify that if $\mathbf{0} \neq \mathbf{z} \in \frac{\mathbf{E}_{\mathbf{c}}}{\mathbf{I} \otimes \mathbf{I}}$ and $\mathbf{q} \in \mathbf{E}_{\mathbf{c}}$, $(\mathbf{z} \mid \mathbf{q}) = 0$, and $\mathbf{d} \in \mathbf{I}$ then

$$H(\mathbf{z}, \mathbf{q}, \mathbf{d}) := o + \left\{ \mathbf{x} \in \mathbf{M} : -(|\mathbf{z}| \otimes \mathbf{c} \mid \mathbf{x})^2 + \left(\mathbf{z} \,\middle|\, \mathbf{x} - \mathbf{d} \otimes \frac{\mathbf{z}}{|\mathbf{z}|} \right)^2 = 1, \right.$$

$$\left. \mathbf{x} + \mathbf{c} \otimes [\mathbf{c} \mid \mathbf{x}) - \frac{\mathbf{z}}{|\mathbf{z}|} \otimes \left[\frac{\mathbf{z}}{|\mathbf{z}|} \,\middle|\, \mathbf{x} \right) = \mathbf{0} \right\} + \mathbf{q}$$

is a world line where $|\mathbf{z}|$ is the magnitude of $\mathbf{z}$ (see 2.9.); a corresponding world line function is

$$r(\mathbf{t}) := o + \left(\frac{\mathbf{c}}{|\mathbf{z}|} \mathrm{sh}\,(|\mathbf{z}| \otimes \mathbf{t}) + \frac{\mathbf{z}}{|\mathbf{z}| \otimes |\mathbf{z}|} \mathrm{ch}\,(|\mathbf{z}| \otimes \mathbf{t}) + \mathbf{d} \otimes \frac{\mathbf{z}}{|\mathbf{z}|} \right) + \mathbf{q} \qquad (\mathbf{t} \in \mathbf{I}).$$

Prove that

(i) $$\mathbf{c} = \dot{r}(\mathbf{0}), \qquad \mathbf{z} = \ddot{r}(\mathbf{0});$$

(ii) $$F(\dot{r}(\mathbf{s}), \dot{r}(\mathbf{t}))_{\mathbf{I}\otimes\mathbf{I}}\ddot{r}(\mathbf{t}) = \ddot{r}(\mathbf{s}) \qquad (\mathbf{t}, \mathbf{s} \in \mathbf{I});$$
(see 1.9.);

(iii) r is rotation free.

⌈Under the canonical identification of the subspaces $\mathbf{E}_{\dot{r}(\mathbf{t})}$ and $\mathbf{E}_{\dot{r}(\mathbf{s})}$ the accelerations at $\mathbf{t}$ and $\mathbf{s}$ coincide; hence this world line can be considered as the history of a uniformly accelerated masspoint.⌋

That is why we call the world line $H(\mathbf{z}, \mathbf{q}, \mathbf{d})$ *uniformly accelerated.*

(iv) If we take an $\mathbf{e} \in \mathbf{I}^+$ and construct the Lorentz reference system whose origin, velocity and space-time unit are o, $\mathbf{c}$ and $\mathbf{e}$, respectively, and the space-time basis is chosen in such a way that $\mathbf{e}_1 = \frac{1}{z}(\mathbf{z}\otimes\mathbf{e}\otimes\mathbf{e})$ where

$$z := |\mathbf{z}|\otimes\mathbf{e} \in \mathbb{R}$$

is the magnitude of $\mathbf{z}$ measured in the given space-time unit, then the world line takes the form

$$\{(x^0, x^1, x^2, x^3) \in \mathbb{R}^4 : -(x^0)^2 + (x^1 - e)^2 = \frac{1}{z^2}, \quad x^2 = b,\ x^3 = c\}$$

where $e := \frac{\mathbf{d}}{\mathbf{e}}$, b and c are some given real numbers; the corresponding world line function becomes

$$\mathbb{R} \longrightarrow \mathbb{R}^4, \quad t \mapsto \left(\frac{1}{z}\operatorname{sh} zt,\ e + \frac{1}{z}\operatorname{ch} zt,\ b,\ c\right).$$

6. Let $o \in M$, $\mathbf{c} \in V(1)$ and $\mathbf{0} \neq \mathbf{b} \in \frac{\mathbf{E_c}}{\mathbf{I}\otimes\mathbf{I}}$. Accept the notation $|\mathbf{b}|$ for the magnitude of $\mathbf{b}$ (see 2.9.). Consider the following observer

$$\operatorname{Dom} \mathbf{U} := o + \{\mathbf{x} \in \mathbf{M} : (\mathbf{b}|\mathbf{x}) > |(|\mathbf{b}|\otimes\mathbf{c}|\mathbf{x})|\},$$

$$\mathbf{U}(o + \mathbf{x}) := \frac{(\mathbf{b}|\mathbf{x})\mathbf{c} - \mathbf{b}\otimes[\mathbf{c}|\mathbf{x})}{\sqrt{(\mathbf{b}|\mathbf{x})^2 - (|\mathbf{b}|\otimes\mathbf{c}|\mathbf{x})^2}}.$$

Verify that

(i) for all $\mathbf{q} \in \mathbf{E_c}$, $(\mathbf{b}|\mathbf{q}) = 0$ and for all $x \in \operatorname{Dom}\mathbf{U}$, $x + \mathbf{q} \in \operatorname{Dom}\mathbf{U}$, we have $\mathbf{U}(x + \mathbf{q}) = \mathbf{U}(x)$;

(ii) the $\mathbf{U}$-lines are

$$H(\lambda\mathbf{b}, \mathbf{q}, \mathbf{0}) \qquad (\mathbf{q} \in \mathbf{E_c},\ (\mathbf{b}|\mathbf{q}) = 0,\ \lambda \in \mathbb{R}^+)$$

(see Exercise 5.);

(iii) $\mathbf{U}$ is regular and the $\mathbf{U}$-surfaces are

$$\{o + \mathbf{x} \in \operatorname{Dom}\mathbf{U} : (\alpha\mathbf{b} + |\mathbf{b}|\otimes\mathbf{c}|\mathbf{x}) = 0\} \qquad (\alpha \in]-1, 1[)$$

hence the instantaneous $\mathbf{U}$-spaces are Euclidean;

(iv) **U** is rotation free;
(v) **U** is rigid;
(vi) **U** restricted to a **U**-time point (**U**-surface) is constant.
Use the results of the previous exercise!

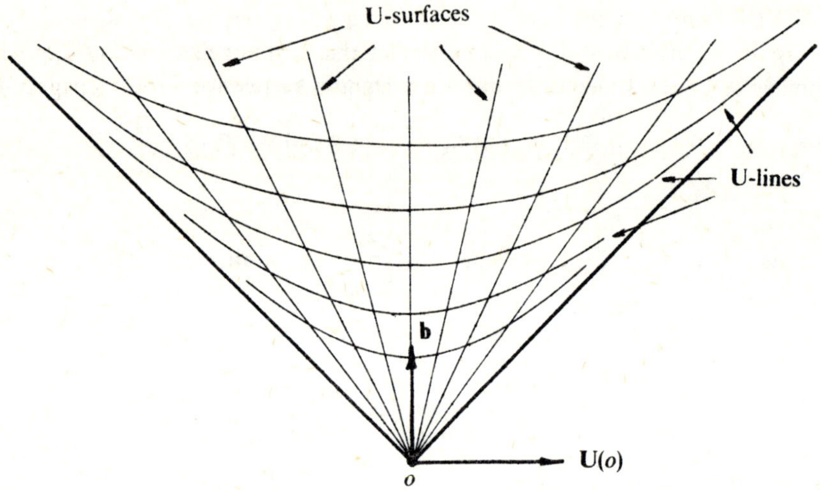

7. Let **U** be the observer given in the previous exercise. Show that the time passed between the two **U**-time points (**U**-surfaces) given by $\alpha_1, \alpha_2 \in]-1, 1[$ along the **U**-line given by $\lambda \in \mathbb{R}^+$ and $\mathbf{q} \in \mathbf{E_c}$, $(\mathbf{b}|\mathbf{q})=0$ equals $\frac{1}{\lambda|\mathbf{b}|}(\operatorname{arth}\alpha_2 - \operatorname{arth}\alpha_1)$, i.e. it depends on λ.

8. Let $o \in M$, $\mathbf{c} \in V(1)$, $\mathbf{0} \neq \mathbf{a} \in \frac{\mathbf{E_c}}{\mathbf{I}\otimes\mathbf{I}}$. Define the observer

$$\mathbf{U}(o+\mathbf{x}) := -\mathbf{a}\otimes[\mathbf{c}|\mathbf{x}) + \mathbf{c}\sqrt{1+(|\mathbf{a}|\otimes\mathbf{c}|\mathbf{x})^2} \qquad (\mathbf{x} \in \mathbf{M}).$$

Verify that
(i) for all $\mathbf{q} \in \mathbf{E_c}$ and for all $x \in M$, $\quad \mathbf{U}(x+\mathbf{q}) = \mathbf{U}(x)$;
(ii) the **U**-lines are

$$H(\mathbf{a}, \mathbf{q}, \mathbf{d}) \qquad (\mathbf{q} \in \mathbf{E_c},\ (\mathbf{a}|\mathbf{q})=0,\ \mathbf{d} \in \mathbf{I})$$

(see Exercise 5.);
(iii) **U** is regular and the **U**-surfaces are

$\{o+\mathbf{x} \in \mathbf{M} : [\mathbf{c}|\mathbf{x}) = \mathbf{0}\}$,

$$\{o+\mathbf{x}\in\mathbf{M} : \lambda[\mathbf{c}|\mathbf{x})>0, \quad \left(\mathbf{a}\,\middle|\,\mathbf{x}-\mathbf{h}\otimes\frac{\mathbf{a}}{|\mathbf{a}|}\right) =$$

$$= \sqrt{1+(|\mathbf{a}|\otimes\mathbf{c}|\mathbf{x})^2} - \operatorname{arcth}\sqrt{1+(|\mathbf{a}|\otimes\mathbf{c}|\mathbf{x})^2}\}$$

$$(\lambda=\pm 1, \quad \mathbf{h}\in\mathbf{I}),$$

hence the instantaneous **U**-spaces are not Euclidean except a single one;

(iv) **U** is rotation free;

(v) the affine planes directed by the subspace spanned by **c** and **a** intersect every instantaneous **U**-space in geodesics; as a consequence, **U** is not rigid;

(vi) the **U**-time point given by $\lambda=\pm 1$, $\mathbf{d}\in\mathbf{I}$ and the **U**-space point given by $\mathbf{h}\in\mathbf{I}$, $\mathbf{q}\in\mathbf{E}_{\mathbf{c}}$, $(\mathbf{a}|\mathbf{q})=0$ intersect each other if and only if $\mathbf{d}>\mathbf{h}$.

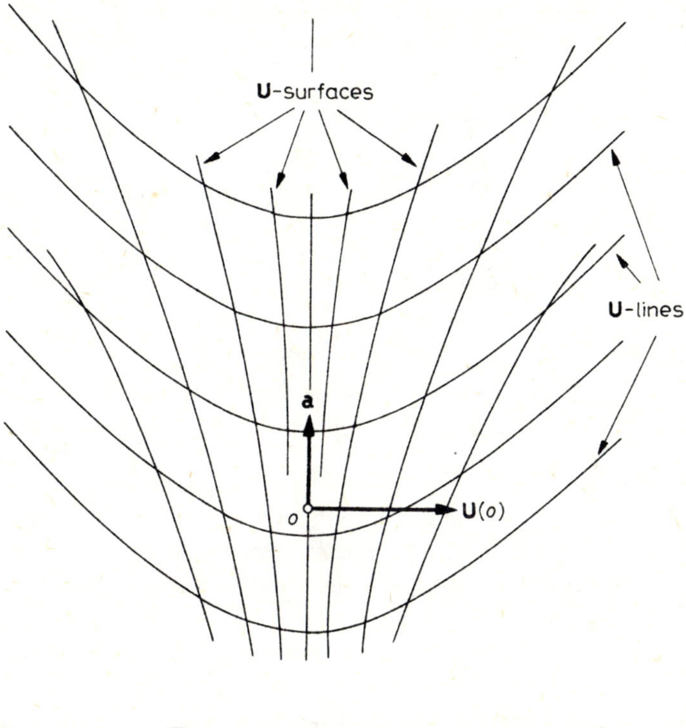

9. Let $\mathbf{c}\in V(1)$, $\mathbf{0}\neq\mathbf{z}\in\frac{\mathbf{E}_{\mathbf{c}}}{\mathbf{I}\otimes\mathbf{I}}$. Put

$$\mathbf{V}_{\mathbf{c},\mathbf{z}}:\mathbf{I}\longrightarrow V(1), \qquad \mathbf{t}\mapsto\mathbf{z}\otimes\mathbf{t}+\mathbf{c}\sqrt{1+(|\mathbf{z}|\otimes\mathbf{t})^2},$$

$$\tau_{\mathbf{c},\mathbf{z}} := \frac{\tau_{\mathbf{c}}}{\sqrt{-(\mathbf{z}|\cdot)^2+(|\mathbf{z}|\otimes\tau_{\mathbf{c}}(\cdot))^2}} : \mathbf{M}\longrightarrow\mathbf{I}.$$

where $\tau_{\mathbf{c}}(\mathbf{x}) := -[\mathbf{c}|\mathbf{x})$ $(\mathbf{x}\in\mathbf{M})$. Recall that $O_o : M\longrightarrow\mathbf{M}, x\mapsto x-o$.

(i) Show that the observer given in Exercise 6. depends only on the direction of **b** (and does not depend on the magnitude of **b**) and

$$\mathbf{U}=\mathbf{V}_{\mathbf{c},\mathbf{b}}\circ\tau_{\mathbf{c},\mathbf{b}}\circ O_o.$$

(ii) The observer given in Exercise 8. has the property

$$\mathbf{U}=\mathbf{V}_{\mathbf{c},\mathbf{a}}\circ\tau_{\mathbf{c}}\circ O_o.$$

(cf. Exercise 6. of Chapter I.).

10. (i) In a convenient Lorentz reference system the observer of Exercise 6. takes the form

$$\{(x^0,x^1,x^2,x^3)\in\mathbb{R}^4 : x^1>|x^0|\}\to\mathbb{R}^4,$$

$$(x^0,x^1,x^2,x^3)\mapsto\frac{1}{\sqrt{-(x^0)^2+(x^1)^2}}(x^1,x^0,0,0).$$

The world lines of the observer are

$$\left\{\left(x^0,\sqrt{\frac{1}{a^2}+(x^0)^2},b,c\right) : \ x^0\in\mathbb{R}\right\}\qquad ((a,b,c)\in\mathbb{R}^+\times\mathbb{R}\times\mathbb{R}).$$

Put

$$C(a,x^0):=\sqrt{\frac{1}{a^2}+(x^0)^2}\qquad(a\in\mathbb{R}^+,\ x^0\in\mathbb{R}),$$

and show that

$$C(a+h,x^0)-C(a,x^0)=$$
$$=\sqrt{1-\left(\frac{\mathrm{d}C(a,x^0)}{\mathrm{d}x^0}\right)^2}\,[C(a+h,0)-C(a,0)]+\operatorname{ordo}(h).$$

⌈The observer **U** is rigid, rotation free, in each of the **U**-time points all masspoints of the observer have the same velocity, the masspoints of the observer are uniformly accelerated in the same direction but the masspoints along the line of the acceleration are differently accelerated.

The last equality shows that also the inertial observer with velocity **c** (which is (1, 0, 0, 0) in the given Lorentz reference system) sees **U** rigid: the distance of near points undergo Lorentz contractions which change according to the instantaneous velocity.⌋

(ii) In a convenient Lorentz reference system the observer of Exercise 8. takes the form

$$(x^0,x^1,x^2,x^3)\mapsto(\sqrt{1+(ax^0)^2},ax^0,0,0)$$

where a is a positive number, it is the magnitude of **a** in the space-time unit of the Lorentz reference system.

Show that the "nonrelativistic limit" (see 8.2.) of this observer is the one given in Exercise 6. of Chapter I.

⌈The masspoints of the observer **U** are uniformly accelerated, **U** is rotation free. The inertial observer with velocity **c** (which is $(1, 0, 0, 0)$ in the given Lorentz reference system) sees that in each of his time points the masspoints of **U** have the same velocity and acceleration and the distance of masspoints do not depend on his time. On the other hand, there is only a single **U**-time point in which all masspoints have the same velocity and acceleration, and **U** is not rigid, i.e. the distances of masspoints depend on **U**-time; evidently they must dilate to compensate the Lorentz contraction.⌋

11. Problem. Can we find a special relativistic observer **U** which we could call *uniformly accelerated*, i.e. which has the properties

(i) **U** is rigid and rotation free,

(ii) **U**-lines are uniformly accelerated (see Exercise 5.),

(iii) at each **U**-time point the velocities of all **U**-lines are equal and the accelerations of all **U**-lines are equal.

Notice that the observer of Exercise 6. satisfies these requirements except that in (iii) regarding the accelerations.

12. Let $o \in M$, $\mathbf{c} \in V(1)$ and suppose $0 \neq \Omega : \mathbf{E_c} \longrightarrow \frac{\mathbf{E_c}}{\mathbf{I}}$ is an antisymmetric linear map, i.e. $\mathbf{t} \otimes \Omega : \mathbf{E_c} \longrightarrow \mathbf{E_c}$ (notation: IV. 1.9.) is antisymmetric with respect to $\gamma_\mathbf{c}$ for all $\mathbf{t} \in \mathbf{I}$ (see I. 5.). Consider the following observer

$$\operatorname{Dom} \mathbf{U} := o + \{\mathbf{x} \in \mathbf{M} \,\vdots\, |\Omega(\mathbf{x}\|_\mathbf{c})|^2 < 1\},$$

$$U(o+\mathbf{x}) := \frac{\mathbf{c} + \Omega(\mathbf{x}\|_\mathbf{c})}{\sqrt{1 - |\Omega(\mathbf{x}\|_\mathbf{c})|^2}}.$$

Prove that

(i) for all $\mathbf{q} \in \operatorname{Ker} \Omega$ and for all $x \in \operatorname{Dom} \mathbf{U}$

$$x + \mathbf{q} \in \operatorname{Dom} \mathbf{U}, \quad \mathbf{U}(x+\mathbf{q}) = \mathbf{U}(x);$$

(ii) for all $x \in \operatorname{Dom} \mathbf{U}$

$$\mathbf{v}_{\mathbf{U}(x),\mathbf{c}} = \Omega((x-o)\|_\mathbf{c})$$

(see 4.2.);

(iii) the **U**-line passing through $x \in \operatorname{Dom} \mathbf{U}$ can be given by the world line function

$$\mathbf{t} \mapsto o - \mathbf{c} \otimes [\mathbf{c} | (x-o)) + \frac{\mathbf{c} \otimes \mathbf{t}}{\sqrt{1 - |\Omega((x-o)\|_\mathbf{c})|^2}} +$$

$$+ \left(\exp \frac{\mathbf{t} \otimes \Omega}{\sqrt{1 - |\Omega((x-o)\|_\mathbf{c})|^2}} \right) (x-o)\|_\mathbf{c}$$

(cf. Exercise 5. of Chapter I.);

(iv) **U** is not regular (hint: consider the vector fields $x \mapsto (x-o)\|_{\mathbf{c}}$ and

$$x \mapsto |\Omega((x-o)\|_{\mathbf{c}})|^2 \mathbf{c} + \Omega((x-o)\|_{\mathbf{c}})$$

and apply Remark 3.2.);

(v) the set J of hyperplanes directed by $\mathbf{E_c}$ is an artificial time for **U**; J is an affine space over **I**;

(vi) identifying $E_{\mathbf{U}}$ (the set of **U**-lines) with $\mathbf{E_c}$ by

$$q \mapsto (q \cap (o + \mathbf{E_c})) - o\,,$$

we have

$$S_{J,\mathbf{U}}(x) = \left(x + \mathbf{E_c}, (\exp\{[\mathbf{c}|x-o)\otimes\Omega\})(x-o)\|_{\mathbf{c}}\right).$$

and $T_t(J) = \mathbf{I}$ for all $t \in J$, $T_q(E_{\mathbf{U}}) = \mathbf{E_c}$ for all $q \in E_{\mathbf{U}}$ (notations: 3.18., 3.19.);

(vii) as a consequence, for all $(t, q) \in \operatorname{Ran} S_{J,\mathbf{U}}$, $g^J_{\mathbf{UU}}[t, q]$ is a symmetric bilinear map $\mathbf{E_c} \times \mathbf{E_c} \longrightarrow \mathbf{I}\otimes\mathbf{I}$, and we can make the identifications

$$g_{JJ}[t, q] = \frac{g_{JJ}[t, q]\,(\mathbf{t}, \mathbf{t})}{\mathbf{t}\otimes\mathbf{t}} \in \mathbb{R}$$

$$g_{J\mathbf{U}}[t, q] = \frac{g_{J\mathbf{U}}[t, q]\,(\mathbf{t}, \cdot\,)}{\mathbf{t}} \in L(\mathbf{E_c}, \mathbf{I}) = \left(\frac{\mathbf{E_c}}{\mathbf{I}}\right)^*$$

where **t** is any non-zero element of **I**. Then we have

$$G_{J,\mathbf{U}}[\mathbf{t}, \mathbf{q}] = \begin{pmatrix} 1 - |\Omega\mathbf{q}|^2 & -2\Omega\mathbf{q} \\ -2\Omega\mathbf{q} & \gamma_{\mathbf{c}} \end{pmatrix};$$

(viii) **U** is rigid.

(ix) Take an $\mathbf{e} \in \mathbf{I}^+$ and construct the Lorentz reference system whose origin, velocity and space-time unit are o, **c** and **e**, respectively, and the space-time basis is chosen in such a way that $\mathbf{e}_3$ spans $\operatorname{Ker}\Omega$, more closely, $\mathbf{e}_3 := \mathbf{e}\otimes j(\mathbf{e}\otimes\Omega)$ (see V. 3.6.).

Show that in this reference system **U** has the form

$$\{(x^0, x^1, x^2, x^3) \in \mathbb{R}^4 : \omega^2((x^1)^2 + (x^2)^2) < 1\} \longrightarrow \mathbb{R}^4,$$

$$(x^0, x^1, x^2, x^3) \mapsto \frac{1}{\sqrt{1 - \omega^2((x^1)^2 + (x^2)^2)}}(1, -\omega x^2, \omega x^1, 0)$$

where ω is the norm of $\mathbf{e}\otimes\Omega$ (cf. Exercise 5. Chapter I.), and the **U**-line passing through $o + \sum_{k=1}^{3} x^k \mathbf{e}_k$ becomes

$$t \mapsto (z(t), x^1 \cos\omega z(t) - x^2 \sin\omega z(t), x^1 \sin\omega z(t) + x^2 \cos\omega z(t), 0)$$

where

$$z(t) := \frac{t}{\sqrt{1-\omega^2((x^1)^2+(x^2)^2)}} \qquad (t \in \mathbb{R}).$$

13. Let $o \in M, \mathbf{c} \in V(1)$ and suppose $0 \neq \Omega : \mathbf{E_c} \to \frac{\mathbf{E_c}}{\mathbf{I}}$ is an antisymmetric linear map (see the previous exercise). Consider the following observer

$$U(o+\mathbf{x}) := \Omega(\mathbf{x}\|_{\mathbf{c}}) + \mathbf{c}\sqrt{1+|\Omega(\mathbf{x}\|_{\mathbf{c}})|^2} \qquad (\mathbf{x} \in \mathbf{M}).$$

Prove that

(i) for all $\mathbf{q} \in \operatorname{Ker} \Omega$ and for all $x \in M$

$$\mathbf{U}(x+\mathbf{q}) = \mathbf{U}(x);$$

(ii) the $\mathbf{U}$-line passing through $x \in M$ can be given by the world line function

$$\mathbf{t} \mapsto o - \mathbf{c} \otimes [\mathbf{c}|x-o) + \mathbf{c} \otimes \mathbf{t}\sqrt{1+|\Omega((x-o)\|_{\mathbf{c}}|^2} + (\exp(\mathbf{t} \otimes \Omega))(x-o)\|_{\mathbf{c}}$$

(cf. Exercise 5. of Chapter I.);

(iii) $\mathbf{U}$ is not regular (hint: consider the vector fields $x \mapsto (x-o)\|_{\mathbf{c}}$ and

$$x \mapsto \frac{|\Omega((x-o)\|_{\mathbf{c}})|^2}{\sqrt{1+|\Omega((x-o)\|_{\mathbf{c}})|^2}}\mathbf{c} + \Omega((x-o)\|_{\mathbf{c}})$$

and apply Remark 3.2.).

(iv) Give an artificial time for $\mathbf{U}$ (cf. the previous exercise) and verify that $\mathbf{U}$ is not rigid.

(v) Take an $\mathbf{e} \in \mathbf{I}^+$ and construct the Lorentz reference system whose origin, velocity and space-time unit are o, $\mathbf{c}$ and $\mathbf{e}$, respectively, and the space-time basis is chosen in such a way that $\mathbf{e}_3$ spans $\operatorname{Ker} \Omega$, more closely, $\mathbf{e}_3 := \mathbf{e} \otimes j(\mathbf{e} \otimes \Omega)$ (see V. 3.6.).

Show that in this references system $\mathbf{U}$ has the form

$$\mathbb{R}^4 \to \mathbb{R}^4, \quad (x^0, x^1, x^2, x^3) \mapsto \left(\sqrt{1+\omega^2((x^1)^2+(x^2)^2)}, -\omega x^2, \omega x^1, 0\right)$$

where ω is the norm of $\mathbf{e} \otimes \Omega$ (cf. Exercise 5. of Chapter I.), and the $\mathbf{U}$-line passing through $o + \sum_{k=1}^{3} x^k \mathbf{e}_k$ becomes

$$t \mapsto \left(t\sqrt{1+\omega^2((x^1)^2+(x^2)^2)}, x^1 \cos \omega t - x^2 \sin \omega t, x^1 \sin \omega t + x^2 \cos \omega t, 0\right).$$

14. ⌈Both observers introduced in Exercises 12. and 13. can be interpreted as some *uniformly rotating observers*. The masspoints of the first one have the same angular velocity relative to the inertial observer with velocity $\mathbf{c}$. The masspoints of the second one have the same angular velocity with respect to their proper times.⌋

Show that if $x-o\in\operatorname{Ker}\Omega$ then the $\mathbf{U}$-line passing through x is $C_{\mathbf{U}}(x)=$ $=x+\mathbf{c}\otimes\mathbf{I}$ for both observers $\mathbf{U}$. As a consequence,

$$\{C_{\mathbf{U}}(x) \,\vdots\, x-o\in\operatorname{Ker}\Omega\}$$

is a one dimensional affine space over $\operatorname{Ker}\Omega$ with the addition $C_{\mathbf{U}}(x)+\mathbf{q}:=$ $=C_{\mathbf{U}}(x+\mathbf{q})$.

⌈This straight line in $\mathbf{U}$-space $E_{\mathbf{U}}$ can be interpreted as the axis of rotation.⌋

15. Let $\mathbf{u}$ and $\mathbf{u}'$ be distinct elements of $V(1)$. Put

$$\cos\Theta(\mathbf{w}):=\frac{(\mathbf{v}_{\mathbf{wu}}\cdot\mathbf{v}_{\mathbf{u}'\mathbf{u}})}{|\mathbf{v}_{\mathbf{wu}}|\,|\mathbf{v}_{\mathbf{u}'\mathbf{u}}|},\qquad (\mathbf{w}\in V(1)\cup V(0),\ \mathbf{w}\neq\mathbf{u}',\ \mathbf{w}\neq\mathbf{u})$$

$$-\cos\Theta'(\mathbf{w}):=\frac{(\mathbf{v}_{\mathbf{wu}'}\cdot\mathbf{v}_{\mathbf{uu}'})}{|\mathbf{v}_{\mathbf{wu}'}|\,|\mathbf{v}_{\mathbf{uu}'}|}$$

($\Theta(\mathbf{w})$ is the angle between the relative velocities $\mathbf{v}_{\mathbf{wu}}$ and $\mathbf{v}_{\mathbf{u}'\mathbf{u}}$, $\Theta'(\mathbf{w})$ is the angle between the relative velocities $\mathbf{v}_{\mathbf{wu}'}$ and $-\mathbf{v}_{\mathbf{uu}'}$). Show that

$$\cos\Theta(\mathbf{w})=\frac{\dfrac{|\mathbf{v}_{\mathbf{wu}'}|}{|\mathbf{v}_{\mathbf{wu}}|}\cos\Theta'(\mathbf{w})+\dfrac{|\mathbf{v}_{\mathbf{u}'\mathbf{u}}|}{|\mathbf{v}_{\mathbf{wu}}|}}{1+|\mathbf{v}_{\mathbf{u}'\mathbf{u}}|\,|\mathbf{v}_{\mathbf{wu}'}|\cos\Theta'(\mathbf{w})}$$

In particular, if $\mathbf{w}\in V(0)$ then $|\mathbf{v}_{\mathbf{wu}}|=|\mathbf{v}_{\mathbf{wu}'}|=1$ (see Proposition 4.6.) and

$$\cos\Theta(\mathbf{w})=\frac{\cos\Theta'(\mathbf{w})+|\mathbf{v}_{\mathbf{u}'\mathbf{u}}|}{1+|\mathbf{v}_{\mathbf{u}'\mathbf{u}}|\cos\Theta'(\mathbf{w})}.$$

⌈This formula is known as the *aberration of light*: two inertial observers with different velocities see the same light beam under different angles with respect to their relative velocities; the relation of angles is given by the above formula. Compare our present result with that of Exercise 9. of Chapter I.⌋

16. Let $\mathbf{u}$ and $\mathbf{u}'$ be distinct elements of $V(1)$ and $\mathbf{w}_1,\mathbf{w}_2\in V(0)$. Put

$$\cos\vartheta:=(\mathbf{v}_{\mathbf{w}_1\mathbf{u}}\cdot\mathbf{v}_{\mathbf{w}_2\mathbf{u}}),$$

$$\cos\vartheta':=(\mathbf{v}_{\mathbf{w}_1\mathbf{u}'}\cdot\mathbf{v}_{\mathbf{w}_2\mathbf{u}'}).$$

Prove that

$$\cos\vartheta'=\lambda\cos\vartheta+1-\lambda$$

where

$$\lambda:=\frac{1+|\mathbf{v}_{\mathbf{uu}'}|\cos\Theta'(\mathbf{w}_1)}{1-|\mathbf{v}_{\mathbf{u}'\mathbf{u}}|\cos\Theta(\mathbf{w}_2)}$$

(the definition of $\cos\Theta(\mathbf{w})$ can be found in the previous exercise).

[ϑ is the angle between two light beams measured by the inertial observer with velocity $\mathbf{u}$, ϑ' is the angle of the same light beams measured by the inertial observer with velocity $\mathbf{u}'$. Hence the above formula can be interpreted as the transformation law of angles under a change of observers.]

17. Derive the splitting formula for 2-tensors and 2-cotensors that are not antisymmetric (see 5.10.), as well as for mixed 2-tensors, i.e. for elements of $\mathbf{M}^*\otimes\mathbf{M}$ and $\mathbf{M}\otimes\mathbf{M}^*$. Use that $A\in\mathbf{M}\otimes\mathbf{M}^*$ can be considered as a linear map $\mathbf{M}\longrightarrow\mathbf{M}$, thus its splitting corresponding to $\mathbf{u}\in V(1)$ is

$$\mathbf{s}_{\mathbf{u}}\circ A\circ\mathbf{s}_{\mathbf{u}}^{-1}=\begin{pmatrix}-(\mathbf{u}\cdot A_{\mathbf{I}}\mathbf{u}) & -A_{\mathbf{I}}^{*}\mathbf{u}\\ A_{\mathbf{I}}\mathbf{u}+(\mathbf{u}\cdot A_{\mathbf{I}}\mathbf{u})\mathbf{u} & (\mathrm{id}_{\mathbf{M}}+\mathbf{u}\otimes\mathbf{u})A|_{\mathbf{E}_{\mathbf{u}}}\end{pmatrix}.$$

Observe that this formula is very simple if A is g-antisymmetric; this case occurs in Theorem 7.20.

18. Define the special relativistic affine reference systems, give a characterization of them and formulate a necessary and sufficient condition that two affine reference system be equivalent (cf. I. 6.4. and Exercise 13. of Chapter I.).

19. Find the explicit form of $z_{\mathbf{c},o}^{-1}$ and $w_{\mathbf{c},o}^{-1}$ (see 7.19. and 7.20.). In particular, give the Poincaré transformations that are images by $z_{\mathbf{c},o}^{-1}$ of the $\mathbf{c}$-split Poincaré transformations having the forms listed in 7.22.

20. Find the $\mathbf{c}$-split Poincaré transformations

$$m_{\mathbf{cc}'oo'}^{-1}(\mathbf{t},\mathbf{q},\mathbf{v},R)m_{\mathbf{cc}'oo'},$$

$$m_{\mathbf{cc}'oo'}^{-1}T(\mathbf{c}')m_{\mathbf{cc}'oo'},\quad m_{\mathbf{cc}'oo'}^{-1}S(\mathbf{c}')m_{\mathbf{cc}'oo'}$$

(see 7.14., 7. 15. and 7.21.).

21. Show that the derived Lie algebra of the Poincaré group equals the Lie algebra of the Poincaré group, i.e. $[\mathbf{L}(\mathscr{P}),\mathbf{L}(\mathscr{P})]=\mathbf{L}(\mathscr{P})$.

22. Using Theorems 7.19. and 7.20. prove that for $\mathbf{u}\in V(1)$

(i) $\{F\in L\colon F_{\mathbf{I}}\mathbf{u}=\mathbf{u}\}$ can be identified with $\mathrm{O}(\gamma_{\mathbf{u}})$ by $F\mapsto F|_{\mathbf{E}_{\mathbf{u}}}$. Furthermore, if $F\mathbf{x}=\mathbf{x}$ all for such Lorentz transformations F then $\mathbf{x}=\mathbf{u}\otimes\mathbf{t}$ for some $\mathbf{t}\in\mathbf{I}$;

(ii) $\{A\in\mathbf{A}(g)\colon A_{\mathbf{I}}\mathbf{u}=\mathbf{0}\}$ can be identified with $\mathbf{A}(\gamma_{\mathbf{u}})$ by $A\mapsto A|_{\mathbf{E}_{\mathbf{u}}}$. Furthermore, if $A\mathbf{x}=\mathbf{0}$ for all such A then $\mathbf{x}=\mathbf{u}\otimes\mathbf{t}$ for some $\mathbf{t}\in\mathbf{I}$.

III. GENERAL RELATIVISTIC SPACE-TIME MODELS

1. Fundamentals

Necessary mathematical tools:

Tensor products and quotients (Chapter IV.)
Theory of smooth manifolds (Sections VII. 1.—VII. 4.)
Foliations (Section VII. 7.)
Connections (Sections VII. 9.)
Pseudo-Riemannian manifolds (Section VII. 10.)

1.1. ⌈The special relativistic space-time models, though they are better from a physical point of view than the nonrelativistic space-time models, suffer from a great insufficiency.

We have seen that a gravitational action can be well described in a nonrelativistic space-time model by a timelike potential which is in fact a scalar (see I. 5.15.). In a special relativistic space-time model such potentials do not exist. To describe gravitational action by a potential (see II. 2.12.) leads to results which do not fit the experimental facts.

That is why we are forced to build up new space-time models. According to the idea of Einstein, space-time models must reflect gravitational actions, *a space-time model must be a model of a gravitational action*; the lack of gravitation is modelled by a special relativistic space-time model.

Here we define the general relativistic space-time models and the corresponding notions of observers etc. and we do not touch the problem how a gravitational action is modelled by a general relativistic space-time model.⌋

1.2. Definition. A *general relativistic space-time model* is a four dimensional oriented and arrow oriented Einsteinian manifold, i.e. it is a triplet $(M, \mathbf{I}, g)$ where

M is an oriented four dimensional smooth manifold,

$\mathbf{I}$ is an oriented one dimensional real vector space and

g is a Einsteinian 2-cotensor field of type $\mathbf{I}\otimes\mathbf{I}$ on M, endowed with an arrow orientation (see VII. 10.10.).

Remark. A special relativistic space-time model is a particular general relativistic space-time model.

We assume in the sequel that a general relativistic space-time model $(M, \mathbf{I}, g)$ *is given.*

1.3. For $x \in M$, $(T_x(M), \mathbf{I}, g[x])$ is a Minkowskian vector space, hence we can speak of *spacelike*, *timelike* and *lightlike* tangent vectors at x. The arrow orientation distinguishes a timelike cone, thus we can say that a timelike or lightlike tangent vector at x is *positive* or *negative* (see II. 1.3.).

The elements of $\mathbf{I}$ and $\mathbf{I}^*$ and of their various tensor products and quotients are called *scalars*. If $\mathbf{A}$ is a space of scalars, *scalar fields*, *vector fields*, *tensor fields*, *covector fields* and *contensor fields* of type $\mathbf{A}$ are defined as in VII. 3.18. and VII. 4.15.

$T_x(M)^*$ can be identified with $\dfrac{T_x(M)}{\mathbf{I}\otimes\mathbf{I}}$ $(x \in M)$ and a covector field of type $\mathbf{A}$ can be viewed as a vector field of type $\dfrac{\mathbf{A}}{\mathbf{I}\otimes\mathbf{I}}$.

1.4. We consider g as a bilinear map on the module of vector fields on M. For $x \in M$ we introduce the notations (cf. II. 1.5.)

$$V(x,0) := \left\{ \mathbf{w} \in \frac{T_x(M)}{\mathbf{I}} : (\mathbf{w} \mid \mathbf{w}) = 0, \quad \mathbf{w} \text{ is positive} \right\},$$

$$V(x,1) := \left\{ \mathbf{u} \in \frac{T_x(M)}{\mathbf{I}} : (\mathbf{u} \mid \mathbf{u}) = -1, \quad \mathbf{u} \text{ is positive} \right\},$$

and for $\mathbf{u} \in V(x,1)$ we put

$$E_{\mathbf{u}}[x] := \{ \mathbf{z} \in T_x(M) : [\mathbf{u} \mid \mathbf{z}) = \mathbf{0} \},$$

$$\gamma_{\mathbf{u}}[x] := g[x]|_{\mathbf{E}_{\mathbf{u}}[x] \times \mathbf{E}_{\mathbf{u}}[x]}.$$

1.5. The set

$$T(M,1) := \bigcup_{x \in M} \{x\} \times V(x,1)$$

can be equipped with a smooth structure in a natural way and then it becomes a submanifold of $T(M)_{\mathbf{I}}$.

$T(M,1)$ is the counterpart of $M \times V(1)$ in the special relativistic space-time model.

1.6. Definition. Two general relativistic space-time models $(M, \mathbf{I}, g)$ and $(M', \mathbf{I}', g')$ are *isomorphic*, if there are

(i) an orientation and arrow preserving diffeomorphism $F : M \longrightarrow M'$,

(ii) an orientation preserving linear bijection $Z : \mathbf{I} \longrightarrow \mathbf{I}'$

in such a way that

$$F^* g' = (Z \otimes Z) \circ g.$$

The pair (F, Z) is an *isomorphism* between the two general relativistic space-time models. If the two models coincide, an isomorphism is called an *automorphism*. An automorphism of the form $(F, \mathrm{id}_{\mathbf{I}})$ is called *strict*.

Remarks. (i) It can be shown that if two special relativistic space-time models are isomorphic as general relativistic space-time models (i.e. by a diffeomorphism) then they are isomorphic as special relativistic space-time models, i.e. by an affine bijection as well.

Indeed, if (F, Z) is a general relativistic isomorphism between the special relativistic space-time models $(M, \mathbf{I}, g)$ and $(M', \mathbf{I}', g')$ then we can find a special relativistic isomorphism (F', Z') such that $(F', Z') \circ (F, Z)$ is a strict general relativistic automorphism of $(M, \mathbf{I}, g)$. One knows that a diffeomorphism of the affine space M preserving g is necessarily a Poincaré transformation ([6], IV. 23.). As a consequence, (F, Z) is special relativistic isomorphism.

(ii) There are a number of non-isomorphic general relativistic space-time models. The classification of general relativistic space-time models and their investigations are mathematical problems having a good and wide literature. The thorough treatment of general relativistic space-time models would require a separate volume, and it lies beyond our present purposes, because we shall apply only nonrelativistic and special relativistic space-time models in describing physical phenomena in the next volumes.

1.7. Definition. A general relativistic space-time model $(M, \mathbf{I}, g)$ is called *flat* if there is a special relativistic space-time model $(M_0, \mathbf{I}_0, g_0)$ and an open subset M' of M_0 such that $(M, \mathbf{I}, g)$ is isomorphic to $(M', \mathbf{I}_0, g_0|_{M'})$.

2. World lines

2.1. Definition. A *light signal* is a connected geodesic whose all tangent vectors are lightlike.

2.2. Definition. A smooth function $r: \mathbf{I} \rightarrowtail M$ having the properties

(i) $\mathrm{Dom}\, r$ is connected,

(ii) $\dot{r}(t) \in V(r(t), 1) \qquad (t \in \mathrm{Dom}\, r)$

is called a *world line function* (notation: VII. 6.1.).

A *world line* is a one dimensional submanifold of M which is the range of a world line function.

A geodesic world line is called *inertial*.

2.3. (i) If r is a world line function then the *acceleration* of r at $\mathbf{t} \in \mathrm{Dom}\, r$, $(\nabla_{\dot{r}}\dot{r})(\mathbf{t})$ is in $\dfrac{\mathbf{E}_{\dot{r}(\mathbf{t})}[r(\mathbf{t})]}{\mathbf{I}\otimes\mathbf{I}}$.

(ii) We can define that a world line is rotation free in the same way as we did it in II. 2.11., taking $\nabla_{\dot{r}}\dot{r}$ instead of $\ddot{r}$.

(iii) Evidently, inertial world lines are rotation free.

2.4. Proposition. A world line function corresponding to a world line is determined uniquely up to a translation of its domain; in other words, if r_1 and r_2 are world line functions, $\operatorname{Ran} r_1 = \operatorname{Ran} r_2$, then there exists an element $\mathbf{t}_0$ of $\mathbf{I}$ such that $\operatorname{Dom} r_2 = \mathbf{t}_0 + \operatorname{Dom} r_1$ and $r_2(\mathbf{t}) = r_1(\mathbf{t} - \mathbf{t}_0)$ $(\mathbf{t} \in \operatorname{Dom} r_2)$ (cf. II. 2.7.).

Definition. If x and y are points of M and r is a world line function such that $x = r(\mathbf{t})$, $y = r(\mathbf{s})$ for some $\mathbf{t}, \mathbf{s} \in \mathbf{I}$ then $\mathbf{s} - \mathbf{t}$ is called the *time passed between x and y along the world line* corresponding to r (cf. II. 2.5. and II. 2.9.).

2.5. Definition. (i) A smooth covector field in M is called a *potential.*

(ii) A smooth map

$$\mathbf{f}: T(M, 1) \rightarrowtail T^*(M)_{\mathbf{I}}$$

for which $\mathbf{f}(x, \mathbf{u}) = (x, \mathbf{f}[x, \mathbf{u}])$ and

$$\mathbf{f}[x, \mathbf{u}] \in \frac{\mathbf{E}_{\mathbf{u}}[x]}{\mathbf{I} \otimes \mathbf{I} \otimes \mathbf{I}} \qquad ((x, \mathbf{u}) \in \operatorname{Dom} \mathbf{f})$$

holds is called a *force field.* The force field $\mathbf{f}$ is said to have a *potential* $\mathbf{K}$ if there exists a covector field $\mathbf{K}: M \rightarrowtail T^*(M)$ such that

$$\mathbf{f}[x, \mathbf{u}] = d\mathbf{K}[x]_{\mathbf{I}}\mathbf{u} \qquad ((x, \mathbf{u}) \in \operatorname{Dom} \mathbf{f})$$

where $d\mathbf{K}$ is the exterior differential of $\mathbf{K}$ and $d\mathbf{K}[x] \in T_x(M)^* \wedge T_x(M)^*$ is considered as a linear map $T_x(M) \longrightarrow T_x(M)^*$, and $d\mathbf{K}[x]_{\mathbf{I}}: \frac{T_x(M)}{\mathbf{I}} \longrightarrow \frac{T_x(M)^*}{\mathbf{I}} = \frac{T_x(M)}{\mathbf{I} \otimes \mathbf{I} \otimes \mathbf{I}}$ (see IV. 1.5. (iv) and IV. 2.10. (ii) and IV. 2.7. (v).)

(iii) A positive element of $\mathbf{I}^*$ is called a *mass.*

2.6. Definition. A world line function r is called *Einsteinian* if there are

(i) a force field $\mathbf{f}$,

(ii) a mass m

such that r satisfies the equation

$$m \otimes \nabla_{\dot{r}}\dot{r} = \mathbf{f} \circ \dot{r}.$$

⌈An inertial world line function is Einsteinian, it corresponds to the zero force field. It describes the history of a masspoint under the graviational action modelled by $(M, \mathbf{I}, g)$.⌋

3. Observers and reference systems

3.1. Definition. An *observer* is a smooth map $\mathbf{U}: M \rightarrowtail T(M, 1) \subset T(M)_{\mathbf{I}}$ for which $\pi_M \circ \mathbf{U} \subset \mathrm{id}_M$ holds; in other words, it is a vector field of cotype **I** such that $\mathbf{U}[x] \in V(x, 1)$ for all $x \in \mathrm{Dom}\,\mathbf{U}$. Depending on whether $\mathrm{Dom}\,\mathbf{U} \neq M$ or $\mathrm{Dom}\,\mathbf{U} = M$, the observer is called *local* or *global*, respectively.

The maximal integral curves of **U** are called **U**-*lines*. The set of **U**-lines is denoted by $E_{\mathbf{U}}$ and is called **U**-*space*. ∎

U can be interpreted as a *velocity field* and $\nabla_{\mathbf{U}}\mathbf{U}$ is the corresponding *acceleration field*.

3.2. We can define regular observers in a similar way as in II. 3.2. and we can prove propositions similar to II. 3.4., II. 3.5., II. 3.18., II. 3.19., II. 3.20. and II. 3.21.

3.3. Splitting of tangent vectors and tensors at $x \in M$ by a $\mathbf{u} \in V(x, 1)$, splitting of vector fields and tensor fields by an observer can be carried out as in the special relativistic case.

3.4. We accept Definitions II. 6.2. and II. 6.7. without modification for a general relativistic space-time model, too.

Two reference systems isomorphic by a strict automorphism of $(M, \mathbf{I}, g)$ are called *equivalent*.

4. The principles of equivalence and relativity and covariance

4.1. ⌈In usual formulation, the principle of equivalence (see [5], 8.2.) says that the inertial mass and the gravitational mass are locally equivalent, which means that a sufficiently small observer moving under the action of a gravitational field does not feel the gravitational action, the observer sees the phenomena proceeding in a small region as if there were no gravitational field. We emphasize that the equivalence between the two masses is valid locally, i.e. for sufficiently small observers.

One hopes perhaps that the convenient mathematical formulation of the principle of equivalence is the following: each point of a general relativistic space-time model $(M, \mathbf{I}, g)$ has an open neighbourhood M' such that $(M', \mathbf{I}, g|_{M'})$ is flat. Unfortunately such a theorem is *not true*. We can say less: for each point x of M there is a reference system such that in the corresponding coordinatization $g[x]$ becomes diagonal and the first order derivative of g at x is zero (see VII. 10.7.).⌋

4.2. ⌈The general principle of relativity and covariance appears usually in connection with gravitation theory in general relativity theory ([5], 9.1.). However, *the principles of relativity and covariance are independent of gravitation theory and they have to be the most fundamental principles of physics.* We can state them as follows.

To describe physical phenomena we must choose a—nonrelativistic or relativistic—space-time model and all physical phenomena must be described in terms of "absolute objects", i.e. by mathematical objects of the space-time model (principle of covariance) and we must have a rule how the "relative objects" connected with observers are derived from the absolute ones (principle of relativity).

We emphasize against the common belief that "nonrelativistic" does not mean "non-covariance"; it means a covariance different from the relativistic one.⌋

PART TWO

MATHEMATICAL TOOLS

The reader is assumed to be familiar with the elements of linear algebra, topology, analysis and group theory.

Further necessary mathematical tools are listed very concisely. Detailed definition and propositions are formulated; proofs of easily verifiable assertions are omitted and proofs easily available in literature are replaced by references.

The material is arranged by its own logical structure and not by didactical requirements. That is why three notions appear advancing their definitions (see IV. 1.20., V. 5.1. and VII. 1.13.).

IV. TENSORIAL OPERATIONS

1. Tensor products

In this section $\mathbb{F}$ denotes the field of complex or real numbers and all vector spaces are given over $\mathbb{F}$.

1.1. Definition. Let $\mathbf{L}$ and $\mathbf{K}$ be vector spaces. A *tensor product* of $\mathbf{L}$ and $\mathbf{K}$ is a pair $(\mathbf{M}, b)$ where

(i) $\mathbf{M}$ is a vector space,

(ii) $b:\mathbf{L}\times\mathbf{K}\longrightarrow\mathbf{M}$ is a bilinear map having the property that

— if $\mathbf{N}$ is a vector space and $c:\mathbf{L}\times\mathbf{K}\longrightarrow\mathbf{N}$ is a bilinear map

— then there exists a unique linear map $L:\mathbf{M}\longrightarrow\mathbf{N}$ such that

$$c=L\circ b$$

1.2. The existence of tensor products is stated in Proposition 1.3.

An easy consequence of the definition (implying the noteworthy fact that $\mathrm{id}_{\mathbf{M}}$ is the only linear map L satisfying $b=L\circ b$) is that if $(\mathbf{M}, b)$ and $(\mathbf{M}', b')$ are tensor products of $\mathbf{L}$ and $\mathbf{K}$ then there exists a unique linear isomorphism $L:\mathbf{M}\longrightarrow\mathbf{M}'$ such that $b'=L\circ b$.

This shows that tensor products of $\mathbf{L}$ and $\mathbf{K}$ are canonically isomorphic, hence we speak of *the* tensor product, and applying a customary abuse of language we denote it by $\mathbf{L}\otimes\mathbf{K}$ writing $(\mathbf{x}, \mathbf{y})\mapsto\mathbf{x}\otimes\mathbf{y}$ for the bilinear map $\mathbf{L}\times\mathbf{K}\longrightarrow\mathbf{L}\otimes\mathbf{K}$.

As a consequence of the definition of tensor products, we can identify the vector space of bilinear maps $\mathbf{L}\times\mathbf{K}\longrightarrow\mathbf{N}$ with the vector space of linear maps $\mathbf{L}\otimes\mathbf{K}\longrightarrow\mathbf{N}$. In other words, a bilinear map $c:\mathbf{L}\times\mathbf{K}\longrightarrow\mathbf{N}$ will be considered as a linear map $c:\mathbf{L}\otimes\mathbf{K}\longrightarrow\mathbf{N}$ in such a way that

$$c(\mathbf{x}\otimes\mathbf{y})=c(\mathbf{x}, \mathbf{y}) \qquad ((\mathbf{x}, \mathbf{y})\in\mathbf{L}\times\mathbf{K}))$$

and vice versa.

1.3. Let $\mathrm{B}(\mathbf{L}^*, \mathbf{K}^*)$ denote the vector space of bilinear forms on $\mathbf{L}^*\times\mathbf{K}^*$ where $\mathbf{L}^*$ and $\mathbf{K}^*$ are the dual vector spaces of $\mathbf{L}$ and $\mathbf{K}$, respectively. Let $(\ |\ )$ denote the bilinear map of duality.

Proposition. $\mathbf{L}\otimes\mathbf{K}$ is the subspace of $B(\mathbf{L}^*, \mathbf{K}^*)$ spanned by the range of the bilinear map

$$\otimes : \mathbf{L}\times\mathbf{K} \longrightarrow B(\mathbf{L}^*, \mathbf{K}^*), \qquad (\mathbf{x}, \mathbf{y}) \mapsto \mathbf{x}\otimes\mathbf{y}$$

defined by

$$(\mathbf{x}\otimes\mathbf{y})(\mathbf{p}, \mathbf{r}) := (\mathbf{p}|\mathbf{x})(\mathbf{r}|\mathbf{y}) \qquad ((\mathbf{p}, \mathbf{r}) \in \mathbf{L}^*\times\mathbf{K}^*).$$

If $\mathbf{L}$ and $\mathbf{K}$ are finite dimensional then

$$\mathbf{L}\otimes\mathbf{K} = B(\mathbf{L}^*, \mathbf{K}^*).$$

([18], section III. 6.).

Remark. For all natural numbers n and m we have

$$\mathbb{F}^n\otimes\mathbb{F}^m = \mathbb{F}^{n\times m}, \qquad (a\otimes b)_{ij} = a_i b_j \qquad \begin{pmatrix} i = 1, \ldots, n \\ j = 1, \ldots, m \end{pmatrix}$$

(where $\mathbb{F}^{n\times m}$ is the set of $n\times m$ matrices with entries from $\mathbb{F}$).

1.4. In some applications we need other realizations of tensor products beside the one given above by the vector space of bilinear forms on duals.

(i) To show that a bilinear map b on $\mathbf{L}\times\mathbf{K}$ realizes $\mathbf{L}\otimes\mathbf{K}$, it suffices to verify that if n is a natural number and $(\mathbf{x}_k, \mathbf{y}_k) \in \mathbf{L}\times\mathbf{K}$ and $\mathbf{x}_k$-s are linearly independent then

$$\sum_{k=1}^{n} b(\mathbf{x}_k, \mathbf{y}_k) = \mathbf{0} \quad \text{implies} \quad \mathbf{y}_k = \mathbf{0} \quad (k = 1, \ldots, n).$$

Indeed, this property assures that if c is another bilinear map on $\mathbf{L}\times\mathbf{K}$ then $b(\mathbf{x}, \mathbf{y}) \mapsto c(\mathbf{x}, \mathbf{y})$ $((\mathbf{x}, \mathbf{y}) \in \mathbf{L}\times\mathbf{K})$ is well defined and can be extended to a linear map L defined on the span of Ran b such that $L\circ b = c$.

(ii) We can establish some canonical isomorphisms between different tensor products. These canonical isomorphisms allow us to *identify* the vector spaces in question. We shall use the symbol $\equiv$ for the canonical identifications.

There is only a single case where we find convenient to denote explicitely the canonical isomorphism and not to establish the identification; this is given in the second item of the following proposition.

1.5. Proposition. Let $\mathbf{L}$, $\mathbf{K}$ and $\mathbf{M}$ be vector spaces. Then

(i) $\mathbb{F}\otimes\mathbf{L} \equiv \mathbf{L}$, $\alpha\otimes\mathbf{x} \equiv \alpha\mathbf{x}$;

(ii) $\mathbf{L}\otimes\mathbf{K} \longrightarrow \mathbf{K}\otimes\mathbf{L}$, $\mathbf{x}\otimes\mathbf{y} \mapsto (\mathbf{x}\otimes\mathbf{y})^* := \mathbf{y}\otimes\mathbf{x}$ gives the canonical isomorphism between the two tensor products;

(iii) $(\mathbf{L}\times\mathbf{K})\otimes\mathbf{M} \equiv (\mathbf{L}\otimes\mathbf{M})\times(\mathbf{K}\otimes\mathbf{M})$,

$$(\mathbf{x}, \mathbf{y})\otimes\mathbf{z} \equiv (\mathbf{x}\otimes\mathbf{z}, \mathbf{y}\otimes\mathbf{z}) \qquad (\mathbf{x}\in\mathbf{L}, \mathbf{y}\in\mathbf{K}, \mathbf{z}\in\mathbf{M});$$

If one of $\mathbf{L}$ and $\mathbf{K}$ is finite dimensional, then

(iv) $\mathbf{L}\otimes\mathbf{K}^* \equiv L(\mathbf{K}, \mathbf{L})$, $(\mathbf{x}\otimes\mathbf{r})\mathbf{y} \equiv (\mathbf{r}|\mathbf{y})\mathbf{x}$ $\quad (\mathbf{x}\in\mathbf{L}, \mathbf{r}\in\mathbf{K}^*, \mathbf{y}\in\mathbf{K})$

where $L(\mathbf{K}, \mathbf{L})$ stands for the vector space of linear maps $\mathbf{K}\longrightarrow\mathbf{L}$;

(v) $\mathbf{L}^*\otimes\mathbf{K}^* \equiv (\mathbf{L}\otimes\mathbf{K})^*$, $(\mathbf{p}\otimes\mathbf{r}|\mathbf{x}\otimes\mathbf{y}) \equiv (\mathbf{p}|\mathbf{x})(\mathbf{r}|\mathbf{y})$ $\quad (\mathbf{p}\in\mathbf{L}^*, \mathbf{r}\in\mathbf{K}^*, \mathbf{y}\in\mathbf{K}, \mathbf{x}\in\mathbf{L})$.

1.6. Remarks. (i) If $\mathbf{L}$ is finite dimensional then the vector space $\mathrm{L}(\mathbf{L})$ of the linear maps $\mathbf{L} \longrightarrow \mathbf{L}$ can be canonically identified with $\mathbf{L}\otimes\mathbf{L}^*$, and $(\mathbf{x}\otimes\mathbf{p})\circ(\mathbf{y}\otimes\mathbf{r}) = (\mathbf{p}|\mathbf{y})\mathbf{x}\otimes\mathbf{r} \quad (\mathbf{x}, \mathbf{y} \in \mathbf{L}, \mathbf{p}, \mathbf{r} \in \mathbf{L}^*)$;

(ii) The map $\mathrm{L}(\mathbf{L}) \equiv \mathbf{L}\otimes\mathbf{L}^* \longrightarrow \mathbf{L}^*\otimes\mathbf{L} \equiv \mathrm{L}(\mathbf{L}^*)$, $A \mapsto A^*$ determined by (ii) is the transposing map, i.e. A^* is the transpose of the linear map A.

(iii) According to the definition of tensor products, the bilinear map of duality on $\mathbf{L}^* \times \mathbf{L}$ induces a linear map

$$\mathrm{Tr}: \mathrm{L}(\mathbf{L}) \equiv \mathbf{L}\otimes\mathbf{L}^* \longrightarrow \mathbb{F} \quad \text{characterized by} \quad \mathbf{x}\otimes\mathbf{p} \mapsto (\mathbf{p}|\mathbf{x}).$$

We deduce from (i) that $\mathrm{Tr}(AB) = \mathrm{Tr}(BA)$ for all $A, B \in \mathrm{L}(\mathbf{L})$.

(iv) If $\mathbf{L}$ and $\mathbf{K}$ are one dimensional, we omit the transposing map and identify $\mathbf{L}\otimes\mathbf{K}$ with $\mathbf{K}\otimes\mathbf{L}$.

1.7. Proposition. If $\{\mathbf{x}_i : i \in I\}$ is a basis in $\mathbf{L}$ and $\{\mathbf{y}_j : j \in J\}$ is a basis in $\mathbf{K}$ then $\{\mathbf{x}_i\otimes\mathbf{y}_j : (i, j) \in I \times J\}$ is a basis in $\mathbf{L}\otimes\mathbf{K}$.

1.8. Proposition. Let $\mathbf{L}$, $\mathbf{K}$ and $\mathbf{L}'$, $\mathbf{K}'$ be vector spaces. Suppose $A: \mathbf{L} \longrightarrow \mathbf{L}'$ and $B: \mathbf{K} \longrightarrow \mathbf{K}'$ are linear maps. Then there is a unique linear map

$$A \circledcirc B: \mathbf{L}\otimes\mathbf{K} \longrightarrow \mathbf{L}'\otimes\mathbf{K}'$$

such that

$$(A \circledcirc B)(\mathbf{x}\otimes\mathbf{y}) = A\mathbf{x}\otimes B\mathbf{y} \qquad (\mathbf{x} \in \mathbf{L}, \mathbf{y} \in \mathbf{K}).$$

Remark. A and B are elements of the vector spaces $\mathrm{L}(\mathbf{L}, \mathbf{L}')$ and $\mathrm{L}(\mathbf{K}, \mathbf{K}')$, respectively. Their tensor product $A\otimes B$ can be identified with $A \circledcirc B$.

1.9. Definition. Let S be a non-void set, let $\mathbf{L}$ and $\mathbf{K}$ be vector spaces. Then for mappings $f: S \longrightarrow \mathbf{L}$, $g: S \longrightarrow \mathbf{K}$, we put

$$f\otimes g: S \longrightarrow \mathbf{L}\otimes\mathbf{K}, \qquad s \mapsto f(s)\otimes g(s).$$

Note the particular case when $\mathbf{x} \in \mathbf{L}$ is considered as a constant map $S \longrightarrow \mathbf{L}$; then $(\mathbf{x}\otimes g)(s) = \mathbf{x}\otimes g(s) \quad (s \in S)$.

1.10. Definition. Let $\mathbf{x}$ be an element of $\mathbf{L}$ and let H be a subset of $\mathbf{K}$. Then we put

$$\mathbf{x}\otimes H := \{\mathbf{x}\otimes\mathbf{y} : \mathbf{y} \in H\}.$$

1.11. For natural numbers $n \geq 2$ the definition of n-fold tensor products of vector spaces is similar to Definition 1.1.: one must consider n-fold linear maps instead of bilinear ones. We can state the existence and the essential uniqueness of n-fold tensor products similarly. We use the notation $\bigotimes_{k=1}^{n} \mathbf{L}_k$ and $\bigotimes_{k=1}^{n} \mathbf{x}_k$ for the n-fold tensor product of the vector spaces $\mathbf{L}_k$ and the vectors $\mathbf{x}_k \in \mathbf{L}_k$ $(k = 1, \ldots, n)$.

We have the canonical identification (in the sense of 1.4.)

$$\left(\bigotimes_{k=1}^{m} \mathbf{L}_k\right) \otimes \left(\bigotimes_{k=m+1}^{n} \mathbf{L}_k\right) \equiv \bigotimes_{k=1}^{n} \mathbf{L}_k ,$$

$$\left(\bigotimes_{k=1}^{m} \mathbf{x}_k\right) \otimes \left(\bigotimes_{k=m+1}^{n} \mathbf{x}_k\right) \equiv \bigotimes_{k=1}^{n} \mathbf{x}_k .$$

1.12. For natural numbers $n \geq 2$ the n-fold tensor product of n copies of the vector space $\mathbf{L}$ is denoted by $\overset{n}{\otimes} \mathbf{L}$; it is a subspace of the vector space $\mathrm{L}^n(\mathbf{L}^*)$ of the n-fold linear forms on $\mathbf{L}^*$. If $\mathbf{L}$ is finite dimensional then $\overset{n}{\otimes} \mathbf{L} = \mathrm{L}^n(\mathbf{L}^*)$.

For convenience we define $\overset{0}{\otimes} \mathbf{L} := \mathbb{F}$, $\overset{1}{\otimes} \mathbf{L} := \mathbf{L}$.

Then we have for all non-negative integers n and m

$$\left(\overset{n}{\otimes} \mathbf{L}\right) \otimes \left(\overset{m}{\otimes} \mathbf{L}\right) \equiv \overset{n+m}{\otimes} \mathbf{L} .$$

Similarly, if $L : \mathbf{L} \longrightarrow \mathbf{L}$ is a linear map, then $\overset{n}{\circledast} L$ denotes the linear map $\overset{n}{\otimes} \mathbf{L} \longrightarrow \overset{n}{\otimes} \mathbf{L}$ defined by 1.8. ($n \geq 2$) and we define $\overset{0}{\circledast} L$ to be the identity map of $\mathbb{F}$, and $\overset{1}{\circledast} L := L$.

1.13. Definition. Let $\mathbf{L}$ be a vector space. An *(anti)symmetric tensor product* of $\mathbf{L}$ is a pair $(\mathbf{M}, b)$ where

(i) $\mathbf{M}$ is a vector space,
(ii) $b : \mathbf{L} \times \mathbf{L} \longrightarrow \mathbf{M}$ is an (anti)symmetric bilinear map having the property that — if $\mathbf{N}$ is a vector space and $c : \mathbf{L} \times \mathbf{L} \longrightarrow \mathbf{N}$ is an (anti)symmetric bilinear map — then there exists a unique linear map $L : \mathbf{M} \longrightarrow \mathbf{N}$ such that

$$c = L \circ b .$$

1.14. For (anti)symmetric tensor products we have results similar to those given in 1.2. The symmetric and antisymmetric tensor products of $\mathbf{L}$ are denoted by $\mathbf{L} \vee \mathbf{L}$ and $\mathbf{L} \wedge \mathbf{L}$, respectively.

1.15. Definition 1.1. implies that there are unique linear maps

$$S : \mathbf{L} \otimes \mathbf{L} \longrightarrow \mathbf{L} \vee \mathbf{L}, \qquad \mathbf{x} \otimes \mathbf{y} \mapsto \mathbf{x} \vee \mathbf{y} ,$$

$$A : \mathbf{L} \otimes \mathbf{L} \longrightarrow \mathbf{L} \wedge \mathbf{L}, \qquad \mathbf{x} \otimes \mathbf{y} \mapsto \mathbf{x} \wedge \mathbf{y} .$$

Furthermore, the bilinear maps

$$\mathbf{L}\times\mathbf{L}\longrightarrow\mathbf{L}\otimes\mathbf{L},\qquad (\mathbf{x},\mathbf{y})\mapsto\frac{1}{2}(\mathbf{x}\otimes\mathbf{y}+\mathbf{y}\otimes\mathbf{x}),$$

$$\mathbf{L}\times\mathbf{L}\longrightarrow\mathbf{L}\otimes\mathbf{L},\qquad (\mathbf{x},\mathbf{y})\mapsto\frac{1}{2}(\mathbf{x}\otimes\mathbf{y}-\mathbf{y}\otimes\mathbf{x})$$

are symmetric and antisymmetric, respectively, hence there are unique linear maps

$$i_S:\mathbf{L}\circledvee\mathbf{L}\longrightarrow\mathbf{L}\otimes\mathbf{L},\qquad \mathbf{x}\circledvee\mathbf{y}\mapsto\frac{1}{2}(\mathbf{x}\otimes\mathbf{y}+\mathbf{y}\otimes\mathbf{x}),$$

$$i_A:\mathbf{L}\circledwedge\mathbf{L}\longrightarrow\mathbf{L}\otimes\mathbf{L},\qquad \mathbf{x}\circledwedge\mathbf{y}\mapsto\frac{1}{2}(\mathbf{x}\otimes\mathbf{y}-\mathbf{y}\otimes\mathbf{x}).$$

Proposition. S and A are surjections, i_S and i_A are injections and

$$S\circ i_S=\mathrm{id}_{\mathbf{L}\circledvee\mathbf{L}},\qquad A\circ i_A=\mathrm{id}_{\mathbf{L}\circledwedge\mathbf{L}}.$$

1.16. With the aid of the injections i_S and i_A we consider $\mathbf{L}\circledvee\mathbf{L}$ and $\mathbf{L}\circledwedge\mathbf{L}$ as subspaces of $\mathbf{L}\otimes\mathbf{L}$. Doing so we have

$$\begin{aligned}\mathbf{x}\circledvee\mathbf{y}&\equiv\frac{1}{2}(\mathbf{x}\otimes\mathbf{y}+\mathbf{y}\otimes\mathbf{x})=S(\mathbf{x}\otimes\mathbf{y}),\\ &\qquad\qquad\qquad\qquad\qquad\qquad(\mathbf{x},\mathbf{y}\in\mathbf{L})\\ \mathbf{x}\circledwedge\mathbf{y}&\equiv\frac{1}{2}(\mathbf{x}\otimes\mathbf{y}-\mathbf{y}\otimes\mathbf{x})=A(\mathbf{x}\otimes\mathbf{y}).\end{aligned}$$

If T is an element of $\mathbf{L}\otimes\mathbf{L}$ then $S(T)$ and $A(T)$ are called the *symmetric part* and the *antisymmetric part* of T, respectively.

Applications require another identification of $\mathbf{L}\circledwedge\mathbf{L}$ with a subspace of $\mathbf{L}\otimes\mathbf{L}$; it is done by $2i_A$. Then the antisymmetric tensor product of $\mathbf{L}$ is denoted by $\mathbf{L}\wedge\mathbf{L}$ and

$$\mathbf{x}\wedge\mathbf{y}:=2A(\mathbf{x}\otimes\mathbf{y})=\mathbf{x}\otimes\mathbf{y}-\mathbf{y}\otimes\mathbf{x}\qquad(\mathbf{x},\mathbf{y}\in\mathbf{L}).$$

Furthermore, $2A(T)$ is called the *antisymmetrization* of $T\in\mathbf{L}\otimes\mathbf{L}$.

1.17. A similar construction leads us to the n-fold symmetric and antisymmetric tensor products $\overset{n}{\circledvee}\mathbf{L}$ and $\overset{n}{\circledwedge}\mathbf{L}$, respectively ($n\geq 2$); they are considered as subspaces of $\overset{n}{\otimes}\mathbf{L}$ by

$$\overset{n}{\underset{k=1}{\circledvee}}\mathbf{x}_k:=\frac{1}{n!}\sum_{\pi\in\mathrm{Perm}\,n}\overset{n}{\underset{k=1}{\otimes}}\mathbf{x}_{\pi(k)}=S\left(\overset{n}{\underset{k=1}{\otimes}}\mathbf{x}_k\right),$$

$$\overset{n}{\underset{k=1}{\circledwedge}}\mathbf{x}_k:=\frac{1}{n!}\sum_{\pi\in\mathrm{Perm}\,n}(\mathrm{sign}\,\pi)\overset{n}{\underset{k=1}{\otimes}}\mathbf{x}_{\pi(k)}=A\left(\overset{n}{\underset{k=1}{\otimes}}\mathbf{x}_k\right),$$

where Perm n denotes the set of permutations of $\{1, 2, \ldots, n\}$. Furthermore we set $\overset{1}{\ovee}\mathbf{L} := \overset{1}{\owedge}\mathbf{L} := \mathbf{L}$, $\overset{0}{\ovee}\mathbf{L} := \overset{0}{\owedge}\mathbf{L} := \mathbb{F}$.

Another identification of the n-fold antisymmetric tensor product of $\mathbf{L}$ with a subspace of $\overset{n}{\otimes}\mathbf{L}$ is denoted by $\overset{n}{\wedge}\mathbf{L}$ and is given by

$$\overset{n}{\underset{k=1}{\wedge}} \mathbf{x}_k := n! A\left(\overset{n}{\underset{k=1}{\otimes}} \mathbf{x}_k\right).$$

1.18. Proposition. For an arbitrary non-negative integer n we have (in the sense of 1.4.), if $\mathbf{L}$ is finite dimensional:

$$\overset{n}{\ovee}\mathbf{L}^* \equiv \left(\overset{n}{\ovee}\mathbf{L}\right)^*, \quad \left(\overset{n}{\underset{k=1}{\ovee}} \mathbf{p}_k \,\middle|\, \overset{n}{\underset{k=1}{\ovee}} \mathbf{x}_k\right) \equiv \frac{1}{n!} \sum_{\pi \in \mathrm{Perm}\, n} \prod_{k=1}^{n} (\mathbf{p}_k | \mathbf{x}_{\pi(k)}),$$

$$\overset{n}{\owedge}\mathbf{L}^* \equiv \left(\overset{n}{\owedge}\mathbf{L}\right)^*, \quad \left(\overset{n}{\underset{k=1}{\owedge}} \mathbf{p}_k \,\middle|\, \overset{n}{\underset{k=1}{\owedge}} \mathbf{x}_k\right) \equiv \frac{1}{n!} \sum_{\pi \in \mathrm{Perm}\, n} \operatorname{sign}\pi \prod_{k=1}^{n} (\mathbf{p}_k | \mathbf{x}_{\pi(k)}).$$

1.19. Proposition. If $\mathbf{L}$ is finite dimensional then

$$\dim\left(\overset{n}{\wedge}\mathbf{L}\right) = \begin{cases} \binom{\dim \mathbf{L}}{n} & \text{if} \quad n \le \dim \mathbf{L}, \\ 0 & \text{if} \quad n > \dim \mathbf{L}. \end{cases}$$

In particular, $\mathbf{L} \wedge \mathbf{L} = 0$ if $\dim \mathbf{L} = 1$.

Furthermore, if $\{\mathbf{x}_i \,\vdots\, i = 1, \ldots, m\}$ is a basis of $\mathbf{L}$ then

$$\{\mathbf{x}_{i_1} \wedge \mathbf{x}_{i_2} \wedge \ldots \wedge \mathbf{x}_{i_n} \,\vdots\, 1 \le i_1 < i_2 < \ldots i_n \le m\}$$

is a basis of $\overset{n}{\wedge}\mathbf{L}$.

1.20. Definition. Let $\mathbf{L}$ and $\mathbf{K}$ be vector spaces of equal finite dimension m and suppose $A : \mathbf{L} \longrightarrow \mathbf{K}$ is a linear map. Then the linear map

$$\det A : \overset{m}{\wedge}\mathbf{K}^* \longrightarrow \overset{m}{\wedge}\mathbf{L}^*, \qquad \alpha \mapsto \alpha \circ \left(\overset{m}{\times} A\right)$$

is called the *determinant* of A. (Recall that in this case $\overset{m}{\wedge}\mathbf{K}^*$ and $\overset{m}{\wedge}\mathbf{L}^*$ are the vector spaces of the m-fold linear antisymmetric forms on $\mathbf{K}$ and $\mathbf{L}$, respectively (see 1.11.)).

Remark. $\overset{m}{\wedge}\, \mathbf{K}^*$ and $\overset{m}{\wedge}\, \mathbf{L}^*$ are one dimensional. Hence we can write (see 2.7. (vi))

$$\det A = \frac{\alpha \circ \left(\overset{m}{\times} A \right)}{\alpha} \in \frac{\overset{m}{\wedge}\, \mathbf{L}^*}{\overset{n}{\wedge}\, \mathbf{K}^*} \qquad (0 \neq \alpha \in \overset{n}{\wedge}\, \mathbf{K}^*).$$

As a consequence, if $\mathbf{L} = \mathbf{K}$, det A is a number from $\mathbb{F}$ which can be shown to equal the usually defined determinant of A.

1.21. Proposition. Let $\mathbf{L}$, $\mathbf{K}$ and A be as before. Then

$$\det A = \frac{\overset{m}{\underset{k=1}{\wedge}}\, A\mathbf{x}_k}{\overset{m}{\underset{k=1}{\wedge}}\, \mathbf{x}_k}$$

for arbitrarily chosen linearly independent vectors $\mathbf{x}_1, \ldots, \mathbf{x}_m$ of $\mathbf{L}$.

Proof. The antisymmetric m-tensor product of m linearly independent vectors is not zero, hence the right hand side of the above equality makes sense and it is an element of

$$\frac{\overset{m}{\wedge}\, \mathbf{K}}{\overset{m}{\wedge}\, \mathbf{L}} \equiv \frac{\left(\overset{m}{\wedge}\, \mathbf{L} \right)^*}{\left(\overset{m}{\wedge}\, \mathbf{K} \right)^*} \equiv \frac{\overset{m}{\wedge}\, \mathbf{L}^*}{\overset{m}{\wedge}\, \mathbf{K}^*}$$

(see 2.8. and 1.18.).

$\overset{m}{\underset{k=1}{\wedge}}\, A\mathbf{x}_k$ is an antisymmetric m-linear form on $\mathbf{K}^*$ and we have for elements $\mathbf{p}_1, \ldots, \mathbf{p}_m$ of $\mathbf{K}^*$

$$\left(\overset{m}{\underset{k=1}{\wedge}}\, A\mathbf{x}_k \right) (\mathbf{p}_1, \ldots, \mathbf{p}_m) = \left(\overset{m}{\underset{k=1}{\wedge}}\, \mathbf{p}_k \right) (A\mathbf{x}_1, \ldots, A\mathbf{x}_m) =$$

$$= \left[(\det A) \otimes \left(\overset{m}{\underset{k=1}{\wedge}}\, \mathbf{p}_k \right) \right] (\mathbf{x}_1, \ldots, \mathbf{x}_m) = \left[(\det A) \otimes \overset{m}{\underset{k=1}{\wedge}}\, \mathbf{x}_k \right] (\mathbf{p}_1, \ldots, \mathbf{p}_m).$$

1.22. Proposition. Let r be a natural number, let $\mathbf{L}_i$, $\mathbf{K}_i$ be vector spaces of equal finite dimension m_i, and suppose $A_i : \mathbf{L}_i \longrightarrow \mathbf{K}_i$ is a linear map ($i = 1, \ldots, r$). Then

$$\det \left(\overset{r}{\underset{i=1}{\times}}\, A_i \right) = \overset{r}{\underset{i=1}{\otimes}} \det A_i\,.$$

Proof. We carry out the proof for $r=2$; the general case can be easily recovered from it.

Let $\{\mathbf{x}_k^i : k=1,\ldots,m_i\}$ be a basis in $\mathbf{L}_i$ $(i=1,2)$. Then $\{(\mathbf{x}_k^1,\mathbf{0}),(\mathbf{0},\mathbf{x}_j^2): k=1,\ldots,m_1, j=1,\ldots,m_2\}$ is a basis in $\mathbf{L}_1\times\mathbf{L}_2$, and

$$\left(\bigwedge_{k=1}^{m_1}(A_1\times A_2)(\mathbf{x}_k^1,\mathbf{0})\right)\wedge\left(\bigwedge_{j=1}^{m_2}(A_1\times A_2)(\mathbf{0},\mathbf{x}_j^2)\right)=$$

$$=\left(\bigwedge_{k=1}^{m_1}(A_1\mathbf{x}_k^1,\mathbf{0})\right)\wedge\left(\bigwedge_{j=1}^{m_2}(\mathbf{0},A_2\mathbf{x}_j^2)\right)=$$

$$=\left[(\det A_1)\otimes\bigwedge_{k=1}^{m_1}(\mathbf{x}_k^1,\mathbf{0})\right]\wedge\left[(\det A_2)\otimes\bigwedge_{j=1}^{m_2}(\mathbf{0},\mathbf{x}_j^2)\right]=$$

$$=(\det A_1)\otimes(\det A_2)\otimes\left[\left(\bigwedge_{k=1}^{m_1}(\mathbf{x}_k^1,\mathbf{0})\right)\wedge\left(\bigwedge_{j=1}^{m_2}(\mathbf{0},\mathbf{x}_j^2)\right)\right].$$

Applying the previous result we get the desired equality.

1.23. Proposition. Let $\mathbf{L}$, $\mathbf{K}$ and $\mathbf{M}$ be vector spaces of equal finite dimension m and let $L:\mathbf{L}\longrightarrow\mathbf{K}$, $K:\mathbf{K}\longrightarrow\mathbf{M}$ be linear maps. Then

$$\det(K\circ L)=(\det K)\otimes(\det L).$$

Proof. Since $\overset{m}{\times}(K\circ L)=\left(\overset{m}{\times}K\right)\circ\left(\overset{m}{\times}L\right)$, we have by Remark 1.20. and 2.7. (ii)

$$\det(K\circ L)=\frac{\alpha\circ\left(\overset{m}{\times}(K\circ L)\right)}{\alpha}=\frac{\alpha\circ\left(\overset{m}{\times}K\right)}{\alpha}\otimes\frac{\left(\alpha\circ\left(\overset{m}{\times}K\right)\right)\circ\left(\overset{m}{\times}L\right)}{\alpha\circ\left(\overset{m}{\times}K\right)}.$$

2. Tensor quotients

In this section $\mathbb{F}$ denotes the field of complex or real numbers and all vector spaces are given over $\mathbb{F}$.

2.1. Definition. Let $\mathbf{L}$ and $\mathbf{A}$ be vector spaces, $\dim\mathbf{A}=1$. A *tensor quotient* of $\mathbf{L}$ by $\mathbf{A}$ is a pair $(\mathbf{M},q)$ where

(i) $\mathbf{M}$ is a vector space,

(ii) $q:\mathbf{L}\times(\mathbf{A}\setminus\{\mathbf{0}\})\longrightarrow\mathbf{M}$ is a *linear quotient map*, i.e.

$$\mathbf{x} \mapsto q(\mathbf{x}, \mathbf{a}) \quad \text{is linear for all} \quad \mathbf{a} \in \mathbf{A} \setminus \{\mathbf{0}\},$$

$$q(\mathbf{x}, \alpha\mathbf{a}) = \frac{1}{\alpha} q(\mathbf{x}, \mathbf{a}) \qquad (\mathbf{x} \in \mathbf{L},\ \alpha \in \mathbb{F} \setminus \{0\},\ \mathbf{a} \in \mathbf{A} \setminus \{\mathbf{0}\}),$$

having the property that

—if $\mathbf{N}$ is a vector space and $r: \mathbf{L} \times (\mathbf{A} \setminus \{\mathbf{0}\}) \longrightarrow \mathbf{N}$ is a linear quotient map

—then there exists a unique linear map $L: \mathbf{M} \longrightarrow \mathbf{N}$ such that

$$r = L \circ q.$$

2.2. The existence of tensor quotients is proved in 2.5.

An easy consequence of the definition (implying the noteworthy fact that $\mathrm{id}_{\mathbf{M}}$ is the only linear map L satisfying $q = L \circ q$) is that if $(\mathbf{M}, q)$ and $(\mathbf{M}', q')$ are tensor quotients of $\mathbf{L}$ by $\mathbf{A}$ then there exists a unique linear isomorphism $L: \mathbf{M} \longrightarrow \mathbf{M}'$ such that $q' = L \circ q$.

This shows that tensor quotients of $\mathbf{L}$ by $\mathbf{A}$ are canonically isomorphic, hence we speak of *the* tensor quotient, and applying an abuse of language we denote it by $\underset{\mathbf{A}}{\overset{\mathbf{L}}{=}}$ writing $(\mathbf{x}, \mathbf{a}) \mapsto \underset{\mathbf{a}}{\overset{\mathbf{x}}{=}}$ for the corresponding linear quotient map.

2.3. Proposition. If $\mathbf{A}$ is a one dimensional vector space then $\underset{\mathbf{A}}{\overset{\mathbf{A}}{=}} = \mathbb{F}$, $\underset{\mathbf{a}}{\overset{\mathbf{b}}{=}}$ is the unique element of $\mathbb{F}$ such that

$$\left(\underset{\mathbf{a}}{\overset{\mathbf{b}}{=}}\right)\mathbf{a} = \mathbf{b} \qquad (\mathbf{b} \in \mathbf{A},\ \mathbf{a} \in \mathbf{A} \setminus \{\mathbf{0}\}).$$

Proof. The above defined map $(\mathbf{b}, \mathbf{a}) \mapsto \underset{\mathbf{a}}{\overset{\mathbf{b}}{=}}$ is evidently linear quotient.

If $r: \mathbf{A} \times (\mathbf{A} \setminus \{\mathbf{0}\}) \longrightarrow \mathbf{N}$ is a linear quotient map then the map

$$L: \mathbb{F} \longrightarrow \mathbf{N}, \qquad \underset{\mathbf{a}}{\overset{\mathbf{b}}{=}} \mapsto r(\mathbf{b}, \mathbf{a})$$

is well defined, for if $\underset{\mathbf{a}}{\overset{\mathbf{b}}{=}} = \underset{\mathbf{a}'}{\overset{\mathbf{b}'}{=}}$ then $\mathbf{b} = \underset{\mathbf{a}'}{\overset{\mathbf{a}}{=}}\mathbf{b}'$, hence $r(\mathbf{b}, \mathbf{a}) = r(\mathbf{b}', \mathbf{a}')$.

Furthermore, using that $(\mathbf{b}, \mathbf{a}) \mapsto \underset{\mathbf{a}}{\overset{\mathbf{b}}{=}}$ is linear quotient, we derive

$$L\left(\alpha \underset{\mathbf{a}}{\overset{\mathbf{b}}{=}}\right) = L\left(\underset{\mathbf{a}}{\overset{\alpha\mathbf{b}}{=}}\right) = r(\alpha\mathbf{b}, \mathbf{a}) = \alpha r(\mathbf{b}, \mathbf{a}) = \alpha L\left(\underset{\mathbf{a}}{\overset{\mathbf{b}}{=}}\right)$$

for all $\alpha \in \mathbb{F}$, $\mathbf{a} \in \mathbf{A} \setminus \{\mathbf{0}\}$, $\mathbf{b} \in \mathbf{A}$, and

$$L\left(\underset{\mathbf{a}}{\overset{\mathbf{b}}{=}} + \underset{\mathbf{a}'}{\overset{\mathbf{b}'}{=}}\right) = L\left(\underset{\mathbf{a}'}{\overset{\left(\underset{\mathbf{a}}{\overset{\mathbf{a}'}{=}}\right)\mathbf{b}}{=\!=\!=\!=}} + \underset{\mathbf{a}'}{\overset{\mathbf{b}'}{=}}\right) = L\left(\underset{\mathbf{a}'}{\overset{\left(\underset{\mathbf{a}}{\overset{\mathbf{a}'}{=}}\right)\mathbf{b}+\mathbf{b}'}{=\!=\!=\!=\!=\!=}}\right) =$$

$$= r\left(\left(\underset{\mathbf{a}}{\overset{\mathbf{a}'}{=}}\right)\mathbf{b}+\mathbf{b}', \mathbf{a}'\right) = r\left(\left(\underset{\mathbf{a}}{\overset{\mathbf{a}'}{=}}\right)\mathbf{b}, \mathbf{a}'\right) + r(\mathbf{b}', \mathbf{a}') = L\left(\underset{\mathbf{a}}{\overset{\mathbf{b}}{=}}\right) + L\left(\underset{\mathbf{a}'}{\overset{\mathbf{b}'}{=}}\right)$$

for all $\mathbf{b}, \mathbf{b}' \in \mathbf{A}$, $\mathbf{a}, \mathbf{a}' \in \mathbf{A} \setminus \{\mathbf{0}\}$; thus L is linear.

The uniqueness of L is evident.

Remark. Note that $\underset{\mathbf{a}}{\overset{\mathbf{0}}{=}} = \mathbf{0}$ for all $\mathbf{a} \in \mathbf{A} \setminus \{\mathbf{0}\}$ and $\dfrac{1}{\left(\underset{\mathbf{a}}{\overset{\mathbf{a}'}{=}}\right)} = \underset{\mathbf{a}}{\overset{\mathbf{a}'}{=}}$ for all $\mathbf{a}, \mathbf{a}' \in \mathbf{A} \setminus \{\mathbf{0}\}$.

2.4. Let $\mathbf{L}$ and $\mathbf{A}$ be vector spaces, $\dim \mathbf{A} = 1$. One easily verifies that the homogeneous relation $\vdash\!\dashv$ on $\mathbf{L} \times (\mathbf{A} \setminus \{\mathbf{0}\})$ given by

$$(\mathbf{x}, \mathbf{a}) \vdash\!\dashv (\mathbf{y}, \mathbf{b})$$

if there exists an

$$\alpha \in \mathbb{F} \setminus \{0\}$$

such that

$$(\alpha\mathbf{x}, \alpha\mathbf{a}) = (\mathbf{y}, \mathbf{b})$$

is an equivalence relation.

Let $\underset{\mathbf{a}}{\overset{\mathbf{x}}{\vdash\!\dashv}}$ denote the equivalence class of $(\mathbf{x}, \mathbf{a}) \in \mathbf{L} \times (\mathbf{A} \setminus \{\mathbf{0}\})$ and put $\underset{\mathbf{A}}{\overset{\mathbf{L}}{\vdash\!\dashv}}$ for the set of equivalence classes.

Proposition. The maps

$$\mathbb{F} \times \underset{\mathbf{A}}{\overset{\mathbf{L}}{\vdash\!\dashv}} \longrightarrow \underset{\mathbf{A}}{\overset{\mathbf{L}}{\vdash\!\dashv}}, \qquad \left(\alpha, \underset{\mathbf{a}}{\overset{\mathbf{x}}{\vdash\!\dashv}}\right) \mapsto \underset{\mathbf{a}}{\overset{\alpha\mathbf{x}}{\vdash\!\dashv}},$$

$$\underset{\mathbf{A}}{\overset{\mathbf{L}}{\vdash\!\dashv}} \times \underset{\mathbf{A}}{\overset{\mathbf{L}}{\vdash\!\dashv}} \longrightarrow \underset{\mathbf{A}}{\overset{\mathbf{L}}{\vdash\!\dashv}}, \qquad \left(\underset{\mathbf{a}}{\overset{\mathbf{x}}{\vdash\!\dashv}}, \underset{\mathbf{b}}{\overset{\mathbf{y}}{\vdash\!\dashv}}\right) \mapsto \underset{\mathbf{b}}{\overset{\underset{\mathbf{a}}{\overset{\mathbf{b}}{=}}\mathbf{x}+\mathbf{y}}{\vdash\!\!=\!\!=\!\!=\!\!\dashv}}$$

are well defined and as multiplication by field elements and addition turn $\underset{\mathbf{A}}{\overset{\mathbf{L}}{\vdash\!\dashv}}$ into a vector space over $\mathbb{F}$.

The zero element of $\underset{\mathbf{A}}{\overset{\mathbf{L}}{\vdash\!\dashv}}$ is $\underset{\mathbf{a}}{\overset{\mathbf{0}}{\vdash\!\dashv}}$ for arbitrary $\mathbf{a} \in \mathbf{A} \setminus \{\mathbf{0}\}$.

2.5. Proposition. We can take

$$\underset{\mathbf{A}}{\overset{\mathbf{L}}{=}} := \underset{\mathbf{A}}{\overset{\mathbf{L}}{\vdash\!\dashv}}, \qquad \underset{\mathbf{a}}{\overset{\mathbf{x}}{=}} := \underset{\mathbf{a}}{\overset{\mathbf{x}}{\vdash\!\dashv}}.$$

Proof. We can argue as we did in 2.3. to verify our assertion.

2.6. Proposition 2.5. demonstrates the existence of tensor quotients. In some applications we need other realizations of tensor quotients beside that one. Furthermore, we can establish some canonical isomorphisms between certain tensor quotients and tensor products. These canonical isomorphisms allow us to *identify* the vector spaces in question. We shall use the symbol $\equiv$ for the canonical identifications (see 1.4.).

2.7. Proposition. Let **L**, **K** and **A**, **B** be vector spaces, dim **A** = dim **B** = 1. Then

(i) $\underset{\mathbb{F}}{\overset{\mathbf{L}}{=}} \equiv \mathbf{L}, \quad \underset{\alpha}{\overset{\mathbf{x}}{=}} \equiv \frac{1}{\alpha}\mathbf{x} \qquad (\mathbf{x}\in\mathbf{L},\quad \alpha\in\mathbb{F}\setminus\{0\});$

(ii) $\underset{\mathbf{A}}{\overset{\mathbf{L}}{=}} \otimes \underset{\mathbf{B}}{\overset{\mathbf{K}}{=}} \equiv \underset{\mathbf{A}\otimes\mathbf{B}}{\overset{\mathbf{L}\otimes\mathbf{K}}{=}} \equiv \underset{\mathbf{A}\otimes\mathbf{B}}{\overset{\mathbf{L}}{=}} \otimes \mathbf{K} \equiv \mathbf{L} \otimes \underset{\mathbf{A}\otimes\mathbf{B}}{\overset{\mathbf{K}}{=}} \equiv \underset{\mathbf{B}}{\overset{\mathbf{L}}{=}} \otimes \underset{\mathbf{A}}{\overset{\mathbf{K}}{=}},$

$$\underset{\mathbf{a}}{\overset{\mathbf{x}}{=}} \otimes \underset{\mathbf{b}}{\overset{\mathbf{y}}{=}} \equiv \underset{\mathbf{a}\otimes\mathbf{b}}{\overset{\mathbf{x}\otimes\mathbf{y}}{=}} \equiv \underset{\mathbf{a}\otimes\mathbf{b}}{\overset{\mathbf{x}}{=}} \otimes \mathbf{y} \equiv \mathbf{x} \otimes \underset{\mathbf{a}\otimes\mathbf{b}}{\overset{\mathbf{y}}{=}} \equiv \underset{\mathbf{b}}{\overset{\mathbf{x}}{=}} \otimes \underset{\mathbf{a}}{\overset{\mathbf{y}}{=}}$$

$$(\mathbf{x}\in\mathbf{L}, \mathbf{y}\in\mathbf{K}, \mathbf{a}\in\mathbf{A}\setminus\{0\};\ \mathbf{b}\in\mathbf{B}\setminus\{0\});$$

(iii) $\underset{\mathbf{B}}{\overset{\left(\underset{\mathbf{A}}{\overset{\mathbf{L}}{=}}\right)}{=}} \equiv \underset{\mathbf{A}\otimes\mathbf{B}}{\overset{\mathbf{L}}{=}}, \qquad \underset{\mathbf{b}}{\overset{\left(\underset{\mathbf{a}}{\overset{\mathbf{x}}{=}}\right)}{=}} \equiv \underset{\mathbf{a}\otimes\mathbf{b}}{\overset{\mathbf{x}}{=}}$

$$(\mathbf{x}\in\mathbf{L}, \mathbf{a}\in\mathbf{A}\setminus\{0\}, \mathbf{b}\in\mathbf{B}\setminus\{0\});$$

(iv) $\underset{\mathbf{A}}{\overset{\mathbf{L}\times\mathbf{K}}{=}} \equiv \underset{\mathbf{A}}{\overset{\mathbf{L}}{=}} \times \underset{\mathbf{A}}{\overset{\mathbf{K}}{=}}, \qquad \underset{\mathbf{a}}{\overset{(\mathbf{x},\mathbf{y})}{=}} \equiv \left(\underset{\mathbf{a}}{\overset{\mathbf{x}}{=}}, \underset{\mathbf{a}}{\overset{\mathbf{y}}{=}}\right)$

$$(\mathbf{x}\in\mathbf{L}, \mathbf{y}\in\mathbf{K}, \mathbf{a}\in\mathbf{A}\setminus\{0\}).$$

(v) $\underset{\mathbf{A}^*}{\overset{\mathbf{L}^*}{=}} \equiv \left(\underset{\mathbf{A}}{\overset{\mathbf{L}}{=}}\right)^*, \qquad \left(\underset{\mathbf{h}}{\overset{\mathbf{p}}{=}} \middle| \underset{\mathbf{a}}{\overset{\mathbf{x}}{=}}\right) \equiv \frac{(\mathbf{p}|\mathbf{x})}{(\mathbf{h}|\mathbf{a})}$

$$(\mathbf{p}\in\mathbf{L}^*, \mathbf{x}\in\mathbf{L}, \mathbf{h}\in\mathbf{A}^*\setminus\{0\}, \mathbf{a}\in\mathbf{A}\setminus\{0\}.$$

(vi) $\underset{\mathbf{A}}{\overset{\mathbf{L}}{=}} \equiv \mathrm{L}(\mathbf{A}, \mathbf{L}), \qquad \left(\underset{\mathbf{a}}{\overset{\mathbf{x}}{=}}\right)\mathbf{b} \equiv \underset{\mathbf{a}}{\overset{\mathbf{b}}{=}}\mathbf{x}$

$$(\mathbf{x}\in\mathbf{L}, \mathbf{a}\in\mathbf{A}\setminus\{0\}, \mathbf{b}\in\mathbf{A}).$$

2.8. Remarks. As particular cases of the above listed equalities we have

$$\text{(i)}\ \mathbf{A} \otimes \frac{\mathbf{L}}{\mathbf{A}} \equiv \frac{\mathbf{A}\otimes\mathbf{L}}{\mathbf{A}} \equiv \mathbf{L}, \qquad \mathbf{b} \otimes \frac{\mathbf{x}}{\mathbf{a}} \equiv \frac{\mathbf{b}\otimes\mathbf{x}}{\mathbf{a}} \equiv \frac{\mathbf{b}}{\mathbf{a}}\mathbf{x}$$

$$(\mathbf{b} \in \mathbf{A}, \mathbf{a} \in \mathbf{A}\setminus\{0\}, \mathbf{x} \in \mathbf{L});$$

$$(ii)\ \frac{\mathbb{F}}{\mathbf{A}} \equiv \mathbf{A}^*, \qquad \left(\frac{\alpha}{\mathbf{a}}\middle|\mathbf{b}\right) \equiv \alpha \frac{\mathbf{b}}{\mathbf{a}} \equiv \frac{\alpha}{\mathbf{a}} \otimes \mathbf{b}$$

$$(\alpha \in \mathbb{F}, \mathbf{b} \in \mathbf{A}, \mathbf{a} \in \mathbf{A}\setminus\{\mathbf{0}\}).$$

(iii) Furthermore, (ii) of Proposition 2.7. suggests us to define

$$(\mathbf{L}\otimes\mathbf{A}) \wedge \left(\frac{\mathbf{L}}{\mathbf{A}}\right) := \mathbf{L} \wedge \mathbf{L}, \qquad (\mathbf{x}\otimes\mathbf{a}) \wedge \left(\frac{\mathbf{y}}{\mathbf{b}}\right) := \left(\frac{\mathbf{a}}{\mathbf{b}}\right) \mathbf{x} \wedge \mathbf{y}$$

$$(\mathbf{x}, \mathbf{y} \in \mathbf{L}, \mathbf{a} \in \mathbf{A}, \mathbf{b} \in \mathbf{A}\setminus\{\mathbf{0}\}).$$

2.9. Proposition. $\mathbf{L} \longrightarrow \frac{\mathbf{L}}{\mathbf{A}}, \mathbf{x} \mapsto \frac{\mathbf{x}}{\mathbf{a}}$ is a linear isomorphism for all $\mathbf{a} \in \mathbf{A}\setminus\{\mathbf{0}\}$.

If $\{\mathbf{x}_i : i \in I\}$ is a basis in $\mathbf{L}$ and $\mathbf{a}$ is a basis in $\mathbf{A}$ (i.e. $\mathbf{a} \in \mathbf{A}\setminus\{\mathbf{0}\}$) then $\left\{\frac{\mathbf{x}_i}{\mathbf{a}} : i \in I\right\}$ is a basis in $\frac{\mathbf{L}}{\mathbf{A}}$.

2.10. Proposition. Let $\mathbf{L}$, $\mathbf{L}'$, $\mathbf{A}$, $\mathbf{A}'$ be vector spaces, $\dim \mathbf{A} = \dim \mathbf{A}' = 1$. Suppose $L : \mathbf{L} \longrightarrow \mathbf{L}'$ and $A : \mathbf{A} \longrightarrow \mathbf{A}'$ are linear maps, $A \neq 0$. Then the map

$$\frac{L}{A} : \frac{\mathbf{L}}{\mathbf{A}} \longrightarrow \frac{\mathbf{L}'}{\mathbf{A}'}, \qquad \frac{\mathbf{x}}{\mathbf{a}} \mapsto \frac{L\mathbf{x}}{A\mathbf{a}}$$

is well defined and linear.

Remarks. (i) L and A are elements of the vector spaces $\mathrm{L}(\mathbf{L}, \mathbf{L}')$ and $\mathrm{L}(\mathbf{A}, \mathbf{A}')$, respectively. One easily shows that the above defined linear map can be taken as the tensor quotient of L by A.

(ii) In the case $\mathbf{A} = \mathbf{A}'$ we shall use the notation

$$L_{\mathbf{A}} := \frac{L}{\mathrm{id}_{\mathbf{A}}}.$$

2.11. Definition. Let S be a non-void set, let $\mathbf{L}$ and $\mathbf{A}$ be vector spaces, $\dim \mathbf{A} = 1$. Then for mappings $f : S \longrightarrow \mathbf{L}$, $g : S \longrightarrow \mathbf{A}\setminus\{0\}$ we put

$$\frac{f}{g} : S \longrightarrow \frac{\mathbf{L}}{\mathbf{A}}, \qquad s \mapsto \frac{f(s)}{g(s)}.$$

Note the particular case when $\mathbf{a} \in \mathbf{A} \setminus \{\mathbf{0}\}$ is considered as a constant map $S \to \mathbf{A}$; then $\frac{f}{\mathbf{a}}(s) = \frac{f(s)}{\mathbf{a}}$ $(s \in S)$.

3. Tensorial operations and orientation

In this section all vector spaces are real and finite dimensional.

3.1. Definition. Two bases $\{\mathbf{x}_i \,\vdots\, i=1, \ldots, n\}$ and $\{\mathbf{x}'_i \,\vdots\, i=1, \ldots, n\}$ of the vector space $\mathbf{L}$ are called *equally oriented* if and only if the linear map L defined by $\mathbf{x}'_i =: L\mathbf{x}_i$ $(i=1, \ldots, n)$ has positive determinant (see 1.20.). An equivalence class of equally oriented bases is called an *orientation* of $\mathbf{L}$. $\mathbf{L}$ is *oriented* if an orientation of $\mathbf{L}$ is given; the bases in the chosen equivalence class are called *positively oriented* the other ones are called *negatively oriented.* (More precisely, an oriented vector space is a pair $(\mathbf{L}, \mathrm{o})$ where $\mathbf{L}$ is a vector space and o is one of the equivalence classes of its bases).

The dual $\mathbf{L}^*$ of an oriented vector space $\mathbf{L}$ is oriented by the dual bases of positively oriented bases of $\mathbf{L}$.

A linear bijection between oriented vector spaces is *orientation preserving* or *orientation reversing* if it sends positively oriented bases into positively oriented ones or into negatively oriented ones, respectively.

3.2. Proposition. If $\{\mathbf{x}_i \,\vdots\, i=1, \ldots, n\}$ and $\{\mathbf{x}'_i \,\vdots\, i=1, \ldots, n\}$ are equally oriented bases in $\mathbf{L}$ and $\{\mathbf{y}_j : j=1, \ldots, m\}$ and $\{\mathbf{y}'_j : j=1, \ldots, m\}$ are equally oriented bases in $\mathbf{K}$ then $\{\mathbf{x}_i \otimes \mathbf{y}_j \,\vdots\, i=1, \ldots, n;\ j=1, \ldots, m\}$ and $\{\mathbf{x}'_i \otimes \mathbf{y}'_j \,\vdots\, i=1, \ldots, n; j=1, \ldots, m\}$ are equally oriented bases in $\mathbf{L} \otimes \mathbf{K}$.

Proof. Let $L : \mathbf{L} \to \mathbf{L}$ and $K : \mathbf{K} \to \mathbf{K}$ be the linear isomorphisms sending the corresponding bases into each other. Then $L \otimes K$ sends the given bases of $\mathbf{L} \otimes \mathbf{K}$ into each other. Since $L \otimes K = (L \otimes \mathrm{id}_{\mathbf{K}})(\mathrm{id}_{\mathbf{L}} \otimes K)$, we end the proof by showing that

$$\det(L \otimes \mathrm{id}_{\mathbf{K}}) = (\det L)^m, \quad \det(\mathrm{id}_{\mathbf{L}} \otimes K) = (\det K)^n.$$

Let $\mathbf{K}_j$ denote the one dimensional subspace of $\mathbf{K}$ spanned by $\mathbf{y}_j$ $(j=1, \ldots, m)$. Then $\mathbf{K}$ can be identified with $\underset{j=1}{\overset{m}{\times}} \mathbf{K}_j$ and so $L \otimes \mathrm{id}_{\mathbf{K}}$ becomes $\underset{j=1}{\overset{m}{\times}} (L \otimes \mathrm{id}_{\mathbf{K}_j})$.

$\mathbf{L} \to \mathbf{L} \otimes \mathbf{K}_j$, $\mathbf{x} \mapsto \mathbf{x} \otimes \mathbf{y}_j$ is a linear isomorphism which turns L into $L \otimes \mathrm{id}_{\mathbf{K}_j}$, hence $\det(L \otimes \mathrm{id}_{\mathbf{K}_j}) = \det L$ $(j=1, \ldots, m)$. Applying Proposition 1.22. we get the desired result.

Remark. A similar result holds for symmetric and antisymmetric tensor products as well; then we take the bases $\{\mathbf{x}_i \vee \mathbf{y}_j : i=1, \ldots, j; j=1, \ldots, n\}$ and $\{\mathbf{x}_i \wedge \mathbf{y}_j : i=1, \ldots, j-1; j=1, \ldots, n\}$.

3.3. Proposition. If $\{\mathbf{x}_i : i=1,\ldots,n\}$ and $\{\mathbf{x}'_i : i=1,\ldots,n\}$ are equally oriented bases in $\mathbf{L}$ and $\mathbf{a}$, $\mathbf{a}'$ are equally oriented bases in $\mathbf{A}$ then $\left\{\frac{\mathbf{x}_i}{\mathbf{a}} : i=1,\ldots,n\right\}$ and $\left\{\frac{\mathbf{x}'_i}{\mathbf{a}'} : i=1,\ldots,n\right\}$ are equally oriented bases in $\frac{\mathbf{L}}{\mathbf{A}}$.

3.4. The previous propositions permit us to introduce a natural orientation on tensor products and tensor quotients of oriented vector spaces.

Definition. If $\mathbf{L}$ and $\mathbf{K}$ are oriented vector spaces then $\mathbf{L}\otimes\mathbf{K}$ is considered oriented by the tensor product of positively oriented bases.

If $\mathbf{L}$ and $\mathbf{A}$ are oriented vector spaces, $\dim \mathbf{A}=1$, then $\frac{\mathbf{L}}{\mathbf{A}}$ is considered oriented by the tensor quotient of positively oriented bases.

Remark. We accept similar definitions for orienting symmetric and antisymmetric tensor products.

3.5. Proposition. An orientation on a one dimensional vector space is equivalent to a total ordering compatible with the linear structure.

Proof. One easily shows that two bases $\mathbf{a}$ and $\mathbf{a}'$ of a one dimensional vector space are equally oriented if and only if $\frac{\mathbf{a}}{\mathbf{a}'}$ is a positive real number. Hence the following construction is well defined.

If an orientation is given, determine the ordering $<$ in such a way that $\mathbf{a}<\mathbf{b}$ if and only if $\mathbf{b}-\mathbf{a}$ is positively oriented.

Conversely, given a total ordering, we define an element to be positively oriented if and only if it is greater then zero.

3.6. Definition. Let $\mathbf{A}$ be a one dimensional oriented vector space. Then we put

$$\mathbf{A}^+ := \{\mathbf{a}\in\mathbf{A} : \mathbf{a}>\mathbf{0}\}, \qquad \mathbf{A}_0^+ := \mathbf{A}^+\cup\{\mathbf{0}\}$$

and we call the elements of the above sets *positive* and *non-negative*, respectively.

Furthermore, the *absolute value* of $\mathbf{a}\in\mathbf{A}$ is

$$|\mathbf{a}| := \begin{cases} \mathbf{a} & \text{if } \mathbf{a}\in\mathbf{A}_0^+ \\ -\mathbf{a} & \text{if } \mathbf{a}\notin\mathbf{A}_0^+ \end{cases}$$

3.7. Proposition. Let $\mathbf{A}$ be a one dimensional vector space. Then the one dimensional vector space $\mathbf{A}\otimes\mathbf{A}$ can be canonically oriented by giving the set of positive elements as follows:

$$(\mathbf{A}\otimes\mathbf{A})^+ := \{\mathbf{a}\otimes\mathbf{a} : \mathbf{a}\in\mathbf{A}\setminus\{\mathbf{0}\}\}.$$

3.8. Proposition. Let $\mathbf{A}$ be an oriented one dimensional vector space. Then

$$\mathbf{A}_0^+ \to (\mathbf{A}\otimes\mathbf{A})_0^+, \quad \mathbf{a}\mapsto\mathbf{a}\otimes\mathbf{a}$$

is a bijection.

Proof. In view of the definition of $(\mathbf{A}\otimes\mathbf{A})_0^+$, it is sufficient to show that the above map is injective.

Suppose $\mathbf{a}\otimes\mathbf{a}=\mathbf{0}$; then evidently $\mathbf{a}=\mathbf{0}$.

Suppose now that $\mathbf{a},\mathbf{b}\in\mathbf{A}^+$ and $\mathbf{a}\otimes\mathbf{a}=\mathbf{b}\otimes\mathbf{b}$. Then

$$1=\frac{\mathbf{a}\otimes\mathbf{a}}{\mathbf{b}\otimes\mathbf{b}}=\frac{\mathbf{a}}{\mathbf{b}}\otimes\frac{\mathbf{a}}{\mathbf{b}}=\left(\frac{\mathbf{a}}{\mathbf{b}}\right)^2$$

(see 2.7. (ii) and 2.3.). Since $\frac{\mathbf{a}}{\mathbf{b}}$ is a positive real number, we deduce that $\mathbf{a}=\mathbf{b}$.

3.9. Definition. The inverse of the map

$$\mathbf{A}_0^+\mapsto(\mathbf{A}\otimes\mathbf{A})_0^+,\qquad \mathbf{a}\mapsto\mathbf{a}\otimes\mathbf{a}$$

is denoted by

$$\sqrt[\otimes]{}:(\mathbf{A}\otimes\mathbf{A})_0^+\longrightarrow\mathbf{A}_0^+$$

and is called the *square root* mapping.

Remark. Note that we have

$$\sqrt[\otimes]{\mathbf{a}\otimes\mathbf{a}}=|\mathbf{a}|\qquad(\mathbf{a}\in\mathbf{A}).$$

4. Tensor roots

4.1. The addition in a one dimensional real vector space is uniquely determined by the multiplication by real numbers. More closely, we have the following.

Proposition. Suppose A is a set and there is a map

$$\mathbb{R}\times A\longrightarrow A,\qquad(\alpha,a)\mapsto\alpha\cdot a$$

with the properties

(i) $1\cdot a=a$ for all $a\in A$,

(ii) $\alpha\cdot(\beta\cdot a)=(\alpha\beta)\cdot a$ for all $\alpha,\beta\in\mathbb{R}, a\in A$,

(iii) there exists a $z\in A$ such that $\mathbb{R}\longrightarrow A$, $\alpha\mapsto\alpha\cdot z$ is a bijection.

Then A is a one dimensional real vector space endowed with the above map as multiplication by real numbers and with the addition

$$\alpha\cdot z+\beta\cdot z:=(\alpha+\beta)\cdot z\qquad(\alpha,\beta\in\mathbb{R}).$$

Remark. It is important that the addition does not depend on z: if z' is another element such that $\{\alpha\cdot z\mid\alpha\in\mathbb{R}\}=A$ then the addition defined by z' coincides with the addition defined by z.

4.2. Definition. Let $\mathbf{A}$ be a one dimensional oriented real vector space. An *n-th tensor root* $(n\in\mathbb{N})$ of $\mathbf{A}$ is a pair $(\mathbf{B},r)$ where

(i) $\mathbf{B}$ is a one dimensional real vector space,

(ii) $r:\mathbf{A}_0^+ \longrightarrow \mathbf{B}$ is an *n-th root map*, i.e.

$$r(\alpha\mathbf{a})=\sqrt[n]{\alpha}\,r(\mathbf{a}) \qquad (\alpha\in\mathbb{R}_0^+,\ \mathbf{a}\in\mathbf{A}_0^+)$$

having the property that

— if $\mathbf{C}$ is a one dimensional real vector space and $s:\mathbf{A}_0^+\longrightarrow\mathbf{C}$ is an n-th root map

— then there exists a unique linear map $L:\mathbf{B}\longrightarrow\mathbf{C}$ such that

$$s=L\circ r\,.$$

Remark. Observe that the range of an n-th root map r is a "half line", i.e. there is a $\mathbf{b}\in\operatorname{Ran} r$ such that $\operatorname{Ran} r=\{\alpha\mathbf{b} : \alpha\in\mathbb{R}_0^+\}$.

4.3. The existence of n-th tensor roots is proved in 4.4. An easy consequence of the definition (implying the noteworthy fact that $\mathrm{id}_{\mathbf{B}}$ is the only linear map L satisfying $r=L\circ r$) is that if $(\mathbf{B}, r)$ and $(\mathbf{B}', r')$ are n-th tensor roots of $\mathbf{A}$ then there exists a unique linear isomorphism $L:\mathbf{B}\longrightarrow\mathbf{B}'$ such that $r'=L\circ r$.

This shows that n-th tensor roots of $\mathbf{A}$ are canonically isomorphic, hence we speak of *the* n-th tensor root, and applying an abuse of language we denote it by $\underset{\otimes}{\sqrt[n]{\mathbf{A}}}$ writing $\mathbf{a}\mapsto\underset{\otimes}{\sqrt[n]{\mathbf{a}}}$ for the corresponding n-th root map.

4.4. Let $\mathbf{A}$ be a one dimensional oriented real vector space, $n\in\mathbb{N}$. Introduce on $\mathbf{A}$ a new multiplication by real numbers:

$$\alpha\cdot\mathbf{a}:=(\operatorname{sign}\alpha)|\alpha|^n\mathbf{a} \qquad (\alpha\in\mathbb{R},\ \mathbf{a}\in\mathbf{A})\,.$$

It is evident that this multiplication by real numbers satisfies the requirements of Proposition 4.1.

Proposition. $\underset{\otimes}{\sqrt[n]{\mathbf{A}}}$ is $\mathbf{A}$ endowed with the linear structure determined by the above defined new multiplication, according to Proposition 4.1. The corresponding n-th root map $\mathbf{a}\mapsto\underset{\otimes}{\sqrt[n]{\mathbf{a}}}$ is the identity of $\mathbf{A}_0^+$.

Proof. It is obvious that the identity of $\mathbf{A}_0^+$ considered as a map from $\mathbf{A}_0^+$ into $\underset{\otimes}{\sqrt[n]{\mathbf{A}}}$ is an injective n-th root map.

Hence if $s:\mathbf{A}_0^+\longrightarrow\mathbf{C}$ is an n-th root map then

$$L_0:\operatorname{Ran}\underset{\otimes}{\sqrt[n]{\ }}\longrightarrow\mathbf{C},\qquad \underset{\otimes}{\sqrt[n]{\mathbf{a}}}\mapsto s(\mathbf{a}) \qquad (\mathbf{a}\in\mathbf{A}_0^+)$$

is well defined and has the property

$$L_0(\alpha\mathbf{b})=\alpha L_0(\mathbf{b}) \qquad (\mathbf{b}\in\operatorname{Ran}\underset{\otimes}{\sqrt[n]{\ }},\ \alpha\in\mathbb{R}_0^+)\,.$$

Putting

$$L(\mathbf{b}):=\begin{cases} L_0(\mathbf{b}) & \text{if} \quad \mathbf{b}\in\operatorname{Ran}\underset{\otimes}{\sqrt[n]{\ }}\,,\\ -L_0(-\mathbf{b}) & \text{if} \quad -\mathbf{b}\in\operatorname{Ran}\underset{\otimes}{\sqrt[n]{\ }}\end{cases}$$

we define a linear map $L: {}_{\otimes}\!\sqrt[n]{\mathbf{A}} \longrightarrow \mathbf{C}$ (see Remark 4.2. and Proposition 4.1.) such that $s = L \circ {}_{\otimes}\!\sqrt[n]{\;}$. The uniquness of L is evident.

4.5. We can establish some canonical isomorphisms between certain tensor products and tensor roots. These canonical isomorphisms allow us to *identify* the vector spaces in question. We shall use the symbol $\equiv$ for the canonical identifications (see 1.4. and 2.6.).

4.6. Proposition. Let **A** and **B** be oriented one dimensional real vector spaces, $n \in \mathbb{N}$. Then

(i) $${}_{\otimes}\!\sqrt[n]{\mathbf{A}\otimes\mathbf{B}} \equiv {}_{\otimes}\!\sqrt[n]{\mathbf{A}} \otimes {}_{\otimes}\!\sqrt[n]{\mathbf{B}}, \qquad {}_{\otimes}\!\sqrt[n]{\mathbf{a}\otimes\mathbf{b}} \equiv {}_{\otimes}\!\sqrt[n]{\mathbf{a}} \otimes {}_{\otimes}\!\sqrt[n]{\mathbf{b}}$$
$$(\mathbf{a} \in \mathbf{A}_0^+,\ \mathbf{b} \in \mathbf{B}_0^+),$$

(ii) $${}_{\otimes}\!\sqrt[n]{\overset{n}{\otimes}\mathbf{A}} \equiv \mathbf{A}, \qquad {}_{\otimes}\!\sqrt[n]{\overset{n}{\underset{k=1}{\otimes}} \mathbf{a}_k} \equiv \left(\prod_{k=1}^{n} \sqrt[n]{\frac{\mathbf{a}_k}{\mathbf{a}}}\right)\mathbf{a}$$
$$(\mathbf{a}_1, \ldots, \mathbf{a}_n \in \mathbf{A}_0^+,\ \mathbf{a} \in \mathbf{A}^+),$$

(iii) $${}_{\otimes}\!\sqrt[n]{\mathbf{A}^*} \equiv \left({}_{\otimes}\!\sqrt[n]{\mathbf{A}}\right)^*, \qquad \left({}_{\otimes}\!\sqrt[n]{\mathbf{z}} \,\middle|\, {}_{\otimes}\!\sqrt[n]{\mathbf{a}}\right) \equiv \sqrt[n]{(\mathbf{z}|\mathbf{a})}$$
$$(\mathbf{z} \in (\mathbf{A}^*)_0^+,\ \mathbf{a} \in \mathbf{A}_0^+).$$

(iv) $${}_{\otimes}\!\sqrt[n]{\mathbb{R}} \equiv \mathbb{R}, \qquad {}_{\otimes}\!\sqrt[n]{\alpha} \equiv \sqrt[n]{\alpha} \qquad (\alpha \in \mathbb{R}_0^+).$$

4.7. Remarks. (i) Similar expressions can be written for relations of tensor quotients and tensor roots because the tensor quotients by a one dimensional vector space **A** can be given by the tensor products of the dual of **A**.

(ii) We easily deduce from (i) and (ii) of Proposition 4.6. that $\overset{n}{\otimes}\, {}_{\otimes}\!\sqrt[n]{\mathbf{A}} \equiv \mathbf{A}$. In particular, the square root mapping introduced in 3.9. coincides with the present one for $n = 2$.

V. PSEUDO-EUCLIDEAN SPACES

1. Bilinear mappings

1.1. Definition. Let $\mathbf{L}_1, \mathbf{L}_2$ and $\mathbf{K}$ be vector spaces over the same field. The *transpose* of a bilinear map $B: \mathbf{L}_1 \times \mathbf{L}_2 \longrightarrow \mathbf{K}$ is the bilinear map

$$B^*: \mathbf{L}_2 \times \mathbf{L}_1 \longrightarrow \mathbf{K}, \qquad (\mathbf{y}, \mathbf{x}) \mapsto B(\mathbf{x}, \mathbf{y}) . \blacksquare$$

Evidently, if $\mathbf{L}_1 = \mathbf{L}_2$, B is symmetric or antisymmetric if and only if $B = B^*$ or $B = -B^*$, respectively.

1.2. If $\mathbf{L}_1, \mathbf{L}_2$ and $\mathbf{K}$ are vector spaces over the same field then we find convenient to write a bilinear map $B: (\mathbf{L}_1 \times \mathbf{L}_2) \times (\mathbf{L}_1 \times \mathbf{L}_2) \longrightarrow \mathbf{K}$ in the form

$$\begin{pmatrix} B_{11} & B_{12} \\ B_{21} & B_{22} \end{pmatrix}$$

where $B_{ik}: \mathbf{L}_i \times \mathbf{L}_k \longrightarrow \mathbf{K}$ ($i, k = 1, 2$) are bilinear maps defined by

$$B_{11}(\mathbf{x}_1, \mathbf{y}_1) := B((\mathbf{x}_1, \mathbf{0}), (\mathbf{y}_1, \mathbf{0})) \qquad (\mathbf{x}_1, \mathbf{y}_1 \in \mathbf{L}_1)$$

etc.

Doing so, we can calculate the value of the bilinear map by the usual matrix multiplication rule:

$$B((\mathbf{x}_1, \mathbf{x}_2), (\mathbf{y}_1, \mathbf{y}_2)) = (\mathbf{x}_1 \mathbf{x}_2) \begin{pmatrix} B_{11} & B_{12} \\ B_{21} & B_{22} \end{pmatrix} \begin{pmatrix} \mathbf{y}_1 \\ \mathbf{y}_2 \end{pmatrix}$$

Evidently, B is symmetric or antisymmetric if and only if $B_{ik} = B^*_{ki}$ or $B_{ik} = -B^*_{ki}$ ($i, k = 1, 2$), respectively.

1.3. Let $\mathbf{L}$ and $\mathbf{K}$ be finite dimensional vector spaces over the same field and suppose $B: \mathbf{L} \times \mathbf{L} \longrightarrow \mathbf{K}$ is a symmetric or antisymmetric bilinear map. Then

$$\operatorname{Ker} B := \{\mathbf{x} \in \mathbf{L} \,\vdots\, B(\mathbf{x}, \cdot) = \mathbf{0}\}$$

is a linear subspace of $\mathbf{L}$.

Definition.

$$\operatorname{rank} B := \dim \mathbf{L} - \dim (\operatorname{Ker} B) .$$

B is called *non-degenerate* if it has maximal rank, i.e. if $\operatorname{Ker} B = \mathbf{0}$.

2. Pseudo-Euclidean vector spaces

2.1. Definition. A *pseudo-Euclidean vector space* is a triplet $(\mathbf{V}, \mathbf{B}, b)$ where

(i) $\mathbf{V}$ is a finite dimensional real vector space,

(ii) $\mathbf{B}$ is a one dimensional real vector space,

(iii) $b: \mathbf{V} \times \mathbf{V} \longrightarrow \mathbf{B} \otimes \mathbf{B}$ is a non-degenerate symmetric bilinear map.

Remarks. (i) Since $\mathbf{B} \otimes \mathbf{B}$ is canonically oriented (see IV. 3.7.), it makes sense to say that $b(\mathbf{x}, \mathbf{y})$ is positive or negative for $\mathbf{x}, \mathbf{y} \in \mathbf{V}$. b is *positive definite* if $b(\mathbf{x}, \mathbf{x}) > \mathbf{0}$ for all non-zero $\mathbf{x} \in \mathbf{V}$.

(ii) The giving of a bilinear map $b: \mathbf{V} \times \mathbf{V} \longrightarrow \mathbf{B} \otimes \mathbf{B}$ is equivalent to the giving of a bilinear form $(\;\mid\;): \dfrac{\mathbf{V}}{\mathbf{B}} \times \dfrac{\mathbf{V}}{\mathbf{B}} \longrightarrow \mathbb{R}$. Indeed, equalities

$$(\mathbf{u} \mid \mathbf{v}) := \frac{b(\mathbf{a} \otimes \mathbf{u}, \mathbf{c} \otimes \mathbf{v})}{\mathbf{a} \otimes \mathbf{c}} \qquad \left(\mathbf{u}, \mathbf{v} \in \frac{\mathbf{V}}{\mathbf{B}},\ \mathbf{a}, \mathbf{c} \in \mathbf{B} \setminus \{0\}\right)$$

$$b(\mathbf{x}, \mathbf{y}) := (\mathbf{a} \otimes \mathbf{c}) \left(\frac{\mathbf{x}}{\mathbf{a}} \,\middle|\, \frac{\mathbf{y}}{\mathbf{c}} \right) \qquad (\mathbf{x}, \mathbf{y} \in \mathbf{V},\ \mathbf{a}, \mathbf{c} \in \mathbf{B} \setminus \{0\})$$

make the desired correspondence.

$(\mathbf{V}, \mathbf{B}, b)$ is a pseudo-Euclidean vector space if and only if $\left(\dfrac{\mathbf{V}}{\mathbf{B}}, \mathbb{R}, (\;\mid\;)\right)$ is a pseudo-Euclidean vector space.

We assume in the sequel that a pseudo-Euclidean vector space $(\mathbf{V}, \mathbf{B}, b)$ *is given.*

2.2. Definition. A basis $\{\mathbf{e}_i : i = 1, \ldots, \dim \mathbf{V}\}$ of $\mathbf{V}$ is called *b-orthogonal* if $b(\mathbf{e}_i, \mathbf{e}_j) = \mathbf{0}$ for $i \neq j$.

A b-orthogonal basis $\{\mathbf{e}_i : i = 1, \ldots, \dim \mathbf{V}\}$ is *normed* to $\mathbf{a} \in \mathbf{B}$ if either $b(\mathbf{e}_i, \mathbf{e}_i) = \mathbf{a} \otimes \mathbf{a}$ or $b(\mathbf{e}_i, \mathbf{e}_i) = -\mathbf{a} \otimes \mathbf{a}$ for all i. If $\mathbf{B} = \mathbb{R}$, a b-orthogonal basis normed to 1 is called *b-orthonormal.*

2.3. We can argue like in the case of real valued bilinear forms ([19], Vol. I., Theorem 2_6.) to have the following.

Proposition. b-orthogonal bases exist in $\mathbf{V}$ and there is a non-negative integer $\operatorname{neg}(b)$ such that for every b-orthogonal basis $\{\mathbf{e}_i : i = 1, \ldots, \dim \mathbf{V}\}$

$\{i : b(\mathbf{e}_i, \mathbf{e}_i) < \mathbf{0}\}$ consists of $\operatorname{neg}(b)$ elements

$\{i : b(\mathbf{e}_i, \mathbf{e}_i) > \mathbf{0}\}$ consists of $\dim \mathbf{V} - \operatorname{neg}(b)$ elements.

Remark. A b-orthogonal basis will be always numbered in such a way that b has negative values on the first $\operatorname{neg}(b)$ elements, i.e.

$$b(\mathbf{e}_i, \mathbf{e}_i) < \mathbf{0} \quad \text{if} \quad i = 1, \ldots, \operatorname{neg}(b)$$

$$b(\mathbf{e}_i, \mathbf{e}_i) > \mathbf{0} \quad \text{if} \quad i = \operatorname{neg}(b) + 1, \ldots, \dim \mathbf{V}.$$

2.4. The bilinear mapping b allows us to identify some vector spaces derived from $\mathbf{V}$ and $\mathbf{B}$ with certain dual vector spaces. We shall use the symbol $\equiv$ for such identifications.

Proposition.

$$\text{(i)}\ \frac{\mathbf{V}}{\mathbf{B}\otimes\mathbf{B}} \equiv \mathbf{V}^*, \qquad \left(\frac{\mathbf{x}}{\mathbf{h}}\,\middle|\,\mathbf{y}\right) \equiv \frac{b(\mathbf{x},\mathbf{y})}{\mathbf{h}}$$

$$(\mathbf{x},\mathbf{y}\in\mathbf{V},\ \mathbf{h}\in\mathbf{B}\otimes\mathbf{B}\setminus\{\mathbf{0}\});$$

$$\text{(ii)}\ \frac{\mathbf{V}}{\mathbf{B}} \equiv \left(\frac{\mathbf{V}}{\mathbf{B}}\right)^*, \qquad \left(\frac{\mathbf{x}}{\mathbf{a}}\,\middle|\,\frac{\mathbf{y}}{\mathbf{c}}\right) \equiv \frac{b(\mathbf{x},\mathbf{y})}{\mathbf{a}\otimes\mathbf{c}} = \left(\frac{\mathbf{x}}{\mathbf{a}}\cdot\frac{\mathbf{y}}{\mathbf{c}}\right)$$

$$(\mathbf{x},\mathbf{y}\in\mathbf{V},\ \mathbf{a},\mathbf{c}\in\mathbf{B}\setminus\{\mathbf{0}\}).$$

Proof. (i) Evidently, given $\frac{\mathbf{x}}{\mathbf{h}}\in\frac{\mathbf{V}}{\mathbf{B}\otimes\mathbf{B}}$, the map $\mathbf{V}\longrightarrow\mathbb{R}$, $\mathbf{y}\mapsto\frac{b(\mathbf{x},\mathbf{y})}{\mathbf{h}}$ is linear and if $\frac{\mathbf{x}}{\mathbf{h}}=\frac{\mathbf{x}'}{\mathbf{h}'}$ then the corresponding maps are identical.

Suppose now that $\frac{\mathbf{x}}{\mathbf{h}}\neq\frac{\mathbf{x}'}{\mathbf{h}'}$ but $\frac{b(\mathbf{x},\mathbf{y})}{\mathbf{h}}=\frac{b(\mathbf{x}',\mathbf{y})}{\mathbf{h}'}$ for all $\mathbf{y}\in\mathbf{V}$. Then

$$\frac{b(\mathbf{x},\mathbf{y})}{\mathbf{h}} = \frac{b\left(\frac{\mathbf{h}}{\mathbf{h}'}\mathbf{x}',\mathbf{y}\right)}{\mathbf{h}} \qquad \text{for all} \quad \mathbf{y}\in\mathbf{V},$$

hence the non-degeneracy of b implies $\mathbf{x}=\frac{\mathbf{h}}{\mathbf{h}'}\mathbf{x}'$ which is a contradiction.

We see that $\frac{\mathbf{V}}{\mathbf{B}\otimes\mathbf{B}}$ is injected in $\mathbf{V}^*$; since both vector spaces have the same finite dimension, this injection is in fact a bijection.

The canonical identification of $\frac{\mathbf{V}}{\mathbf{B}\otimes\mathbf{B}}$ with $\mathbf{V}^*$ by the aid of b is linear, due to the bilinearity of b.

(ii) The demonstration is similar to the previous one.

Remark. If $\mathbf{V}$ is oriented, then $\mathbf{V}^*$ and $\frac{\mathbf{V}}{\mathbf{B}\otimes\mathbf{B}}$ are oriented as well (see IV. 3.1. and IV. 3.3.). Under the identification $\mathbf{V}^*\equiv\frac{\mathbf{V}}{\mathbf{B}\otimes\mathbf{B}}$, their orientations coincide if neg(b) is even and are opposite if neg(b) is odd.

2.5. Since $\mathbf{B}^*\otimes\mathbf{B}^*\equiv(\mathbf{B}\otimes\mathbf{B})^*\equiv\frac{\mathbb{R}}{\mathbf{B}\otimes\mathbf{B}}$ (see (IV. 1.5. (v) and IV. 2.8. (ii)) the previous result allows us to define a bilinear map

$$b^{(*)}: \mathbf{V}^* \times \mathbf{V}^* \longrightarrow \mathbf{B}^* \otimes \mathbf{B}^*, \qquad (\mathbf{p}, \mathbf{r}) \mapsto \frac{b(\mathbf{h}\otimes\mathbf{p}, \mathbf{k}\otimes\mathbf{r})}{\mathbf{h}\otimes\mathbf{k}}$$

with arbitrary non-zero $\mathbf{h}, \mathbf{k} \in \mathbf{B}\otimes\mathbf{B}$. Then $(\mathbf{V}^*, \mathbf{B}^*, b^{(*)})$ is a pseudo-Euclidean vector space, $\operatorname{neg} b^{(*)} = \operatorname{neg} b$.

2.6. Antisymmetric tensor product of different vector spaces cannot be defined in general. Proposition 2.4. permit us to introduce the useful notion of antisymmetric tensor product of $\mathbf{V}$ and $\mathbf{V}^*$. Proposition IV. 2.7. gives us

$$\mathbf{V}\otimes\mathbf{V}^* \equiv \mathbf{V}\otimes\frac{\mathbf{V}}{\mathbf{B}\otimes\mathbf{B}} \equiv \frac{\mathbf{V}}{\mathbf{B}}\otimes\frac{\mathbf{V}}{\mathbf{B}},$$

which suggests the following

Definition.

$$\mathbf{V} \wedge \mathbf{V}^* := \frac{\mathbf{V}}{\mathbf{B}} \wedge \frac{\mathbf{V}}{\mathbf{B}}, \qquad \mathbf{x} \wedge \mathbf{p} := \left(\frac{\mathbf{x}}{\mathbf{a}}\right) \wedge (\mathbf{a}\otimes\mathbf{p})$$

$$(\mathbf{x} \in \mathbf{V},\ \mathbf{p} \in \mathbf{V}^*,\ \mathbf{a} \in \mathbf{B}\setminus\{\mathbf{0}\}).$$

Remark. $\mathbf{V} \wedge \mathbf{V}^*$ is considered as a subspace of $\mathbf{V}\otimes\mathbf{V}^*$ (see IV. 1.16.). $\mathbf{V}\otimes\mathbf{V}^*$ is the vector space of linear maps $\mathbf{V} \longrightarrow \mathbf{V}$ (see IV. 1.5. (iv)). We easily obtain that $\mathbf{x} \wedge \mathbf{p} \in \mathbf{V} \wedge \mathbf{V}^*$ acts as a linear map in the following way:

$$(\mathbf{x} \wedge \mathbf{p})\mathbf{y} = (\mathbf{p}|\mathbf{y})\mathbf{x} - b(\mathbf{x}, \mathbf{y})\otimes\mathbf{p} \qquad (\mathbf{y} \in \mathbf{V}).$$

2.7. Proposition.

$$(\ \nmid\ ) : (\mathbf{V} \wedge \mathbf{V}^*) \times (\mathbf{V} \wedge \mathbf{V}^*) \longrightarrow \mathbb{R}, \qquad (A, B) \mapsto -\frac{1}{2}\operatorname{Tr}(AB)$$

(notation: IV. 1.6. (iii)) is a non-degenerate symmetric bilinear form, which is positive definite if and only if b is (positive or negative) definite.

Proof. $(\ \nmid\ )$ is clearly bilinear and IV. 1.6. (i), (iii) show that it is symmetric. It is a simple fact that

$$(\mathbf{x} \wedge \mathbf{p} \nmid \mathbf{y} \wedge \mathbf{r}) = -(\mathbf{p}|\mathbf{y})(\mathbf{r}|\mathbf{x}) + b^{(*)}(\mathbf{p}, \mathbf{r})\otimes b(\mathbf{x}, \mathbf{y})$$

$$(\mathbf{x}, \mathbf{y} \in \mathbf{V},\ \mathbf{p}, \mathbf{r} \in \mathbf{V}^*).$$

Let $\{\mathbf{e}_k : k = 1, \ldots, \dim \mathbf{V}\}$ be a b-orthogonal basis of $\mathbf{V}$, normed to $\mathbf{a} \in \mathbf{B}$. Then $\left\{\mathbf{e}_i \wedge \frac{\mathbf{e}_k}{\mathbf{a}\otimes\mathbf{a}} : i < k\right\}$ is a basis in $\mathbf{V} \wedge \mathbf{V}^*$ (see Proposition IV. 1.19.) and

$$\left(\mathbf{e}_i \wedge \frac{\mathbf{e}_k}{\mathbf{a}\otimes\mathbf{a}} \,\middle|\, \mathbf{e}_j \wedge \frac{\mathbf{e}_m}{\mathbf{a}\otimes\mathbf{a}}\right) = \frac{b(\mathbf{e}_k, \mathbf{e}_m)}{\mathbf{a}\otimes\mathbf{a}}\,\frac{b(\mathbf{e}_i, \mathbf{e}_j)}{\mathbf{a}\otimes\mathbf{a}}$$

(since $i < k$ and $j < m$, the term $-\frac{b(\mathbf{e}_k, \mathbf{e}_j)}{\mathbf{a}\otimes\mathbf{a}}\,\frac{b(\mathbf{e}_i, \mathbf{e}_m)}{\mathbf{a}\otimes\mathbf{a}}$ is necessarily zero) and this proves our assertion.

2.8. Definition. Let L be an element of $\mathrm{L}(\mathbf{V})$. The *b-adjoint* of L is the linear map $L^*:\mathbf{V}\to\mathbf{V}$ given by

$$b(\mathbf{x}, L^*\mathbf{y})=b(L\mathbf{x},\mathbf{y}) \qquad (\mathbf{x},\mathbf{y}\in\mathbf{V}).$$

2.9. Proposition. If we make the identification $\mathbf{V}^*=\frac{\mathbf{V}}{\mathbf{B}\otimes\mathbf{B}}$, then

$$L^*=\mathrm{id}_{\mathbf{B}\otimes\mathbf{B}}\otimes L^* \quad \text{or, equivalently,} \quad L^*=\frac{L^*}{\mathrm{id}_{\mathbf{B}\otimes\mathbf{B}}}.$$

As a consequence,

$$(L^*)^*=L_{\mathbf{B}\otimes\mathbf{B}} \quad\text{and}\quad (L^*)^*_{\mathbf{B}}=L_{\mathbf{B}}$$

(notations: IV. 1.8. and Remark IV. 2.10. (ii)).

Proof. Let $\mathbf{x}\in\mathbf{V}$, $\mathbf{p}\in\mathbf{V}^*$. Then one easily finds that

$$(\mathbf{x}\otimes\mathbf{p})^*=\mathbf{p}\otimes b(\mathbf{x},\cdot).$$

Thus we have for all $\mathbf{r}\in\mathbf{V}^*$, $\mathbf{h}\in\mathbf{B}\otimes\mathbf{B}\setminus\{0\}$

$$\frac{(\mathbf{x}\otimes\mathbf{p})^*}{\mathrm{id}_{\mathbf{B}\otimes\mathbf{B}}}(\mathbf{r})=\frac{\mathbf{p}\otimes b(\mathbf{x},\cdot)}{\mathrm{id}_{\mathbf{B}\otimes\mathbf{B}}}\left(\frac{\mathbf{h}\otimes\mathbf{r}}{\mathbf{h}}\right)\equiv\mathbf{p}\otimes\frac{b(\mathbf{x},\mathbf{h}\otimes\mathbf{r})}{\mathbf{h}}=$$
$$=\mathbf{p}(\mathbf{x}|\mathbf{r})=(\mathbf{p}\otimes\mathbf{x})(\mathbf{r})=(\mathbf{x}\otimes\mathbf{p})^*(\mathbf{r})$$

which proves the assertion for linear transformations of the form $\mathbf{x}\otimes\mathbf{p}$. Since such linear transformations span $\mathrm{L}(\mathbf{V})$ and the correspondence $L\mapsto L^*$ is evidently linear, the proof is complete.

Remark. This result has the following immediate consequences:

$$(LK)^*=K^*L^*, \qquad (L^*)^*=L \qquad (L,K\in\mathrm{L}(\mathbf{V}))$$

which can be verified directly as well.

2.10. Definition.

$$\mathrm{O}(b):=\{L\in\mathrm{GL}(\mathbf{V})\,\vdots\, L^*=L^{-1}\},$$
$$\mathbf{A}(b):=\{L\in\mathrm{L}(\mathbf{V})\,\vdots\, L^*=-L\}.$$

The elements of $\mathrm{O}(b)$ and $\mathbf{A}(b)$ are called *b-orthogonal* and *b-antisymmetric* maps, respectively.

Proposition.

(i) $L\in\mathrm{O}(b)$ if and only if

$$b(L\mathbf{x},L\mathbf{y})=b(\mathbf{x},\mathbf{y}) \qquad (\mathbf{x},\mathbf{y}\in\mathbf{V})$$

(ii) $A\in\mathbf{A}(b)$ if and only if

$$b(\mathbf{x},A\mathbf{y})=-b(A\mathbf{x},\mathbf{y}) \qquad (\mathbf{x},\mathbf{y}\in\mathbf{V})$$

and if and only if $b(\mathbf{x},A\mathbf{x})=0$ $(\mathbf{x}\in\mathbf{V})$.

Remarks. (i) We say that b-orthogonal maps *preserve* the bilinear map b.

(ii) If $(\mathbf{V}', \mathbf{B}', b')$ is another pseudo-Euclidean vector space, a linear map $L:\mathbf{V} \longrightarrow \mathbf{V}'$ is called *b-b'-orthogonal*, if $b'(L\mathbf{x}, L\mathbf{y}) = (\mathbf{h}\otimes\mathbf{h})\otimes b(\mathbf{x}, \mathbf{y})$ for some non-zero $\mathbf{h} \in \frac{\mathbf{B}'}{\mathbf{B}}$ and for all $\mathbf{x}, \mathbf{y} \in \mathbf{V}$.

2.11. Proposition.

$$\mathbf{V} \wedge \mathbf{V}^* \equiv \mathsf{A}(b)\,.$$

Proof. It is a simple task to show that $\mathsf{A}(b)$ is a linear subspace of $\mathrm{L}(\mathbf{V})$. We easily deduce from Remark 2.6. that the elements of $\mathbf{V} \wedge \mathbf{V}^*$ are b-antisymmetric maps, i.e. $\mathbf{V} \wedge \mathbf{V}^*$ is a linear subspace of $\mathsf{A}(b)$. Furthermore we know that

$$\dim(\mathbf{V} \wedge \mathbf{V}^*) = \binom{\dim \mathbf{V}}{2}.$$

A b-orthogonal basis establishes a linear bijection between $\mathsf{A}(b)$ and the vector space of antisymmetric $(\dim \mathbf{V}) \times (\dim \mathbf{V})$ matrices whose dimension is $\binom{\dim \mathbf{V}}{2}$. Hence, the dimension of $\mathsf{A}(b)$ equals the dimension of $\mathbf{V} \wedge \mathbf{V}^*$ which implies that the two vector spaces coincide.

Remark. In view of Proposition 2.7., $(\mathsf{A}(b), \mathbb{R}, (\;|\;))$ is a pseudo-Euclidean vector space. We can write

$$(A|B) = \frac{1}{2}\operatorname{Tr}(A^* B) \qquad (A, B \in \mathsf{A}(b))\,.$$

and $(\;|\;)$ makes the identification $\mathsf{A}(b)^* \equiv \mathsf{A}(b)$.

2.12. Proposition.

(i) $|\det L| = 1 \qquad (L \in \mathrm{O}(b))$;

(ii) $\operatorname{Tr} A = 0 \qquad (A \in \mathsf{A}(b))$.

Proof. (i) Put $n := \dim \mathbf{V}$ and let $\{\mathbf{e}_k \,\vdots\, k = 1, \ldots, n\}$ be a b-orthogonal basis of $\mathbf{V}$, normed to $\mathbf{a} \in \mathbf{B}$. Then the definitions of antisymmetric tensor products and determinants yield for $L \in \mathrm{O}(b)$

$$\left(\bigwedge_{k=1}^{n} \frac{\mathbf{e}_k}{\mathbf{a}\otimes\mathbf{a}}\right)(\mathbf{e}_1, \ldots, \mathbf{e}_n) = \left(\bigwedge_{k=1}^{n} \frac{L\mathbf{e}_k}{\mathbf{a}\otimes\mathbf{a}}\right)(L\mathbf{e}_1, \ldots, L\mathbf{e}_n) =$$

$$= (\det L)\left(\bigwedge_{k=1}^{n} \frac{L\mathbf{e}_k}{\mathbf{a}\otimes\mathbf{a}}\right)(\mathbf{e}_1, \ldots, \mathbf{e}_n) =$$

$$= (\det L)\left(\bigwedge_{k=1}^{n} \mathbf{e}_k\right)\left(\frac{L\mathbf{e}_k}{\mathbf{a}\otimes\mathbf{a}}, \ldots, \frac{L\mathbf{e}_n}{\mathbf{a}\otimes\mathbf{a}}\right) =$$

$$=(\det L)\left(\overset{n}{\underset{k=1}{\wedge}} \frac{\mathbf{e}_k}{\mathbf{a}\otimes\mathbf{a}}\right)(L\mathbf{e}_1, \ldots, L\mathbf{e}_n)=$$

$$=(\det L)^2\left(\overset{n}{\underset{k=1}{\wedge}} \frac{\mathbf{e}_k}{\mathbf{a}\otimes\mathbf{a}}\right)(\mathbf{e}_1, \ldots, \mathbf{e}_n).$$

(ii) Since Tr is a linear map, it is enough to show that $\mathrm{Tr}(\mathbf{x}\wedge\mathbf{p})=0$ for $\mathbf{x}\in\mathbf{V}, \mathbf{p}\in\mathbf{V}^*$, and this is evident from Remark 2.6.

2.13. Proposition. Let L be an element of $\mathrm{O}(b)$. Then $L[\mathbf{A}(b)]L^{-1}=\mathbf{A}(b)$ and

$$(LAL^{-1}|LBL^{-1})=(A|B) \qquad (A, B\in\mathbf{A}(b)).$$

Proof. If $A\in\mathbf{A}(b)$ then $(LAL^{-1})^*=(L^{-1})^*A^*L^*=-LAL^{-1}$, hence LAL^{-1} is in $\mathbf{A}(b)$, too.

The equality concerning the bilinear map $(\ |\ )$ can be verified by taking the basis elements of $\mathbf{V}\wedge\mathbf{V}^*=\mathbf{A}(b)$ used in the proof of Proposition 2.7. and by noticing that

$$L\left(\mathbf{x}\wedge\frac{\mathbf{y}}{\mathbf{h}}\right)L^{-1}=L\mathbf{x}\wedge\frac{L\mathbf{y}}{\mathbf{h}} \qquad (\mathbf{x},\mathbf{y}\in\mathbf{V},\ \mathbf{h}\in\mathbf{B}\otimes\mathbf{B}\setminus\{0\}).$$

2.14. Proposition. Put $n:=\dim\mathbf{V}$ and let $\{\mathbf{e}_k : k=1, \ldots, n\}$ and $\{\mathbf{e}'_k : k=1, \ldots, n\}$ be equally oriented b-orthogonal bases of $\mathbf{V}$, normed to $\mathbf{a}\in\mathbf{B}$ and $\mathbf{a}'\in\mathbf{B}$, respectively. Suppose $\mathbf{a}$ and $\mathbf{a}'$ are equally oriented, i.e. $\frac{\mathbf{a}}{\mathbf{a}'}$ is a positive real number. Then

$$\frac{\overset{n}{\underset{k=1}{\wedge}} \mathbf{e}_k}{\overset{n}{\otimes}\mathbf{a}}=\frac{\overset{n}{\underset{k=1}{\wedge}} \mathbf{e}'_k}{\overset{n}{\otimes}\mathbf{a}'}.$$

Proof. The linear map $\mathbf{V}\longrightarrow\mathbf{V}$ defined by

$$L\mathbf{e}_k:=\frac{\mathbf{a}}{\mathbf{a}'}\mathbf{e}'_k \qquad (k=1, \ldots, n)$$

is clearly b-orthogonal and $\det L=1$ (see Proposition 2.12. and Definition IV. 3.1.).

Applying Proposition IV. 1.21. and the usual rules of manipulating tensor products and quotients (see for instance IV. 2.7. (ii)) we get the desired result.

2.15. Definition. Suppose $\mathbf{V}$ and $\mathbf{B}$ oriented. Let $\{\mathbf{e}_k : k=1, \ldots, n\}$ be a positively oriented b-orthogonal basis of $\mathbf{V}$, normed to $\mathbf{a}\in\mathbf{B}^+$. Then the antisymmetric maps

$$\mu^*: \overset{n}{\times}\mathbf{V}^*\longrightarrow\overset{n}{\otimes}\mathbf{B}^*, \quad (\mathbf{p}_1, \ldots, \mathbf{p}_n)\mapsto\left[\left(\overset{n}{\underset{k=1}{\wedge}}\mathbf{e}_k\right)(\mathbf{p}_1, \ldots, \mathbf{p}_n)\right]\overset{n}{\otimes}\frac{1}{\mathbf{a}}$$

$$\mu\colon \overset{n}{\times}\,\mathbf{V} \longrightarrow \overset{n}{\otimes}\,\mathbf{B}\,, \quad (\mathbf{x}_1, \ldots, \mathbf{x}_n) \mapsto \left[\left(\bigwedge_{k=1}^{n} \frac{\mathbf{e}_k}{\mathbf{a}\otimes\mathbf{a}}\right)(\mathbf{x}_1, \ldots, \mathbf{x}_n)\right] \overset{n}{\otimes}\,\mathbf{a}$$

are called the *canonical volume forms* on $\mathbf{V}^*$ and $\mathbf{V}$, respectively.

Remarks. (i) Proposition 2.14. says that the definition of μ^* and μ is independent of bases.

(ii) μ^* and μ determine a $\overset{n}{\otimes}\,\mathbf{B}^*$ valued and a $\overset{n}{\otimes}\,\mathbf{B}$ valued measure on the parallelepipedons, thus also on the Borel subsets of $\mathbf{V}^*$ and $\mathbf{V}$, respectively; they are called the *canonical measure* on $\mathbf{V}^*$ and $\mathbf{V}$.

(iii) The canonical measures on $\mathbf{V}$ and $\mathbf{V}^*$ are invariant under translations, under b-orthogonal and $b^{(*)}$-orthogonal transformations, respectively.

(iv) One can show that

$$|\mu(\mathbf{x}_1, \ldots, \mathbf{x}_n)| = \sqrt[\otimes]{|\det\{b(\mathbf{x}_i, \mathbf{x}_j)\}_{i,j=1,\ldots,n}|}$$

where the determinant of a matrix with entries from $\mathbf{B}\otimes\mathbf{B}$ is defined as the determinant of a real matrix, with tensor products instead of products.

2.16. Proposition. If we put $n := \dim \mathbf{V}$, we have (in the sense of IV. 1.4.)

$$\text{(i)}\ \overset{n}{\wedge}\,\mathbf{V}^* \equiv \overset{n}{\otimes}\,\mathbf{B}^*, \qquad \bigwedge_{k=1}^{n} \mathbf{p}_k \equiv \mu^*(\mathbf{p}_1, \ldots, \mathbf{p}_n)$$

$$\text{(ii)}\ \overset{n}{\wedge}\,\mathbf{V} \equiv \overset{n}{\otimes}\,\mathbf{B}, \qquad \bigwedge_{k=1}^{n} \mathbf{x}_k \equiv \mu(\mathbf{x}_1, \ldots, \mathbf{x}_n).$$

Proof. The definition of antisymmetric tensor products (see IV. 1.13.) implies the existence of two linear maps sending $\bigwedge_{k=1}^{n} \mathbf{p}_k$ into $\mu^*(\mathbf{p}_1, \ldots, \mathbf{p}_n)$ and $\bigwedge_{k=1}^{n} \mathbf{x}_k$ into $\mu(\mathbf{x}_1, \ldots, \mathbf{x}_n)$, respectively. These linear maps are bijections because all vector spaces in question have one dimension and $\mu^* \neq 0$, $\mu \neq 0$.

3. Euclidean vector spaces

3.1. Definition. A pseudo-Euclidean vector space is called *Euclidean* if the corresponding bilinear map b is positive definite, or, equivalently, $\operatorname{neg}(b) = 0$ (see 2.3.). ▮

We assume in the sequel that a Euclidean vector space $(\mathbf{E}, \mathbf{D}, \gamma)$ *is given, and* $\mathbf{E}$ *and* $\mathbf{D}$ *are oriented*, $\dim \mathbf{E} = 3$.

Then Proposition 2.4. says that $\mathbf{E}^* \equiv \dfrac{\mathbf{E}}{\mathbf{D}\otimes\mathbf{D}}$, and according to 2.5., also $(\mathbf{E}^*, \mathbf{D}^*, \gamma^{(*)})$ is a Euclidean vector space.

3.2. Proposition. (Cauchy-Schwartz inequality)

$$|\gamma(\mathbf{x},\mathbf{y})| \leq \sqrt[\otimes]{\gamma(\mathbf{x},\mathbf{x})} \otimes \sqrt[\otimes]{\gamma(\mathbf{y},\mathbf{y})} \qquad (\mathbf{x},\mathbf{y} \in \mathbf{E})$$

and equality holds if and only if **x** and **y** are parallel (notations: IV. 3.6. and IV. 3.9.).

Remarks. (i) Consequently, $\mathbf{x} \mapsto \sqrt[\otimes]{\gamma(\mathbf{x},\mathbf{x})}$ can be considered as a **D** valued Euclidean norm on **E**.

(ii) Similar assertions are true regarding $\gamma^{(*)}$.

3.3. Proposition. The bilinear form $(\,|\,)$ on $\frac{\mathbf{E}}{\mathbf{D}}$ (see Remark 2.1. (ii)) is an inner product, hence $\mathbf{s} \mapsto |\mathbf{s}| := \sqrt{(\mathbf{s}|\mathbf{s})}$ is a norm on $\frac{\mathbf{E}}{\mathbf{D}}$.

3.4. Proposition. The bilinear form $\mathbf{(\,|\,)}$ on $\mathbf{A}(\gamma) \equiv \mathbf{E} \wedge \mathbf{E}^* \equiv \frac{\mathbf{E}}{\mathbf{D}} \wedge \frac{\mathbf{E}}{\mathbf{D}}$ (see 2.7. and 2.11.) is an inner product, hence $A \mapsto |A| := \sqrt{\mathbf{(}A\mathbf{|}A\mathbf{)}}$ is a norm on $\mathbf{A}(\gamma)$. ∎

Recall that $\mathbf{A}(\gamma)$ is three dimensional (see 2.11.).

3.5. Proposition. Let A be a non-zero element of $\mathbf{A}(\gamma)$. Then the kernel of A is one dimensional.

Proof. Since A is not zero, there are elements $\mathbf{x}_1$ and $\mathbf{x}_2$ of **E** such that $\gamma(\mathbf{x}_1, A\mathbf{x}_2) \neq \mathbf{0}$. Evidently, $\mathbf{x}_1$ and $\mathbf{x}_2$ are not parallel (see 2.13. (ii)), and moreover, $A\mathbf{x}_1$ and $A\mathbf{x}_2$ cannot be parallel, in particular, neither of them is zero. Thus Ran A is at least two dimensional (i.e. Ker A is at most one dimensional).

Take a γ-orthogonal basis $\{\mathbf{e}_1, \mathbf{e}_2, \mathbf{e}_3\}$ in **E**, normed to $\mathbf{d} \in \mathbf{D}^+$. Then

$$A \equiv \alpha_1 \frac{\mathbf{e}_2}{\mathbf{d}} \wedge \frac{\mathbf{e}_3}{\mathbf{d}} + \alpha_2 \frac{\mathbf{e}_3}{\mathbf{d}} \wedge \frac{\mathbf{e}_1}{\mathbf{d}} + \alpha_3 \frac{\mathbf{e}_1}{\mathbf{d}} \wedge \frac{\mathbf{e}_2}{\mathbf{d}}$$

for some real numbers $\alpha_1, \alpha_2, \alpha_3$ (see 2.11.). We easily find that A maps the non-zero vector $\sum_{i=1}^{3} \alpha_i \mathbf{e}_i$ into zero.

3.6. Let A be a non-zero element of $\mathbf{A}(\gamma)$. If **x** is an element γ-orthogonal to Ker A then $A\mathbf{x}$ is an element γ-orthogonal to both Ker A and **x**.

Moreover we find that $A^2\mathbf{x}$ is a vector opposite to **x**. Using this fact we easily show that if **y** is another vector γ-orthogonal to Ker A then $\{\mathbf{x}, A\mathbf{x}\}$ and $\{\mathbf{y}, A\mathbf{y}\}$ are equally oriented bases in the plane γ-orthogonal to Ker A.

Definition. Let $j(A)$ be the element of $\frac{\mathbf{E}}{\mathbf{D}}$ that

(i) $j(A) \in \frac{\mathrm{Ker}\,A}{\mathbf{D}}$,

(ii) $(j(A)|j(A)) = \mathbf{(}A\mathbf{|}A\mathbf{)}$,

(iii) $\mathbf{d} \otimes j(A), \mathbf{x}, A\mathbf{x}$ is a positively oriented basis for arbitrary $\mathbf{d} \in \mathbf{D}^+$ and non-zero **x** γ-orthogonal to Ker A.

3.7. Proposition. The map $j:\mathbf{A}(\gamma) \longrightarrow \frac{\mathbf{E}}{\mathbf{D}}$ is an inner product preserving and orientation reversing linear bijection.

Proof. We have to show first that j is linear. To this end it suffices to demonstrate that $j(A_1)+j(A_2)$ is in $\frac{\operatorname{Ker}(A_1+A_2)}{\mathbf{D}}$.

Observe that $\left\{j(A_2), j(A_1), \frac{A_2}{\mathrm{id}_{\mathbf{D}}} j(A_1)\right\}$

and $\left\{j(A_1), j(A_2), \frac{A_1}{\mathrm{id}_{\mathbf{D}}} j(A_2)\right\}$

(notation: IV. 2.10.) are positively oriented bases in $\frac{\mathbf{E}}{\mathbf{D}}$, and both $j(A_1)$ and $j(A_2)$ are orthogonal to

$$\frac{A_2}{\mathrm{id}_{\mathbf{D}}} j(A_1) \qquad \text{and} \qquad \frac{A_1}{\mathrm{id}_{\mathbf{D}}} j(A_2).$$

As a consequence, these vectors have opposite directions. It is left to the reader to verify that they are of equal norm.

As a result we have what we want to prove:

$$\frac{A_1+A_2}{\mathrm{id}_{\mathbf{D}}}\big(j(A_1)+j(A_2)\big) = \mathbf{0}.$$

Consider $\mathbf{A}(\gamma)$ as $\frac{\mathbf{E}}{\mathbf{D}} \wedge \frac{\mathbf{E}}{\mathbf{D}}$. If $\{\mathbf{a}_1, \mathbf{a}_2, \mathbf{a}_3\}$ is positively oriented orthonormal basis in $\frac{\mathbf{E}}{\mathbf{D}}$ then

$$j(\mathbf{a}_2 \wedge \mathbf{a}_3) = -\mathbf{a}_1,\; j(\mathbf{a}_3 \wedge \mathbf{a}_1) = -\mathbf{a}_2,\; j(\mathbf{a}_1 \wedge \mathbf{a}_2) = -\mathbf{a}_3,$$

which proves that j reverses orientation (see Remark 2.4.).

The other assertions of the proposition are evident.

3.8. Let us introduce the notation

$$[A, B] := AB - BA \qquad (A, B \in \mathbf{A}(\gamma))$$

(cf. VIII. 4.1. and VIII. 4.12.).

Proposition. Let A and B elements of $\mathbf{A}(\gamma)$. Then

(i) $[A, B]$ is an element of $\mathbf{A}(\gamma)$, orthogonal to both A and B and

$$|[A, B]|^2 = |A|^2|B|^2 - (A|B)^2;$$

consequently, $[A, B]$ is zero if and only if A and B are parallel;

(ii) $([A, B]|C) = ([C, A]|B) = ([B, C]|A)$ for all $C \in \mathbf{A}(\gamma)$.

Proof. The proof of both assertions is based on the fact that the inner product on $\mathbf{A}(\gamma)$ is given by the aid of Tr (see 2.7.) and the order of composition of linear maps can be changed under Tr (see IV. 1.6. (iii)). The equality of (i) can be demonstrated most easily by identifying $\mathbf{A}(\gamma)$ with $\frac{\mathbf{E}}{\mathbf{D}} \wedge \frac{\mathbf{E}}{\mathbf{D}}$.

3.9. Proposition. We have

$$[C, [A, B]] = (C \mid B)A - (C \mid A)B$$

for all $A, B, C \in \mathbf{A}(\gamma)$.

Proof. Since the equality is linear in all its entries, it suffices to take arbitrarily an A, a B orthogonal to A, and A, B, $[A, B]$ in the role of C. For $C = [A, B]$ the equality is trivial.

Furthermore, $[A, [A, B]]$ is orthogonal to both A and $[A, B]$, hence $[A, [A, B]] = \beta B$ for some $\beta \in \mathbb{R}$. Take the inner product of both sides by B, apply 3.8. (ii) to have $-|[A, B]|^2 = \beta |B|^2$ which implies $\beta = -|A|^2$.

Similar considerations for $[B, [A, B]]$ yield the desired result.

3.10. Definition. The maps

$$\frac{\mathbf{E}}{\mathbf{D}} \times \frac{\mathbf{E}}{\mathbf{D}} \longrightarrow \frac{\mathbf{E}}{\mathbf{D}}, \qquad (\mathbf{a}, \mathbf{b}) \mapsto \mathbf{a} \times \mathbf{b} := -j(\mathbf{a} \wedge \mathbf{b}),$$

$$\mathbf{E} \times \mathbf{E}^* \longrightarrow \frac{\mathbf{E}}{\mathbf{D}}, \qquad (\mathbf{q}, \mathbf{p}) \mapsto \mathbf{q} \times \mathbf{p} := -j(\mathbf{q} \wedge \mathbf{p})$$

are called *vectorial products.*

Proposition.

(i) $j([A, B]) = j(A) \times j(B) \qquad (A, B \in \mathbf{A}(\gamma))$;

(ii) $A_{\mathbf{D}}\mathbf{a} = j(A) \times \mathbf{a} \qquad \left(A \in \mathbf{A}(\gamma),\ \mathbf{a} \in \frac{\mathbf{E}}{\mathbf{D}}\right)$

and consequently,

$$A_{\mathbf{D}} j(B) = j([A, B]) \qquad (A, B \in \mathbf{A}(\gamma)).$$

Proof. Use a positively oriented orthonormal basis in $\frac{\mathbf{E}}{\mathbf{D}}$ and the relations given in the proof of 3.7.

Remark. Propositions 3.7., 3.8. and 3.9. result in

$$|\mathbf{a} \times \mathbf{b}|^2 = |\mathbf{a}|^2 |\mathbf{b}|^2 - (\mathbf{a} \mid \mathbf{b})^2,$$

$$(\mathbf{a} \times \mathbf{b} \mid \mathbf{c}) = (\mathbf{c} \times \mathbf{a} \mid \mathbf{b}) = (\mathbf{b} \times \mathbf{c} \mid \mathbf{a}),$$

$$\mathbf{c} \times (\mathbf{a} \times \mathbf{b}) = (\mathbf{c} \mid \mathbf{b})\mathbf{a} - (\mathbf{c} \mid \mathbf{a})\mathbf{b}$$

for all $\mathbf{a}, \mathbf{b}, \mathbf{c} \in \frac{\mathbf{E}}{\mathbf{D}}$.

4. Minkowskian vector spaces

4.1. Definition. A pseudo-Euclidean vector space is called *Minkowskian* if its dimension is greater than one and neg $(b)=1$ for the corresponding bilinear map b (see 2.3.).

We assume in the sequel that a Minkowskian vector space $(\mathbf{M}, \mathbf{I}, g)$ *is given, and* $\mathbf{M}$ *and* $\mathbf{I}$ *are oriented.*

Then Proposition 2.4. says that $\mathbf{M}^* \equiv \dfrac{\mathbf{M}}{\mathbf{I}\otimes\mathbf{I}}$.

4.2. Let us introduce the notation

$$T := \{\mathbf{x} \in \mathbf{M} : g(\mathbf{x}, \mathbf{x}) < \mathbf{0}\}.$$

T is an open subset of $\mathbf{M}$ as a preimage of an open subset by a continuous function.

Proposition. Let $\mathbf{x}$ be an element of T. Suppose $\mathbf{y}$ is a non-zero element of $\mathbf{M}$ and $g(\mathbf{x}, \mathbf{y}) = \mathbf{0}$. Then $g(\mathbf{y}, \mathbf{y}) > \mathbf{0}$.

Proof. Apply Proposition 2.3.

4.3. Proposition. Let $\mathbf{u}, \mathbf{v}, \mathbf{w} \in \dfrac{\mathbf{M}}{\mathbf{I}}$, $(\mathbf{u}|\mathbf{u}) = (\mathbf{v}|\mathbf{v}) = (\mathbf{w}|\mathbf{w}) = -1$ (notation: Remark 2.1. (ii)). Then

$$(\mathbf{u}|\mathbf{v})(\mathbf{v}|\mathbf{w})(\mathbf{w}|\mathbf{u}) \leq -\frac{(\mathbf{u}|\mathbf{v})^2 + (\mathbf{v}|\mathbf{w})^2 + (\mathbf{w}|\mathbf{u})^2}{3} < 0,$$

and equality holds if and only if $\mathbf{u}, \mathbf{v}, \mathbf{w}$ are parallel.

As a consequence,

$$g(\mathbf{x}, \mathbf{y}) \otimes g(\mathbf{y}, \mathbf{z}) \otimes g(\mathbf{z}, \mathbf{x}) \leq$$

$$\leq \frac{g(\mathbf{x}, \mathbf{y}) \otimes g(\mathbf{x}, \mathbf{y}) \otimes g(\mathbf{z}, \mathbf{z}) + g(\mathbf{y}, \mathbf{z}) \otimes g(\mathbf{y}, \mathbf{z}) \otimes g(\mathbf{x}, \mathbf{x}) + g(\mathbf{z}, \mathbf{x}) \otimes g(\mathbf{z}, \mathbf{x}) \otimes g(\mathbf{y}, \mathbf{y})}{3} < \mathbf{0}$$

for all $\mathbf{x}, \mathbf{y}, \mathbf{z} \in T$, and equality holds if and only if $\mathbf{x}, \mathbf{y}, \mathbf{z}$ are parallel.

Proof. $\mathbf{z}_\mathbf{w} := (\mathbf{v}|\mathbf{w})\mathbf{u} - (\mathbf{w}|\mathbf{u})\mathbf{v}$ is $(\ |\ )$-orthogonal to $\mathbf{w}$, hence we deduce from the previous proposition that $(\mathbf{z}_\mathbf{w}|\mathbf{z}_\mathbf{w}) \geq 0$ and equality holds if and only if $\mathbf{z}_\mathbf{w} = 0$, which involves

$$(\mathbf{u}|\mathbf{v})(\mathbf{v}|\mathbf{w})(\mathbf{w}|\mathbf{u}) \leq -\frac{(\mathbf{v}|\mathbf{w})^2 + (\mathbf{w}|\mathbf{u})^2}{2} < 0$$

and equality holds if and only if $\mathbf{u}$ and $\mathbf{v}$ are parallel.

Construct $\mathbf{z}_\mathbf{u}$ and $\mathbf{z}_\mathbf{v}$ in a similar way, sum up the corresponding inequalities to get the desired result.

4.4. Proposition. If $\mathbf{u}, \mathbf{v} \in \frac{\mathbf{M}}{\mathbf{I}}$ and $(\mathbf{u} \dagger \mathbf{u}) = (\mathbf{v} \dagger \mathbf{v}) = -1$ then

$$|(\mathbf{u} \dagger \mathbf{v})| \geq 1$$

and equality holds if and only if $\mathbf{u} = \mathbf{v}$ or $\mathbf{u} = -\mathbf{v}$.

As a consequence, we have the "reversed" Cauchy-Schwartz inequality

$$|g(\mathbf{x}, \mathbf{y})| \geq {}_{\otimes}\!\sqrt{-g(\mathbf{x}, \mathbf{x})} \otimes {}_{\otimes}\!\sqrt{-g(\mathbf{y}, \mathbf{y})} \qquad (\mathbf{x}, \mathbf{y} \in T)$$

where equality holds if and only if $\mathbf{x} = \alpha\mathbf{y}$ for some $\alpha \in \mathbb{R}$ (notations: IV. 3.6. and IV. 3.9.).

Proof. Take $\mathbf{w} := \mathbf{u}$ in the previous inequality.

4.5. Definition. We say that the elements $\mathbf{x}$ and $\mathbf{y}$ of T have the *same arrow* if $g(\mathbf{x}, \mathbf{y}) < \mathbf{0}$.

Proposition. Having the same arrow is an equivalence relation on T and there are two equivalence classes (called *arrow classes*).

Proof. The relation having the same arrow is evidently reflexive and symmetric. Suppose now that $\mathbf{x}$ and $\mathbf{y}$ have the same arrow and $\mathbf{y}$ and $\mathbf{z}$ have the same arrow. Then Proposition 4.3. implies that $\mathbf{x}$ and $\mathbf{z}$ have the same arrow as well, hence the relation is transitive.

Let $\mathbf{x}$ be an element of T. It is obvious that $\mathbf{x}$ and $-\mathbf{x}$ have not the same arrow. On the other hand, Proposition 4.4. shows that either $\mathbf{y}$ and $\mathbf{x}$ or $\mathbf{y}$ and $-\mathbf{x}$ have the same arrow for all $\mathbf{y} \in T$, which means that there are exactly two arrow classes.

4.6. Proposition. The arrow classes of T are open and convex.

Proof. Let $\mathbf{x} \in T$. The arrow class of $\mathbf{x}$ is the intersection of T with $\{\mathbf{y} \in \mathbf{M} : g(\mathbf{x}, \mathbf{y}) < \mathbf{0}\}$ which is an open subset of $\mathbf{M}$.

Suppose now that $\mathbf{x}$ and $\mathbf{y}$ have the same arrow and α, β are non-negative real numbers, $\alpha + \beta = 1$. Then

$$g(\alpha\mathbf{x} + \beta\mathbf{y}, \alpha\mathbf{x} + \beta\mathbf{y}) = \alpha^2 g(\mathbf{x}, \mathbf{x}) + 2\alpha\beta g(\mathbf{x}, \mathbf{y}) + \beta^2 g(\mathbf{y}, \mathbf{y}) < \mathbf{0},$$

hence $\alpha\mathbf{x} + \beta\mathbf{y}$ is in T, and

$$g(\alpha\mathbf{x} + \beta\mathbf{y}, \mathbf{x}) = \alpha g(\mathbf{x}, \mathbf{x}) + \beta g(\mathbf{y}, \mathbf{x}) < \mathbf{0},$$

thus $\alpha\mathbf{x} + \beta\mathbf{y}$ and $\mathbf{x}$ have the same arrow.

4.7. Proposition. ("reversed" triangle inequality). Let $\mathbf{x}$, $\mathbf{y}$ be elements of T having the same arrow. Then

$${}_{\otimes}\!\sqrt{-g(\mathbf{x} + \mathbf{y}, \mathbf{x} + \mathbf{y})} \geq {}_{\otimes}\!\sqrt{-g(\mathbf{x}, \mathbf{x})} + {}_{\otimes}\!\sqrt{-g(\mathbf{y}, \mathbf{y})}\,.$$

Proof. Proposition 4.4. results in

$$-g(\mathbf{x}+\mathbf{y},\mathbf{x}+\mathbf{y})=-g(\mathbf{x},\mathbf{x})-g(\mathbf{y},\mathbf{y})-2g(\mathbf{x},\mathbf{y})\geq$$

$$\geq -g(\mathbf{x},\mathbf{x})-g(\mathbf{y},\mathbf{y})+2\,{}_{\otimes}\!\!\sqrt{-g(\mathbf{x},\mathbf{x})}\otimes{}_{\otimes}\!\!\sqrt{-g(\mathbf{y},\mathbf{y})}=$$

$$=\left({}_{\otimes}\!\!\sqrt{-g(\mathbf{x},\mathbf{x})}+{}_{\otimes}\!\!\sqrt{-g(\mathbf{y},\mathbf{y})}\right)\otimes\left({}_{\otimes}\!\!\sqrt{-g(\mathbf{x},\mathbf{x})}+{}_{\otimes}\!\!\sqrt{-g(\mathbf{y},\mathbf{y})}\right)$$

4.8. Definition. We say that $(\mathbf{M}, \mathbf{I}, g)$ is *arrow oriented* or an *arrow orientation is associated* with g if we pick out one of the arrow classes of T.

A linear isomorphism between arrow oriented Minkowskian vector spaces is called *arrow preserving* or *arrow reversing* if it maps the chosen arrow classes into each other or into the opposite ones, respectively.

5. Pseudo-Euclidean affine spaces

5.1. Definition. A *pseudo-Euclidean affine space* is a triplet $(V, \mathbf{B}, b)$ where
(i) V is an affine space over the finite dimensional real vector space $\mathbf{V}$;
(ii) $(\mathbf{V}, \mathbf{B}, b)$ is a pseudo-Euclidean vector space.

We assume in the sequel that a pseudo-Euclidean affine space $(V, \mathbf{B}, b)$ is given.

5.2. Let $\mathbf{K}: V \rightarrowtail \mathbf{V}^*$ be a differentiable mapping. Then the linear map $\mathrm{D}\mathbf{K}[x]: \mathbf{V} \longrightarrow \mathbf{V}^*$ ($x \in \mathrm{Dom}\,\mathbf{K}$) can be viewed as an element of $\mathbf{V}^*\otimes\mathbf{V}^*$ (see IV. 1.5. (iv)). Furthermore, we have that $\mathbf{V}^*\otimes\mathbf{V}^* \equiv \dfrac{\mathbf{V}}{\mathbf{B}\otimes\mathbf{B}}\otimes\mathbf{V}^* \equiv \dfrac{\mathbf{V}\otimes\mathbf{V}^*}{\mathbf{B}\otimes\mathbf{B}}$ (see 2.4. and IV. 2.7. (ii)).

Definition. The map

$$\mathrm{d}\mathbf{K}: V \rightarrowtail \mathbf{V}^* \wedge \mathbf{V}^*, \qquad x \mapsto 2A(\mathrm{D}\mathbf{K}[x]) \qquad (x \in \mathrm{Dom}\,\mathbf{K})$$

(see IV. 1.16.) is called the *exterior derivative* of $\mathbf{K}$.

The map

$$\delta\mathbf{K}: V \rightarrowtail \frac{\mathbb{R}}{\mathbf{B}\otimes\mathbf{B}}, \qquad x \mapsto \mathrm{Tr}_{\mathbf{B}\otimes\mathbf{B}}(\mathrm{D}\mathbf{K}[x]) \qquad (x \in \mathrm{Dom}\,\mathbf{K})$$

(see IV. 1.6. (iii) and IV. 2.10. (ii)) is called the *divergence* of $\mathbf{K}$.

Remark. Notice that the definition of the exterior derivative is independent of b, it can be introduced in any affine space. On the other hand, the definition of divergence depends on how $\mathbf{V}^*$ is identified with $\dfrac{\mathbf{V}}{\mathbf{B}\otimes\mathbf{B}}$, i.e. depends on b.

5.3. Let $\mathbf{Z}$ be a finite dimensional real vector space and suppose $F: V \rightarrowtail \mathbf{Z}$ is a twice differentiable map. If $\mathbf{r} \in \mathbf{Z}^*$ then $\mathbf{r} \circ F: V \rightarrowtail \mathbb{R}$ is a twice differentiable map, and $\mathrm{D}(\mathbf{r} \circ F): V \rightarrowtail \mathbf{V}^*$ is a differentiable map. Hence we can take $\delta\mathrm{D}(\mathbf{r} \circ F): V \rightarrowtail \dfrac{\mathbb{R}}{\mathbf{B} \otimes \mathbf{B}}$.

Fix an $x \in \operatorname{Dom} F$. Then

$$\mathbf{Z}^* \longrightarrow \frac{\mathbb{R}}{\mathbf{B} \otimes \mathbf{B}}, \qquad \mathbf{r} \mapsto \delta\mathrm{D}(\mathbf{r} \circ F)[x]$$

is a linear map, i.e. it is an element of $\dfrac{\mathbb{R}}{\mathbf{B} \otimes \mathbf{B}} \otimes \mathbf{Z} \equiv \dfrac{\mathbf{Z}}{\mathbf{B} \otimes \mathbf{B}}$, which we denote by $\Delta F[x]$.

Definition. The map

$$V \rightarrowtail \frac{\mathbf{Z}}{\mathbf{B} \otimes \mathbf{B}}, \qquad x \mapsto \Delta F[x] \qquad (x \in \operatorname{Dom} F)$$

is called the *Laplacian* of $F: V \rightarrowtail \mathbf{Z}$.

5.4. Since the canonical measure μ on $\mathbf{V}$ is translation invariant, we can define a canonical measure on V:

$$E \mapsto \mu(O_x[E]) \qquad (E \in B(V))$$

where $B(V)$ denotes the algebra of Borel sets of V and x is an arbitrarily chosen element of V (see VI. 1.1.). The translation invariance of μ involves that the definition is independent of $x \in V$.

VI. AFFINE SPACES

1. Fundamentals

1.1. Definition. An *affine space* is a triplet $(L, \mathbf{L}, -)$ where

(i) L is a non-void set,

(ii) $\mathbf{L}$ is a finite dimensional real or complex vector space,

(iii) $-$ is a map from $L \times L$ into $\mathbf{L}$, denoted by

$$(x, y) \mapsto x - y,$$

having the properties

a) for every $y \in L$ the map $O_y : L \longrightarrow \mathbf{L}$, $\quad x \mapsto x - y$ is bijective,

b) $(x-y)+(y-z)+(z-x)=\mathbf{0} \quad (x, y, z \in L)$.

1.2. Given $y \in L$, the inverse of the map O_y is denoted by $\mathbf{L} \longrightarrow L$, $\mathbf{x} \mapsto y+\mathbf{x}$. Hence, by definition,

$$y+(x-y)=x \qquad (x, y \in L).$$

Furthermore we have

$$x-y=\mathbf{0} \quad \text{if and only if} \quad x=y \qquad (x, y \in L),$$
$$(x+\mathbf{x})+\mathbf{y}=x+(\mathbf{x}+\mathbf{y}) \qquad (x \in L,\ \mathbf{x}, \mathbf{y} \in \mathbf{L}).$$

1.3. The properties listed above allow us to give another equivalent, less transparent but useful definition of affine spaces. A triplet $(L, \mathbf{L}, +)$ is an affine space if

(i) L is a non-void set,

(ii) $\mathbf{L}$ is a finite dimensional real or complex vector space,

(iii) $+$ is a map from $L \times \mathbf{L}$ into L, denoted by

$$(x, \mathbf{y}) \mapsto x+\mathbf{y}$$

having the properties

a) $(x+\mathbf{x})+\mathbf{y}=x+(\mathbf{x}+\mathbf{y}) \quad (x \in L, \mathbf{x}, \mathbf{y} \in \mathbf{L})$,

b) for every $x \in L$ the map $\mathbf{L} \longrightarrow L$, $\mathbf{y} \mapsto x+\mathbf{y}$ is bijective.

1.4. As it is usual in mathematics, we shall denote an affine space by a single letter; we say that L is an affine space over the vector space $\mathbf{L}$ and we shall call the map — *subtraction.*

The *dimension* of an affine space L is, by definition, that of the underlying vector space $\mathbf{L}$. If $\mathbf{L}$ is oriented, then L is called *oriented.*

1.5. Proposition. Let n be a natural number and take $x_i \in L, \alpha_i \in \mathbb{R}_0^+$ $(i=1, \dots, n)$ such that $\sum_{i=1}^{n} \alpha_i = 1$. Then there is a unique element x of L such that $\sum_{i=1}^{n} \alpha_i(x_i - x) = \mathbf{0}$.

Definition. x is called the *convex combination* of x_i-s with coefficients α_i, and is denoted by $\sum_{i=1}^{n} \alpha_i x_i$.

1.6. Proposition. There is a (necessarily unique) Hausdorff topology on an affine space L such that O_y is a homeomorphism for all $y \in L$.

Proof. $\mathbf{L}$ is of finite dimension, hence it has a unique Hausdorff topology compatible with the linear structure ([8], Theorem 3.2.). Picking out an arbitrary element z of L, let us equip L with the topology induced by O_z, i.e. the collection of open subsets of L should be $\{O_z^{-1}[U] \,\vdots\, U \text{ is open in } \mathbf{L}\}$. Since O_z is bijective, it becomes a homeomorphism. Evidently, the map $O_y \circ O_z^{-1}$: $\mathbf{L} \to \mathbf{L}, \mathbf{x} \mapsto \mathbf{x} + (y - z)$ is a homeomorphism for all $y \in L$, hence $O_y = (O_y \circ O_z^{-1}) \circ O_z$ is a homeomorphism as well.

1.7. Definition. A non-void subset S of an affine space L is called an *affine subspace* if there is a linear subspace $\mathbf{S}$ of $\mathbf{L}$ such that for some (hence for all) $y \in S$ $\{x - y \,\vdots\, x \in S\} = \mathbf{S}$, or equivalently, $y + \mathbf{S} = S$.

S is called *directed* by $\mathbf{S}$ and the *dimension* of S is that of $\mathbf{S}$.

One dimensional and two dimensional affine subspaces are called *straight lines* and *planes*, respectively. *Hyperplanes* are affine subspaces having dimension of L but one.

Two affine subspaces of an affine space are called *parallel* if they are directed by the same linear subspace.

1.8. Proposition. An affine subspace S directed by $\mathbf{S}$ is an affine space over $\mathbf{S}$ with the subtraction inherited from L.

Points of L are zero dimensional affine subspaces.

If $\mathbf{S}$ is a linear subspace of $\mathbf{L}$ and $x \in L$ then $x + \mathbf{S} := \{x + \mathbf{s} \,\vdots\, \mathbf{s} \in \mathbf{S}\}$ is the unique affine subspace directed by $\mathbf{S}$ and containing x.

1.9. Proposition. Let n be a natural number and suppose L_i is an affine space over $\mathbf{L}_i$ $(i=1, \dots, n)$. Then $\times_{i=1}^{n} L_i$ is an affine space over $\times_{i=1}^{n} \mathbf{L}_i$ with the subtraction

$$(x_1, \dots, x_n) - (y_1, \dots, y_n) := (x_1 - y_1, \dots, x_n - y_n).$$

1.10. Definition. Let L and K be affine spaces over $\mathbf{L}$ and $\mathbf{K}$, respectively. A map $F: L \longrightarrow K$ is called *affine* if there is a (necessarily unique) linear map $\mathrm{D}F: \mathbf{L} \longrightarrow \mathbf{K}$ such that

$$F(y)-F(x)=\mathrm{D}F(y-x) \qquad (x, y \in L)$$

or, equivalently,

$$F(x+\mathbf{x})-F(x)=\mathrm{D}F(\mathbf{x}) \qquad (x \in L, \mathbf{x} \in \mathbf{L}).$$

If L and K are oriented and F is bijective, F is *orientation preserving* or *orientation reversing* if $\mathrm{D}F$ is such.

1.11. Proposition. Let $F: L \longrightarrow K$ be an affine map. Then

(i) F is injective or surjective if and only if $\mathrm{D}F$ is such,

(ii) $\mathrm{D}F=0$ if and only if F is a constant map;

(iii) the restriction of F to an affine subspace S of L is an affine map and

$$\mathrm{D}(F|_S)=(\mathrm{D}F)|_{\mathbf{S}};$$

(iv) $\operatorname{Ran} F$ is an affine subspace of K, directed by $\operatorname{Ran} \mathrm{D}F$;

(v) if Z is an affine subspace of K, directed by $\mathbf{Z}$, the $\overset{-1}{F}[Z]$ is an affine subspace of L, directed by $\overset{-1}{\mathrm{D}F}[\mathbf{Z}]$.

(vi) F preserves convex combinations.

1.12. Proposition. Let F and G be affine maps such that $F \circ G$ exists. Then $F \circ G$ is an affine map and

$$\mathrm{D}(F \circ G)=\mathrm{D}F \circ \mathrm{D}G.$$

1.13. Proposition. Let n be a natural number. If $F_i: L_i \longrightarrow K_i$ is an affine map for all $i=1, \ldots, n$ then $\underset{i=1}{\overset{n}{\times}} F_i$ is an affine map and

$$\mathrm{D}\left(\underset{i=1}{\overset{n}{\times}} F_i\right)=\underset{i=1}{\overset{n}{\times}} \mathrm{D}F_i.$$

If $F: L \longrightarrow K_i$ is an affine map for all $i=1, \ldots, n$ then $(F_i)_{i=1,\ldots,n}$ is an affine map and

$$\mathrm{D}(F_1, \ldots, F_n)=(\mathrm{D}F_1, \ldots, \mathrm{D}F_n).$$

1.14. Proposition. Let $F, G: L \longrightarrow K$ be affine maps. Then

$$F-G: L \longrightarrow \mathbf{K}, \qquad x \mapsto F(x)-G(x)$$

is an affine map (see 1.15.) and

$$\mathrm{D}(F-G)=\mathrm{D}F-\mathrm{D}G.$$

1.15. Proposition. If $\mathbf{L}$ is a finite dimensional real or complex vector space then $\mathbf{L}$ is an affine space over itself by the vector subtraction. Linear subspaces of $\mathbf{L}$ as well as the so called linear manifolds—sets of the form $\mathbf{x}+\mathbf{S}$ where $\mathbf{x}\in\mathbf{L}$ and $\mathbf{S}$ is a linear subspace of $\mathbf{L}$—are the affine subspaces of $\mathbf{L}$. If $\mathbf{K}$ is another vector space, a map $F:\mathbf{L}\longrightarrow\mathbf{K}$ is affine if and only if there is an $\mathbf{a}\in\mathbf{K}$ such that $F(\mathbf{x})=DF(\mathbf{x})+\mathbf{a}$ $(\mathbf{x}\in\mathbf{L})$.

1.16. Proposition. Let L be an affine space over $\mathbf{L}$.
For all $y\in L$ the map O_y is affine,

$$\mathrm{D}O_y=\mathrm{id}_{\mathbf{L}}.$$

For all $\mathbf{x}\in\mathbf{L}$ the map

$$T_{\mathbf{x}}:L\longrightarrow L,\qquad x\mapsto x+\mathbf{x},$$

(the so called *translation by* $\mathbf{x}$) is affine,

$$\mathrm{D}T_{\mathbf{x}}=\mathrm{id}_{\mathbf{L}}.$$

Moreover, if $F:L\longrightarrow L$ is an affine map such that $\mathrm{D}F=\mathrm{id}_{\mathbf{L}}$, then $F=T_{\mathbf{x}}$ for some $\mathbf{x}\in\mathbf{L}$.

2. Differentiation

2.1. Definition. Let L and K be affine spaces. A map $F:L\rightarrowtail K$ is said *differentiable* at an interior point x of $\mathrm{Dom}\,F$ if there is a (necessarily unique) linear map $\mathrm{D}F[x]:\mathbf{L}\longrightarrow\mathbf{K}$ and there is a neighbourhood $U(x)\subset\mathrm{Dom}\,F$ of x such that

$$F(y)-F(x)=\mathrm{D}F[x](y-x)+\mathrm{ordo}\,(y-x)\qquad (y\in U(x)),$$

or, equivalently, there is a neighbourhood V of zero of $\mathbf{L}$ such that $x+V\subset\subset\mathrm{Dom}\,F$ and

$$F(x+\mathbf{x})-F(x)=\mathrm{D}F[x]\mathbf{x}+\mathrm{ordo}\,(\mathbf{x})\qquad (\mathbf{x}\in V)$$

where $\lim\limits_{\mathbf{x}\to\mathbf{0}}\dfrac{\mathrm{ordo}\,(\mathbf{x})}{|\mathbf{x}|}=\mathbf{0}$ for some (hence for all) norm $|\ \ |$ on $\mathbf{L}$.

$\mathrm{D}F[x]$ is called the *derivative* of F at x.

F is *differentiable on a subset* H of its domain if it is differentiable at every point of H. F is *differentiable* if it is differentiable on its domain (which is necessarily open in this case). F is *continuously differentiable* if it is differentiable and the map $\mathrm{Dom}\,F\longrightarrow\mathrm{L}(\mathbf{L},\mathbf{K})$, $x\mapsto\mathrm{D}F[x]$ is continuous.

2.2. Definition. A differentiable map F between oriented affine spaces of equal dimension is *orientation preserving* or *orientation reversing* if $\mathrm{D}F[x]$ is such for all $x\in\mathrm{Dom}\,F$.

2.3. Let n be a natural number and let K and L_i be affine spaces, $i=1, \ldots, n$. Suppose $F: \mathop{\times}\limits_{i=1}^{n} L_i \rightarrowtail K$ is differentiable at $(x_1, \ldots, x_n)$. Pick out a natural number $k \le n$. Then the map

$$L_k \rightarrowtail K, \qquad y \mapsto F(x_1, \ldots, x_{k-1}, y, x_{k+1} \ldots, x_n)$$

is differentiable at x_k; its derivative is denoted by

$$\mathrm{D}_k F[x_1, \ldots, x_n] \quad \text{or} \quad \mathrm{D}^{(k)} F[x_1, \ldots, x_n].$$

2.4. Proposition. (i) An affine map is differentiable, its derivative is the same at all points of its domain and equals the linear map appearing in the definition of affine maps;

(ii) a differentiable map is continuous;

(iii) if F and G are differentiable maps then $F \circ G$ is differentiable and

$$\mathrm{D}(F \circ G)[x] = \mathrm{D}F[G(x)] \circ \mathrm{D}G[x] \qquad (x \in \mathrm{Dom}\, F \circ G)$$

(iv) if F and G are differentiable maps then $F - G$ is differentiable and

$$\mathrm{D}(F-G)[x] = \mathrm{D}F[x] - \mathrm{D}G[x] \qquad (x \in \mathrm{Dom}\, F \cap \mathrm{Dom}\, G).$$

2.5. Definition. Let L and K be affine spaces. If $F: L \rightarrowtail K$ is a differentiable map and its derivative map $L \rightarrowtail \mathrm{L}(\mathbf{L}, \mathbf{K}), x \mapsto \mathrm{D}F[x]$ is differentiable then F is called *twice differentiable.*

Differentiability of higher order is defined similarly. An infinitely many times differentiable map is called *smooth.*

The second derivative of F at x is denoted by $\mathrm{D}^2F[x]$; it is an element of $\mathrm{L}(\mathbf{L}, \mathrm{L}(\mathbf{L}, \mathbf{K}))$ which is canonically isomorphic to the vector space $\mathrm{L}^2(\mathbf{L}, \mathbf{K})$ of bilinear maps $\mathbf{L} \times \mathbf{L} \longrightarrow \mathbf{K}$. Similarly, $\mathrm{D}^n F[x]$, the n-th derivative of F at x can be considered as an element of $\mathrm{L}^n(\mathbf{L}, \mathbf{K})$, the n-fold linear maps $\mathop{\times}\limits^{n} \mathbf{L} \longrightarrow \mathbf{K}$; a well-known theorem asserts that $\mathrm{D}^n F[x]$ is symmetric if F is n times continuously differentiable.

2.6. If A is a one dimensional affine space over $\mathbf{A}$ and $F: A \rightarrowtail K$ is differentiable at $a \in \mathrm{Dom}\, F$ then $\mathrm{D}F[a]$ can be considered as an element of $\dfrac{\mathbf{K}}{\mathbf{A}}$ (see IV. 2.7. (vi)). Similarly, if F is n times differentiable at a then $\mathrm{D}^n F[a]$ is an element of $\dfrac{\mathbf{K}}{\overset{n}{\otimes} \mathbf{A}}$.

Note that in this case the differentiability of F at a is equivalent to the fact that there exists

$$\lim_{\mathbf{h}\to 0} \frac{F(a+\mathbf{h})-F(a)}{\mathbf{h}}$$

which equals $DF[a]$.

For such an F we shall use the notations $\dot{F}(a):=DF[a]$, $\ddot{F}(a):=D^2F[a]$ etc.

2.7. Definition. Let A and K be affine spaces, dim $A=1$. Suppose A is real and oriented. A map $F: A \rightarrowtail K$ is called *differentiable on the right* at an interior point a of Dom F if there exists

$$D^+F[a]:=\lim_{\substack{\mathbf{h}\to 0\\ \mathbf{h}>0}} \frac{F(a+\mathbf{h})-F(a)}{\mathbf{h}}$$

which is called the *right derivative* of F at a.

The differentiability on the left and the left derivative $D^-F[a]$ is defined similarly.

2.8. Definition. Let A and K be as before. A map $F: A \rightarrowtail K$ is called *piecewise differentiable* if it is

(i) continuous,

(ii) differentiable with the possible exception of finitely many points where F is differentiable both on the right and on the left.

F is *piecewise twice differentiable* if

(i) it is piecewise differentiable,

(ii) it is twice differentiable where it is differentiable

(iii) if a is a point where F is not differentiable then there exist

$$(D^+)^2F[a]:=\lim_{\substack{\mathbf{h}\to 0\\ \mathbf{h}>0}} \frac{DF[a+\mathbf{h}]-D^+F[a]}{\mathbf{h}},$$

$$(D^-)^2F[a]:=\lim_{\substack{\mathbf{h}\to 0\\ \mathbf{h}<0}} \frac{DF[a+\mathbf{h}]-D^-F[a]}{\mathbf{h}}.$$

3. Vector fields and differential equations

3.1. Definition. Let L be a real affine space over $\mathbf{L}$ and suppose $\mathbf{A}$ is a one dimensional real vector space. Maps

$$L \rightarrowtail \mathbf{A}\otimes\left(\overset{p}{\otimes}\mathbf{L}\right) \quad \text{and} \quad L \rightarrowtail \frac{\overset{p}{\otimes}\mathbf{L}}{\mathbf{A}}$$

are called *p-tensor fields of type* $\mathbf{A}$ and *p-tensor fields of cotype* $\mathbf{A}$, respectively.

If **L** is replaced with **L***, we get *p-cotensor fields of type (cotype)* **A**.

If $\mathbf{A}=\mathbb{R}$ we omit the terms "of type $\mathbb{R}$" and "of cotype $\mathbb{R}$".

Instead of 1-tensor field and 1-cotensor field we use the name *vector field* and *covector field*, respectively.

3.2. Definition. Let L be an affine space and suppose **X** is a differentiable vector field of cotype **A** in L. Suppose A is an affine space over **A**. Then a *solution* of the first order differential equation

$$\dot{x}=\mathbf{X}\circ x$$

is a differentiable map $r: A \rightarrowtail L$ such that

(i) Dom r is connected;

(ii) Ran $r \subset$ Dom **X**;

(iii) $\dot{r}(a)=\mathbf{X}(r(a)) \quad (a \in \mathrm{Dom}\, r)$.

The range of a solution is called an *integral curve* of **X**. An integral curve is *maximal* if it is not contained properly in an integral curve.

3.3. Definition. Let **X** be as before. For a given $(a_0, x_0) \in A \times \mathrm{Dom}\,\mathbf{X}$ a *solution of the initial value problem*

$$\dot{x}=\mathbf{X}\circ x,$$

$$x(a_0)=x_0$$

is a map r such that

(i) r is a solution of the corresponding differential equation;

(ii) a_0 is an accumulation point of Dom r;

(iii) the limit of r at a_0 exists and equals x_0.

The range of the initial value problem is called the integral curve of **X** *passing through* (a_0, x_0).

3.4. The existence and local uniqueness theorem on solutions of initial value problems and on maximal integral curves, well-known for differential equations in finite dimensional vector spaces, are valid for the present case, too. They will be formulated in a more general context (see VII. 6.5.).

3.5. Definition. Let **X** be a differentiable vector field of cotype **A** in the affine space L. A solution of the corresponding differential equation is called *global* if its domain is the whole space A.

3.6. It may happen that **X** is not everywhere defined still it has global solutions. Conversely, an everywhere defined **X** can have maximal integral curves which are not ranges of global solutions.

Proposition. Let **X** be an everywhere defined differentiable vector field of cotype **A** in the affine space L. Suppose there are elements z in L, $\mathbf{a}\neq 0$ in **A**, α in $\mathbb{R}$ and a norm $\|\ \|$ on $\dfrac{\mathbf{L}}{\mathbf{A}}$ such that

$$\|\mathbf{X}(x)\| \le \left\|\frac{x-z}{\mathbf{a}}\right\| + \alpha \qquad (x \in L)\,.$$

Then all maximal integral curves of $\mathbf{X}$ are ranges of global solutions ([9], Vol. II. Theorem 4).

Remark. The norm $\|\ \|$ and the element z of L are not important in this condition. If the above inequality holds then a similar inequality is valid for any other norms and elements of L with some other elements of $\mathbf{A}$ and $\mathbb{R}$.

3.7. Definition. Let $\mathbf{Z}: L \times \frac{\mathbf{L}}{\mathbf{A}} \rightarrowtail \frac{\mathbf{L}}{\mathbf{A}\otimes\mathbf{A}}$ be a differentiable mapping. A solution of the second order differential equation

$$\ddot{x} = \mathbf{Z} \circ (x, \dot{x})$$

is a twice differentiable map $r: A \rightarrowtail L$ such that

(i) $\mathrm{Dom}\, r$ is connected;

(ii) $(r(a), \dot{r}(a)) \in \mathrm{Dom}\, \mathbf{Z} \quad (a \in \mathrm{Dom}\, r)$;

(iii) $\ddot{r}(a) = \mathbf{Z}(r(a), \dot{r}(a)) \quad (a \in \mathrm{Dom}\, r)$.

3.8. Definition. Let $\mathbf{Z}$ be as before. For a given $(a_0, x_0, \mathbf{v}_0) \in A \times \mathrm{Dom}\, \mathbf{Z}$ a *solution of the initial value problem*

$$\ddot{x} = \mathbf{Z} \circ (x, \dot{x}),$$

$$x(a_0) = x_0, \qquad \dot{x}(a_0) = \mathbf{v}_0$$

is a map r such that

(i) r is a solution of the corresponding second order differential equation;

(ii) a_0 is an accumulation point of $\mathrm{Dom}\, r$;

(iii) the limits of r and $\dot{r}$ at a_0 exist and equal x_0 and $\mathbf{v}_0$, respectively.

3.9. A second order differential equation can be considered as a first order one on a larger space. Let $\mathbf{Z}: L \times \frac{\mathbf{L}}{\mathbf{A}} \rightarrowtail \frac{\mathbf{L}}{\mathbf{A}\otimes\mathbf{A}}$ be a differentiable map. Then

$$\mathbf{S}: L \times \frac{\mathbf{L}}{\mathbf{A}} \rightarrowtail \frac{\mathbf{L}}{\mathbf{A}} \times \frac{\mathbf{L}}{\mathbf{A}\otimes\mathbf{A}}, \qquad (x, \mathbf{v}) \mapsto (\mathbf{v}, \mathbf{Z}(x, \mathbf{v}))$$

is a differentiable vector field of cotype $\mathbf{A}$ in $L \times \frac{\mathbf{L}}{\mathbf{A}}$.

If r is a solution of the initial value problem defined in 3.8 then $(r, \dot{r})$ is a solution of the initial value problem

$$\dot{s} = \mathbf{S} \circ s, \qquad s(a_0) = (x_0, \mathbf{v}_0)\,.$$

Conversely, if R is a solution of the above initial value problem then $\mathrm{pr}_L \circ R$ is a solution of the initial value problem defined in 3.8. (pr_L is the canonical projection $L \times \frac{\mathbf{L}}{\mathbf{A}} \longrightarrow L$, $(x, \mathbf{v}) \mapsto x$).

VII. SMOOTH MANIFOLDS

1. Fundamentals

1.1. Definition. Let M be a topological space. A pair (U, u) is called a *chart* in M if

(i) U is an open subset of M,

(ii) $u: U \longrightarrow \operatorname{Ran} u \subset \mathbb{R}^m$ is a homeomorphism—called a *coordinatization*—for some non-negative integer m which is called the *dimension* of the chart.

The chart is called *global* or *local* according to whether $U = M$ or $U \neq M$.

1.2. Definition. Two charts (U_1, u_1) and (U_2, u_2) in M are *compatible* if $u_1 \circ u_2^{-1}$ and $u_2 \circ u_1^{-1}$ are smooth (see VI. 2.5.) (the void map is smooth by definition!).

1.3. Definition. An *atlas* on a topological space M is a family $\{(U_i, u_i) : i \in I\}$ of compatible charts such that $M = \bigcup_{i \in I} U_i$. It is *maximal* if any chart compatible with all members of the atlas is itself a member of the atlas. A maximal atlas is called a *smooth structure*, too.

1.4. Definition. A *smooth manifold* is a pair $(M, \mathscr{A})$ where

(i) M is a topological space,

(ii) $\mathscr{A}$ is a maximal atlas on M.

The smooth manifold is called *m dimensional* if all its charts are m dimensional.

1.5. Remarks. (i) For the sake of brevity, we shall use the term *manifold* instead of smooth manifold.

(ii) As usual in mathematics, a manifold will be denoted by a single letter, i.e. instead of $(M, \mathscr{A})$ we write only M, and a chart in M will always mean a chart of the given maximal atlas.

(iii) An atlas determines a unique maximal atlas containing the given one: it is the atlas containing all the charts compatible with the charts of the given atlas. Hence we can define a smooth structure by giving arbitrarily an atlas.

(iv) An m dimensional affine space (in particular a vector space) is considered as an m dimensional manifold determined by a global chart given by an optional affine isomorphism onto $\mathbb{R}^m$.

1.6. The following simple fact is well-known: if $F:\mathbb{R}^m \rightarrowtail \mathbb{R}^n$ is a differentiable injection whose inverse is also differentiable, then $m=n$. Hence we have the following nice result which allows us to consider only manifolds that have fixed dimension.

Proposition. Let M be a manifold. Then there are uniquely determined disjoint open subsets M_m $(m=0, 1, 2, \ldots)$ of M such that

(i) $M = \bigcup_{m=0}^{\infty} M_m$.

(ii) M_m is either void or an m dimensional manifold (as a submanifold of M, see 1.7.).

1.7. Definition. Let N be an open subset of the manifold M. If $\{(U_i, u_i) \vdots i \in I\}$ is a (maximal) atlas on M we put $J := \{i \in I \vdots N \cap U_i \neq \emptyset\}$. Then $\{(N \cap U_i, u_i|_N) \vdots i \in J\}$ is a (maximal) atlas turning N into a manifold which is called an *open submanifold* of M.

1.8. Definition. Let M and N be manifolds with maximal atlases $\{(U_i, u_i) \vdots i \in I\}$ and $\{(V_j, v_j) \vdots j \in J\}$, respectively. If $M \times N$ is equipped with the product topology, then $\{(U_i \times V_j, u_i \times v_j) \vdots (i, j) \in I \times J\}$ is a (not necessarily maximal) atlas turning $M \times N$ into a smooth manifold which is called the *Cartesian product* of M and N.

1.9. Definition. Let M and N be manifolds. A map $F: M \rightarrowtail N$ is called

(i) *differentiable* at the interior point x of Dom F,

(ii) *s times (continuously) differentiable* (s is either in $\mathbb{N}_0$ or is infinity)

if for all charts (U, u) in M and (V, v) in N the map $v \circ F \circ u^{-1}$ (the so called *local representative* of F with respect to (U, u) and (V, v)) is

(i) differentiable at $u(x)$,

(ii) s times (continuously) differentiable, respectively.

If s is infinity, we say that F is *smooth*.

Remarks. (i) It suffices that the listed differentiability properties hold for charts (U, u) and (V, v) of arbitrarily given (not necessarily maximal) atlases of M and N, respectively.

(ii) If F is differentiable then Dom F is necessarily open.

(iii) Zero times continuously differentiable maps are the continuous ones.

(iv) As a consequence of (i) and Remark 1.5. (iv), the present definition of differentiability for maps between affine spaces coincides with the definitions given in VI. 2.1. and VI. 2.5.

(v) The composition of differentiable maps is differentiable.

1.10. Definition. Let M and N be manifolds. A map $F: M \longrightarrow N$ is called a *diffeomorphism* if

(i) F is bijective,

(ii) both F and F^{-1} are smooth.

A map $F: M \rightarrowtail N$ is a *little diffeomorphism* if Dom F and Ran F are open and F is a diffeomorphism between Dom F and Ran F.

Diff(M) denotes the set of diffeomorphisms $M \longrightarrow M$.

Remark. If (U, u) is a chart in M then u is a little diffeomorphism.

1.11. Definition. An atlas $\{(U_i, u_i) : i \in I\}$ on a manifold is called *oriented* if all the functions $u_i \circ u_j^{-1}$ $(i, j \in I)$ are orientation preserving, i.e. they have positive Jacobian at every point of their domain (the void map is orientation preserving by definition). A manifold M is *orientable* if it has an oriented atlas. Two oriented atlases on M are *equally oriented* if their union is oriented. An equivalence class of equally oriented atlases is called an *orientation* of M. M is *oriented* if an orientation of M is given. The atlases in the chosen equivalence class are called *positively oriented.*

Proposition. A connected manifold has at most two orientations ([11], Theorem I. 4.1.).

1.12. Definition. A differentiable map F between oriented manifolds of equal dimensions is *orientation preserving* or *orientation reversing* if for all positively oriented charts (U, u) in Dom F and (V, v) in Ran F the map $v \circ F \circ u^{-1}$ is orientation preserving or orientation reversing, respectively.

1.13. Definition. Let N be a subset of a manifold M. Consider N equipped with the relative topology induced by M and suppose there is given a smooth structure on N. N is called a *submanifold* of M if the canonical injection $i: N \longrightarrow M$ is smooth and D$i[x]$ (see 2.9.) is injective for all $x \in N$.

Proposition. A subset N is an n dimensional submanifold of the m dimensional manifold M if and only if for all $x \in N$ there is a chart (U, u) in M such that $x \in U$ and $u^{-1}[\mathbb{R}^n \times \{0\}] = U \cap N$ (where we have taken $\mathbb{R}^m = \mathbb{R}^n \times \mathbb{R}^{m-n}$ and 0 is the zero element in $\mathbb{R}^{m-n}$) and $(U \cap N, u|_N)$ is a chart in N ([2], Proposition III. 5.9.).

2. Tangent space

In this section M denotes an m dimensional manifold.

2.1. If $x \in M$ and $\mathbf{L}$ is a finite dimensional real vector space, $\mathrm{D}(x, \mathbf{L})$ stands for the set of maps $M \rightarrowtail \mathbf{L}$ which are defined on a neighbourhood of x and are differentiable at x. We write $\mathrm{D}(x) := \mathrm{D}(x, \mathbb{R})$. $\mathrm{C}^1(x)$ is the subset of $\mathrm{D}(x)$ consisting of the functions continuously differentiable in a neighbourhood of x.

$\mathrm{C}(x)$ denotes the set of real valued functions defined on a neighbourhood of x and continuous at x.

The real numbers, considered as constant functions, are elements of $\mathrm{D}(x)$.

If $F, G \in \mathrm{D}(x, \mathbf{L})$ and $f, g \in \mathrm{D}(x)$ then $fF + gG$ is in $\mathrm{D}(x, \mathbf{L})$.

Moreover, if $F \in \mathrm{D}(x, \mathbf{L})$, $F(x) = 0$ and $f \in \mathrm{C}(x)$ then fF is in $\mathrm{D}(x, \mathbf{L})$.

2.2. Definition. Let **K** be a finite dimensional real vector space and suppose $\phi: M \rightarrowtail \mathbf{K}$ is a little diffeomorphism, $x \in \operatorname{Dom} \phi$. Then we put for any finite dimensional real vector space **L**

$$\frac{\mathrm{d}}{\mathrm{d}\phi}[x]: \mathrm{D}(x, \mathbf{L}) \longrightarrow \mathrm{L}(\mathbf{K}, \mathbf{L}), \qquad F \mapsto \mathrm{D}(F \circ \phi^{-1})[\phi(x)]$$

(see VI. 2.1.) and we shall find convenient to write

$$\frac{\mathrm{d}F}{\mathrm{d}\phi}[x] := \left(\frac{\mathrm{d}}{\mathrm{d}\phi}[x]\right)(F).$$

Proposition.

(i) $\dfrac{\mathrm{d}(\alpha F + \beta G)}{d\phi}[x] = \alpha \dfrac{\mathrm{d}F}{\mathrm{d}\phi}[x] + \beta \dfrac{\mathrm{d}G}{\mathrm{d}\phi}[x]$

for all $\alpha, \beta \in \mathbb{R}$, $F, G \in \mathrm{D}(x, \mathbf{L})$,

(ii) $\dfrac{\mathrm{d}(fF)}{\mathrm{d}\phi}[x] = F(x) \otimes \dfrac{\mathrm{d}f}{\mathrm{d}\phi}[x] + f(x)\dfrac{\mathrm{d}F}{\mathrm{d}\phi}[x]$

for all $f \in \mathrm{D}(x)$, $F \in \mathrm{D}(x, \mathbf{L})$ (we have used IV. 1.5. (iv) and that $\dfrac{\mathrm{d}f}{\mathrm{d}\phi}[x]$ is in $\mathbf{K}^*$ by definition),

(iii) $\dfrac{\mathrm{d}(fF)}{\mathrm{d}\phi}[x] = f(x)\dfrac{\mathrm{d}F}{\mathrm{d}\phi}[x]$ if $f \in \mathrm{C}(x)$, $F \in \mathrm{D}(x, \mathbf{L})$ and $F(x) = 0$;

(iv) $\dfrac{\mathrm{d}F}{\mathrm{d}\phi}[x] = 0$ if $F \in \mathrm{D}(x, \mathbf{L})$ is constant in a neighbourhood of x.

2.3. Proposition. If **K** and **K**′ are finite dimensional real vector spaces and $\phi: M \rightarrowtail \mathbf{K}$, $\phi': M \rightarrowtail \mathbf{K}'$ are little diffeomorphisms, $x \in \operatorname{Dom} \phi \cap \operatorname{Dom} \phi'$, then

$$\frac{\mathrm{d}F}{\mathrm{d}\phi}[x] = \frac{\mathrm{d}F}{\mathrm{d}\phi'}[x] \circ \frac{\mathrm{d}\phi'}{\mathrm{d}\phi}[x] \qquad (F \in \mathrm{D}(x, \mathbf{L})).$$

Remark. The above equality is written symbolically in the form

$$\frac{\mathrm{d}}{\mathrm{d}\phi}[x] = \frac{\mathrm{d}}{\mathrm{d}\phi'}[x] \circ \frac{\mathrm{d}\phi'}{\mathrm{d}\phi}[x].$$

2.4. Definition. Let $\phi: M \rightarrowtail \mathbb{R}^m$ be a little diffeomorphism. Then $\phi = (\phi^1, \phi^2, \dots, \phi^m)$ and we put for $x \in \operatorname{Dom} \phi$

$$\frac{\partial}{\partial \phi^k}[x]: \mathrm{D}(x, \mathbf{L}) \longrightarrow \mathbf{L}, \qquad F \mapsto \left(\frac{\mathrm{d}F}{\mathrm{d}\phi}[x]\right)\mathbf{e}_k$$

where $\{\mathbf{e}_k : k=1, \ldots, m\}$ is the standard basis of $\mathbb{R}^m$. We shall find convenient to write

$$\frac{\partial F}{\partial \phi^k}[x] := \left(\frac{\partial}{\partial \phi^k}[x]\right)(F).$$

Proposition.

(i) $$\frac{\partial(\alpha F+\beta G)}{\partial \phi^k}[x] = \alpha \frac{\partial F}{\partial \phi^k}[x] + \beta \frac{\partial F}{\partial \phi^k}[x]$$

for all $\alpha, \beta \in \mathbb{R}$, $F, G \in \mathrm{D}(x, \mathbf{L})$.

(ii) $$\frac{\partial(fF)}{\partial \phi^k}[x] = \frac{\partial f}{\partial \phi^k}[x] F(x) + f(x) \frac{\partial F}{\partial \phi^k}[x]$$

for all $f \in \mathrm{D}(x)$ and $F \in \mathrm{D}(x, \mathbf{L})$.

(iii) $$\frac{\partial(fF)}{\partial \phi^k}[x] = f(x) \frac{\partial F}{\partial \phi^k}[x] \qquad \text{if } f \in \mathrm{C}(x),\ F \in \mathrm{D}(x, \mathbf{L}) \text{ and } F(x)=0.$$

(iv) $$\frac{\partial F}{\partial \phi^k}[x]=0 \qquad \text{if } F \in \mathrm{D}(x, \mathbf{L}) \text{ is constant in a neighbourhood of } x.$$

Furthermore, if $\phi' : M \rightarrowtail \mathbb{R}^m$ is another little diffeomorphism, $x \in \operatorname{Dom} \phi \cap \operatorname{Dom} \phi'$ then

$$\frac{\partial}{\partial \phi^k}[x] = \sum_{i=1}^{m} \frac{\partial}{\partial \phi'^i}[x] \frac{\partial \phi'^i}{\partial \phi^k}[x] \qquad (k=1, \ldots, m).$$

2.5. Remarks. (i) Observe that in the above case $\left(\frac{\partial \phi'^i}{\partial \phi^k}[x]\right)_{i,k=1,\ldots,m}$ is the Jacobian matrix at $\phi(x)$ of $\phi' \circ \phi^{-1} : \mathbb{R}^m \rightarrowtail \mathbb{R}^m$.

(ii) Note the interesting particular case when $M=\mathbb{R}^m$. Then

$$\frac{\mathrm{d}F}{\mathrm{d}(\mathrm{id}_{\mathbb{R}^m})}[x] = \mathrm{D}F[x], \qquad \frac{\partial F}{\partial \mathrm{pr}^k}[x] = \partial_k F[x]$$

$$(x \in \mathbb{R}^m, \qquad k=1, \ldots, m)$$

where pr^k is the k-th canonical projection, i.e. $\mathrm{id}_{\mathbb{R}^m} = (\mathrm{pr}^1, \ldots, \mathrm{pr}^m)$.

2.6. Definition. Let $\mathbf{K}$ be a finite dimensional real vector space. A map $\mathbf{v} : \mathrm{D}(x) \longrightarrow \mathbf{K}^*$ is called a $\mathbf{K}$-*derivation at* $x \in M$ if

(i) $\mathbf{v}(\alpha f + \beta g) = \alpha \mathbf{v}(f) + \beta \mathbf{v}(g) \qquad (\alpha, \beta \in \mathbb{R},\ f, g \in \mathrm{D}(x))$,

(ii) $\mathbf{v}(fg) = g(x)\mathbf{v}(f) + f(x)\mathbf{v}(g) \qquad (f, g \in \mathrm{D}(x))$,

(iii) $\mathbf{v}(fg) = f(x)\mathbf{v}(g) \qquad (f \in \mathrm{C}(x),\ g \in \mathrm{D}(x),\ g(x)=0)$,

(iv) $\mathbf{v}(f)=0$ if $f \in \mathrm{D}(x)$ is constant in a neighbourhood of x.

$\mathbb{R}$-derivations at x are called *derivations* at x.

Remarks. (i) Let (U, u) be a chart, $x \in U$. Then $\dfrac{\mathrm{d}}{\mathrm{d}u}[x] : \mathrm{D}(x) \longrightarrow \mathbb{R}^m$ is an $\mathbb{R}^m$-derivation at x and $\dfrac{\partial}{\partial u^k}[x] : \mathrm{D}(x) \longrightarrow \mathbb{R}$ $(k=1, \ldots, m)$ are derivations at x.

(ii) If f and g are differentiable functions equal in a neighbourhood of x then (iv) of Definition 2.6. implies that $\mathbf{v}(f)=\mathbf{v}(g)$ for all $\mathbf{K}$-derivations $\mathbf{v}$ at x.

2.7. Definition. A derivation at $x \in M$ restricted to the subset $\mathrm{C}^1(x)$ of $\mathrm{D}(x)$ is called a *tangent vector at x.*

The set of tangent vectors at x is denoted by $T_x(M)$ and is called the *tangent space* of M at x.

Proposition. Under the pointwise defined operations, $T_x(M)$ is an m dimensional real vector space.

If (U, u) is a chart, $x \in U$, then

$$T_x(M) \longrightarrow \mathbb{R}^m, \quad \mathbf{v} \longmapsto \mathbf{v}_m(u) := (\mathbf{v}(u^1), \ldots, \mathbf{v}(u^m))$$

is a linear bijection, in other words, if $u^k := \mathrm{pr}^k \circ u$ then

$$\left\{\frac{\partial}{\partial u^k}[x] : k=1, \ldots, m\right\}$$

is a basis in $T_x(M)$.

Proof. We can suppose without loss of generality that $u(x)=0$. Then $\dfrac{\partial}{\partial u^k}[x]$ $(k=1, \ldots, m)$ are linearly independent tangent vectors for $\dfrac{\partial u^i}{\partial u^k}[x]=\delta^i_k$ (Kronecker symbol).

Take an $f: U \rightarrowtail \mathbb{R}$ which is continuously differentiable in a neighbourhood of x, and put $\hat{f} := f \circ u^{-1}$. Then we obtain from Lagrange's mean value theorem that there exists a function $\hat{\vartheta} : \mathbb{R}^m \rightarrowtail \mathbb{R}$ whose limit at zero is zero, and

$$\hat{f}(\xi)=\hat{f}(0)+\sum_{k=1}^{m} \partial_k \hat{f}(\hat{\vartheta}(\xi))\xi^k \qquad (\xi \in u[\mathrm{Dom}\, f])$$

which results in

$$f=f(x)+\sum_{k=1}^{m} \frac{\partial f}{\partial u^k}[\vartheta(\cdot)]u^k$$

where $\vartheta := \hat{\vartheta} \circ u$.

Now, if $\mathbf{v}$ is a derivation at x we get by Definition 2.6. that

$$\mathbf{v}(f)=\sum_{k=1}^{m} \frac{\partial f}{\partial u^k}[x]\mathbf{v}(u^k)$$

which says that

$$\mathbf{v} = \sum_{k=1}^{m} \mathbf{v}(u^k) \frac{\partial}{\partial u^k}[x].$$

2.8. Remarks. (i) If $x \in M$ and $r : \mathbb{R} \rightarrowtail M$ is a differentiable map, $r(0) = x$ (r is an "arc passing through x") then $f \mapsto \mathrm{D}(f \circ r)[0]$ is a derivation at x.

In the case when M is a surface in $\mathbb{R}^n$, one defines $\dot{r}(0)$ (see VI. 2.6.) as the tangent vector of r at x. Then $\mathrm{D}(f \circ r)[0] = \mathrm{D}f[x]\dot{r}(0)$ and we see that if s is another arc passing through x then the derivations at x corresponding to r and s are the same if and only if $\dot{r}(0) = \dot{s}(0)$.

That is why we defined in general a tangent vector as a derivation.

(ii) If (U, u) and (U', u') are charts, $x \in U \cap U'$ then 2.6. shows that

$$\frac{\partial}{\partial u^k}[x] = \sum_{i=1}^{m} \frac{\partial u'^i}{\partial u^k}[x] \frac{\partial}{\partial u'^i}[x]$$

As a consequence, an orientation of M determines an orientation of $T_x(M)$ for all $x \in M$: the one defined by the basis $\left\{\frac{\partial}{\partial u^k}[x] : k = 1, \ldots, m\right\}$ where (U, u) is any positively oriented chart such that $x \in U$.

(iii) If M is a real affine space over $\mathbf{M}$ then for all $x \in M$ we have a linear bijection

$$\mathbf{M} \longrightarrow T_x(M), \qquad \mathbf{x} \mapsto \mathbf{v_x}$$

defined by

$$\mathbf{v_x}(f) := \mathrm{D}f[x]\mathbf{x} \qquad (f \in \mathrm{C}^1(x)).$$

We use this bijection to *identify* $T_x(M)$ with $\mathbf{M}$ for all $x \in M$.

(iv) If M and N are manifolds, $x \in M$ and $y \in N$, then the map

$$T_{(x,y)}(M \times N) \longrightarrow T_x(M) \times T_y(N), \qquad \mathbf{v} \mapsto (\mathbf{v}_M, \mathbf{v}_N)$$

defined by

$$\mathbf{v}_M(f) := \mathbf{v}(f \otimes 1) \qquad (f \in \mathrm{C}^1(x)),$$

$$\mathbf{v}_N(g) := \mathbf{v}(1 \otimes g) \qquad (g \in \mathrm{C}^1(y))$$

is a linear bijection (we can show this e.g. by using a basis induced by a chart $(U \times V, u \times v)$) that allows us to identify $T_{(x,y)}(M \times N)$ with $T_x(M) \times T_y(N)$.

2.9. Proposition. Let N be another manifold and suppose $F : M \rightarrowtail N$ is a continuously differentiable map. Then for $x \in \mathrm{Dom}\, F$ the map

$$\mathrm{D}F[x] : T_x(M) \longrightarrow T_{F(x)}(N)$$

defined by

$$(\mathrm{D}F[x]\mathbf{v})(g) := \mathbf{v}(g \circ F) \qquad (\mathbf{v} \in T_x(M), g \in \mathrm{C}^1(F(x)))$$

is linear.

Definition. $DF[x]$ is called the *derivative* of F at x.

Remarks. (i) If (U, u) and (V, v) are charts in M and N such that $x \in U$, $F(x) \in V$, then the matrix of $DF[x]$ in the bases induced by u and v (see Proposition 2.7.) is

$$\left(DF[x]\frac{\partial}{\partial u^k}[x]\right)v^i \qquad \begin{pmatrix} k=1, \ldots, \dim M \\ i=1, \ldots, \dim N \end{pmatrix}$$

which equals the Jacobian matrix (the derivative) of $v \circ F \circ u^{-1}$ at $u(x)$.

(ii) If F and G are composable continuously differentiable maps then

$$D(G \circ F)[x] = DG[F(x)] \circ DF[x].$$

(iii) If M and N are affine spaces then $DF[x]$ defined here and in VI. 2.1. are the same. Indeed, by Remark 2.8. (iii) we have

$$(DF[x]\mathbf{v})(g) = Dg[F(x)](DF[x]\mathbf{v}) = D(g \circ F)[x]\mathbf{v} = \mathbf{v}(g \circ F).$$

(iv) If (U, u) is a chart in M then Remark (2.8.) (iii) yields that $Du[x]\mathbf{v}$ can be applied to the functions $\mathrm{pr}^k : \mathbb{R}^m \longrightarrow \mathbb{R}$ resulting in $\mathbf{v}(\mathrm{pr}^k \circ u)$ $(k = 1, \ldots, m)$, hence

$$Du[x] : T_x(M) \longrightarrow T_{u(x)}(\mathbb{R}^m) = \mathbb{R}^m, \qquad \mathbf{v} \mapsto \mathbf{v}_m(u).$$

(notation: 2.7.).

2.10. Definition. The set

$$T(M) := \bigcup_{x \in M} \{x\} \times T_x(M)$$

is called the *tangent bundle* of M. The map

$$\Pi_M : T(M) \longrightarrow M, \qquad (x, \mathbf{v}) \mapsto x$$

is the *canonical projection* of the tangent bundle.

2.11. Definition. Let $\{(U_i, u_i) : i \in I\}$ be the maximal atlas of M. Then $T(M)$ is equipped with the topology generated by the family of sets

$$\left\{ \bigcup_{x \in U_i} \{x\} \times Du_i[x]^{-1}[G] : G \text{ is an open subset of } \mathbb{R}^m, i \in I \right\}.$$

$T(M)$ is considered as a manifold equipped with the following smooth structure.

Proposition. If (U, u) is a chart of M then $(\overset{-1}{\Pi}_M[U], \phi_u)$ is a chart of $T(M)$ where

$$\phi_u : \overset{-1}{\Pi}_M[U] \longrightarrow \mathbb{R}^m \times \mathbb{R}^m, \quad (x, \mathbf{v}) \mapsto (u(x), \mathbf{v}_m(u)) = (u(x), Du[x]\mathbf{v}).$$

If $\{(U_i, u_i) : i \in I\}$ is the maximal atlas of M then $\{(\overset{-1}{\Pi}_M[U_i], \phi_{u_i}) : i \in I\}$ is an atlas of $T(M)$.

If $T(M)$ is considered as a smooth manifold determined by this atlas, then the canonical projection Π_M is smooth.

2.12. Remark. If M is an affine space over $\mathbf{M}$ then, by Remark 2.8. (iii), $T(M)=M\times\mathbf{M}$. The topology and the smooth structure of $T(M)$ is the product topology and the product smooth structure (see Definition 1.8.) corresponding to $M\times\mathbf{M}$.

2.13. Proposition. Let N be another manifold and suppose $F:M\rightarrowtail N$ is continuously differentiable. Then the map

$$\mathrm{D}F:T(M)\rightarrowtail T(N),\qquad (x,\mathbf{v})\mapsto(F(x),\mathrm{D}F[x]\mathbf{v})$$

$$(x\in\operatorname{Dom}F)$$

is differentiable and has the property

$$\Pi_N\circ\mathrm{D}F=F\circ\Pi_M.$$

If F and G are composable continuously differentiable maps then $F\circ G$ is continuously differentiable and

$$\mathrm{D}(F\circ G)=\mathrm{D}F\circ\mathrm{D}G.$$

Definition. $\mathrm{D}F$ is called the *derivative* of F.

2.14. Warning. The symbol $\mathrm{D}F$ is used in two different meanings: $\mathrm{D}F(x,\mathbf{v})=(F(x),\mathrm{D}F[x]\mathbf{v})$.

$\mathrm{D}F(\cdot)$ can be nicely defined as a map from $T(M)$ into $T(N)$, it can be well handled in general contexts.

On the other hand, $\mathrm{D}F[\cdot]$ cannot be defined generally as a map, though it is the essential part of $\mathrm{D}F(\cdot)$ and in simple special cases as affine spaces $\mathrm{D}F[\cdot]$ itself is a map (see VI. 2.5.) and there it would be an unnecessary complication to involve $\mathrm{D}F(\cdot)$ instead of $\mathrm{D}F[\cdot]$.

2.15. Definition. The dual $T_x(M)^*$ of $T_x(M)$ is called the *cotangent space of M* at $x\in M$. The set

$$T^*(M):=\bigcup_{x\in M}\{x\}\times T_x(M)^*$$

is the *cotangent bundle* of M. The map

$$(\Pi^*)_M:T^*(M)\longrightarrow M,\qquad (x,\mathbf{p})\longrightarrow x$$

is the *canonical projection* of the cotangent bundle.

2.16. Definition. Let $\{(U_i,u_i)\,\vdots\, i\in I\}$ be the maximal atlas of M. Then $T^*(M)$ is equipped with the topology generated by the family of sets

$$\left\{\bigcup_{x\in U_i}\{x\}\times \mathrm{D}u_i[x]^*[G]\,\vdots\, G\text{ is an open subset of }\mathbb{R}^m, i\in I\right\}.$$

$T^*(M)$ is considered as a manifold equipped with the following smooth structure.

Proposition. If (U, u) is a chart of M then $(\overset{-1}{(\Pi^*)_M}[U], \phi_u)$ is a chart of $T^*(M)$ where

$$\phi_u : \overset{-1}{(\Pi^*)_M}[U] \longrightarrow \mathbb{R}^m \times \mathbb{R}^m, \quad (x, \mathbf{p}) \mapsto (u(x), (\mathrm{D}u[x]^*)^{-1}\mathbf{p}).$$

If $\{(U_i, u_i) : i \in I\}$ is the maximal atlas of M then $\{(\overset{-1}{(\Pi^*)_M}[U_i], \phi_{u_i}) : i \in I\}$ is an atlas of $T^*(M)$.

If $T^*(M)$ is considered as a smooth manifold determined by this atlas, then the canonical projection $(\Pi^*)_M$ is smooth.

2.17. Remark. If M is an affine space over $\mathbf{M}$ then, by Remark 2.8. (iii), $T^*(M) = M \times \mathbf{M}^*$. The topology and the smooth structure of $T^*(M)$ is the product topology and the product smooth structure (see Definition 1.8.) corresponding to $M \times \mathbf{M}^*$.

2.18. Let $f \in \mathrm{C}^1(x)$. Then $\mathrm{D}f[x] : T_x(M) \longrightarrow T_{f(x)}(\mathbb{R}) \equiv \mathbb{R}$ (see Remark 2.8. (iii)) is a linear map, i.e. it is an element of $T_x(M)^*$ $(x \in M)$. We introduce the notation

$$\mathrm{d}f[x] := \mathrm{D}f[x]$$

and we call it the *differential of f at $x \in M$.*

Notice that—similarly to Remark 2.9. (iii)—we have

$$(\mathrm{d}f[x] \mid \mathbf{v}) = \mathbf{v}(f) \qquad (\mathbf{v} \in T_x(M)).$$

2.19. Proposition. Let (U, u) be a chart in M, $x \in U$. Then $\{\mathrm{d}u^k[x] : k = 1, \ldots, m\}$ is the dual basis of $\left\{\frac{\partial}{\partial u^k}[x] : k = 1, \ldots, m\right\}$, i.e.

$$\left(\mathrm{d}u^k[x] \,\middle|\, \frac{\partial}{\partial u^i}[x]\right) = \delta_i^k \qquad (i, k = 1, \ldots, m)$$

and we have for all $f \in \mathrm{C}^1(x)$

$$\mathrm{d}f[x] = \sum_{k=1}^{m} \frac{\partial f}{\partial u^k}[x]\mathrm{d}u^k[x].$$

Remark. As a consequence, if (U', u') is another chart, $x \in U'$ then

$$\mathrm{d}u^k[x] = \sum_{i=1}^{m} \frac{\partial u^k}{\partial u'^i}[x]\mathrm{d}u'^i[x].$$

2.20. Proposition. Let N be another manifold and suppose $F : M \rightarrowtail N$ is a little diffeomorphism. Then

$$\mathrm{D}F^{*-1} : T^*(M) \rightarrowtail T^*(N), \qquad (x, \mathbf{p}) \mapsto (F(x), (\mathrm{D}F[x]^*)^{-1}\mathbf{p})$$

is a little diffeomorphism.

3. Tensor fields

In this section M denotes an m dimensional manifold whose topology is of Hausdorff type and satisfies the second axiom of countability.

r and p denote non-negative integers.

$\mathrm{D}(M)$ stands for the algebra of the smooth functions $M \longrightarrow \mathbb{R}$.

3.1. Definition.

$$T^r(M) := \bigcup_{x \in M} \{x\} \times \overset{r}{\otimes} T_x(M)$$

is called the *r-tangent bundle* of M (see IV. 1.12.). The map

$$\Pi_M : T^r(M) \longrightarrow M, \qquad (x, t) \mapsto x$$

is the *canonical projection* of the r-tangent bundle.

The *symmetric* and *antisymmetric* r-tangent bundles $S^r(M)$ and $A^r(M)$ are defined similarly, with $\overset{r}{\vee}$ and $\overset{r}{\wedge}$ instead of $\overset{r}{\otimes}$.

Remarks. (i) We have the equalities (see IV. 1.12. and IV. 1.17.)

$$T^0(M) = S^0(M) = A^0(M) = M \times \mathbb{R},$$

$$T^1(M) = S^1(M) = A^1(M) = T(M).$$

(ii) The canonical projection is denoted by the same symbol for all r; similar notations appear later, too (see 3.3.). Hopefully, they do not lead to confusions.

(iii) $S^r(M)$ and $A^r(M)$ are subsets of $T^r(M)$ (see IV. 1.16.).

(iv) If M is an affine space over $\mathbf{M}$ then $T^r(M) = M \times \overset{r}{\otimes}\mathbf{M}$, $S^r(M) = M \times \overset{r}{\vee}\mathbf{M}$, $A^r(M) = M \times \overset{r}{\wedge}\mathbf{M}$.

3.2. A basis of $T_x(M)$ determines a basis of $\overset{r}{\otimes} T_x(M)$, $\overset{r}{\vee} T_x(M)$ and $\overset{r}{\wedge} T_x(M)$ (see IV. 1.7. and IV. 1.19.). Thus if (U, u) is a chart in M, $x \in U$, then u induces a basis in all these tensor products (see 2.7.).

The r-tangent bundles for $r \geq 2$ are equipped with a topology and a smooth structure in a similar way as it is done in 2.11. for $r = 1$. Then all the canonical projections become smooth maps.

3.3. Let N be another manifold and suppose $F : M \rightarrowtail N$ is a continuously differentiable map. Then we put for all r (see IV. 1.12.)

$$\mathrm{D}F : T^r(M) \rightarrowtail T^r(N), \qquad (x, t) \mapsto (F(x), \ (\overset{r}{\otimes} \mathrm{D}F[x])t)$$

$$(x \in \operatorname{Dom} F)$$

and $\mathrm{D}F$ is considered also as a map $S^r(M) \rightarrowtail S^r(N)$ and $A^r(M) \rightarrowtail A^r(N)$ defined by the same formula.

3.4. Definition. A smooth map $T\colon M \rightarrowtail T^r(M)$ for which $\Pi_M \circ T \subset \mathrm{id}_M$ holds is called an *r-tensor field*. If its range is contained in $S^r(M)$ or $A^r(M)$, it is called *symmetric* or *antisymmetric*, respectively.

In the case $r=1$ we use the name *vector field*.

Everywhere defined tensor fields are called *global*.

The set of global (symmetric or antisymmetric) r-tensor fields is denoted by $\mathrm{T}^r\mathrm{F}(M)$ ($\mathrm{S}^r\mathrm{F}(M)$, $\mathrm{A}^r\mathrm{F}(M)$, respectively) and we shall use the notation $\mathrm{V}(M) := \mathrm{T}^1\mathrm{F}(M)$, too.

Remarks. (i) The value of T at $x \in \mathrm{Dom}\, T$ is a pair whose first member is x itself and the second one is an element of $\overset{r}{\otimes}\, T_x(M)$; we shall use the notation

$$T(x) = (x, T[x])$$

and we warn the reader as we did it in 2.14.

(ii) 0-tensor fields can be identified with smooth functions $M \rightarrowtail \mathbb{R}$, hence $\mathrm{T}^0\mathrm{F}(M) = \mathrm{D}(M)$.

(iii) An r-tensor field in an affine space M can be identified with a smooth map $M \rightarrowtail \overset{r}{\otimes}\mathbf{M}$.

3.5. Definition.

(i) Let T_1 and T_2 be r-tensor fields. Their *sum* is the r-tensor field

$$(T_1 + T_2)(x) := (x, T_1[x] + T_2[x]) \qquad (x \in \mathrm{Dom}\, T_1 \cap \mathrm{Dom}\, T_2).$$

(ii) Let S and T be a p-tensor field and an r-tensor field. Their *tensor product* is the $(p+r)$-tensor field

$$(S \otimes T)(x) := (x, S[x] \otimes T[x]) \qquad (x \in \mathrm{Dom}\, S \cap \mathrm{Dom}\, T).$$

Remark. If f is a 0-tensor field, i.e. a smooth real valued function, we write fT instead of $f \otimes T$ for all tensor fields T.

3.6. Proposition. The global (symmetric, antisymmetric) r-tensor fields form a module over $\mathrm{D}(M)$ by the operations given in the previous definition. In particular, they are real vector spaces because $\mathbb{R}$ can be considered as the subset of $\mathrm{D}(M)$ consisting of the constant functions.

3.7. Proposition. The dual of the module $\mathrm{T}^r\mathrm{F}(M)$ can be identified with the $\mathrm{D}(M)$-module

$$\{C : T^r(M) \longrightarrow \mathbb{R} \,\vdots\, C \text{ is smooth, } C(x, \cdot) \text{ is linear for all } x \in M\}$$

by the formula

$$(C \mid T) := C \circ T.$$

([2], Proposition III. 7.2.).

Similar assertions are true for $\mathrm{S}^r\mathrm{F}(M)$ and $\mathrm{A}^r\mathrm{F}(M)$, too.

3.8. Definition. Let N be another manifold and let $F: M \rightarrowtail N$ be a smooth map. An r-tensor field S in N is called *F-related* to an r-tensor field T in M if

$$\mathrm{D}F \circ T \subset S \circ F.$$

If $M = N$ and T is F-related to itself then T is called *F-invariant*.

Remark. A function $g: N \rightarrowtail \mathbb{R}$ is F-related to a function $f: M \rightarrowtail \mathbb{R}$ if $f|_{\mathrm{Dom}\, f \cap \mathrm{Dom}\, F} \subset g \circ F$.

Hence a constant function on N is F-related to the same constant function on M.

3.9. Proposition. Let $F: M \rightarrowtail N$ be a little diffeomorphism. Then there is a unique r-tensor field defined on $F[\mathrm{Dom}\, T] \subset N$ and F-related to the r-tensor field T in M, namely

$$F_*(T) := \mathrm{D}F \circ T \circ F^{-1}.$$

Remark. $F_*(f) = f \circ F^{-1}$ for $f \in \mathrm{D}(M)$.

3.10. Proposition. Let $F: M \longrightarrow N$ be a diffeomorphism. Then the map $T \mapsto F_*(T)$ has the properties

(i) $F_*(T_1 + T_2) = F_*(T_1) + F_*(T_2)$,

(ii) $F_*(S \otimes T) = F_*(S) \otimes F_*(T)$.

In particular, $F_*: \mathrm{T}^r\mathrm{F}(M) \longrightarrow \mathrm{T}^r\mathrm{F}(N)$ is an $\mathbb{R}$-linear isomorphism, and then

$$(F_*)^{-1} = (F^{-1})_*.$$

Furthermore, if G is another diffeomorphism composable with F then

$$(F \circ G)_* = F_* \circ G_*.$$

3.11. (i) If (U, u) is a chart in M, then

$$x \mapsto \frac{\partial}{\partial u^k}(x) := \left(x, \frac{\partial}{\partial u^k}[x]\right) \qquad (k = 1, \ldots, m)$$

are vector fields defined on U.

(ii) Let $\mathbf{X}$ be a vector field and let (U, u) be a chart in M, $U \subset \mathrm{Dom}\, \mathbf{X}$. Then Proposition 2.7. yields that for $x \in U$ there are real numbers $X^k(u(x))$ $(k = 1, \ldots, m)$ such that

$$\mathbf{X}(x) = \left(x, \sum_{k=1}^{m} X^k(u(x)) \frac{\partial}{\partial u^k}[x]\right) = \sum_{k=1}^{m} X^k(u(x)) \frac{\partial}{\partial u^k}(x)$$

in other words there are smooth functions $X^k: \mathrm{Ran}\, u \longrightarrow \mathbb{R}$ $(k = 1, \ldots, m)$ such that

$$\mathbf{X}|_U = (X^k \circ u) \frac{\partial}{\partial u^k}$$

where we used the so called summation convention.

The right side of the above equality is called the *local representative* of $\mathbf{X}$ in (U, u) and X^k $(k=1, \ldots, m)$ are the *local components* of $\mathbf{X}$ with respect to (U, u).

If (U', u') is another chart in M meeting (U, u) then the local components of $\mathbf{X}$ with respect to $(U \cap U', u'|_{U \cap U'})$ are

$$X'^k = X^i \frac{\partial u'^k}{\partial u^i} \qquad (k=1, \ldots, m).$$

(iii) The local components of a vector field $\mathbf{X}$ with respect to a chart (U, u) are exactly the components of the vector field $u_*(\mathbf{X}|_U)$ in $\mathbb{R}^m$, i.e.

$$(u_*(\mathbf{X}|_U))\,[\xi] = (X^1(\xi), X^2(\xi), \ldots, X^m(\xi))$$
$$(\xi \in \operatorname{Ran} u).$$

(iv) Similar observations can be made for other tensor fields as well. For instance, the local representative of a 2-tensor field T in the chart (U, u), $U \subset \operatorname{Dom} T$, is

$$T|_U = (T^{ik} \circ u) \frac{\partial}{\partial u^i} \otimes \frac{\partial}{\partial u^k},$$

where $T^{ik}: \operatorname{Ran} u \longrightarrow \mathbb{R}$ $(i, k=1, \ldots, m)$ are smooth functions.

3.12. Definition. A *derivation* of $D(M)$ is a map $L: D(M) \longrightarrow D(M)$ with the properties

(i) L is $\mathbb{R}$-linear,

(ii) $L(fg) = L(f)g + fL(g) \qquad (f, g \in D(M))$.

Proposition. The derivations of $D(M)$ form a module over $D(M)$ defined by the pointwise operations. In particular, it is a real vector space because $\mathbb{R}$ can be considered as the subset of $D(M)$ consisting of the constant functions.

3.13. Proposition. Let L and K be derivations of $D(M)$. Their *commutator*

$$[L, K] := L \circ K - K \circ L$$

is also a derivation of $D(M)$.

The map $(L, K) \mapsto [L, K]$ is a Lie product, i.e. it is antisymmetric, bilinear and satisfies the Jacobi identity:

$$[[L, K], J] + [[J, L], K] + [[K, J], L] = 0$$

for all derivations L, K, J of $D(M)$.

Furthermore, for all $f \in D(M)$ we have

$$[L, fK] = L(f)K + f[L, K].$$

3.14. Proposition. Let $\mathbf{X}$ be a global vector field. A derivation of $D(M)$ can be associated with $\mathbf{X}$ by

$$f \mapsto (x \mapsto (\mathrm{d}f[x] \mid \mathbf{X}[x])).$$

This correspondence between the module of global vector fields and the module of derivations of D(M) is an isomorphism ([2], Theorem III. 6.5.).

Remarks. (i) According to the previously given isomorphism we shall identify the modules in question. Moreover, we shall use the same symbol for a vector field (a map $M \longrightarrow T(M)$) and for the corresponding derivation of D(M) (a map D(M) $\longrightarrow$ D(M)). If any danger of confusion arises, we always make clear which of the two possibilities should be taken.

(ii) As a consequence of the given isomorphism, the commutator of global vector fields is defined. If $\mathbf{X}, \mathbf{Y} \in \mathrm{V}(M)$ and (U, u) is a chart in M then the local representative of $[\mathbf{X}, \mathbf{Y}]$ in (U, u) is

$$\left((X^k \circ u)\frac{\partial(Y^i \circ u)}{\partial u^k} - (Y^k \circ u)\frac{\partial(X^i \circ u)}{\partial u^k}\right)\frac{\partial}{\partial u^i}$$

where X^k and Y^k ($k = 1, \ldots, m$) are the local components of $\mathbf{X}$ and $\mathbf{Y}$, respectively, in (U, u).

(iii) If M is an affine space, the commutator of the vector fields $\mathbf{X}, \mathbf{Y} : M \longrightarrow \mathbf{M}$ (see Remark 3.4. (iii)) is $\mathrm{D}Y[\cdot]X - \mathrm{D}X[\cdot]Y$.

3.15. Proposition. Let N be another manifold. Suppose $F : M \longrightarrow N$ is a smooth map.

(i) If $\mathbf{Y} \in \mathrm{V}(N)$ is F-related to $\mathbf{X} \in \mathrm{V}(M)$ and they are considered as derivations of D(N) and D(M), respectively, then

$$\mathbf{X}(g \circ F) = \mathbf{Y}(g) \circ F \qquad (g \in \mathrm{D}(N)).$$

(ii) If $\mathbf{Y}_1, \mathbf{Y}_2 \in \mathrm{V}(N)$ are F-related to $\mathbf{X}_1, \mathbf{X}_2 \in \mathrm{V}(M)$, respectively, then $[\mathbf{Y}_1, \mathbf{Y}_2]$ is F-related to $[\mathbf{X}_1, \mathbf{X}_2]$.

In particular, if F is a diffeomorphism,

$$F_*[\mathbf{X}_1, \mathbf{X}_2] = [F_*(\mathbf{X}_1), F_*(\mathbf{X}_2)].$$

([2], Proposition III. 6.14.).

In particular, if $M = N$, the commutator of F-invariant vector fields is F-invariant as well.

3.16. We can define s times continuously differentiable r-tensor fields by an evident modification of Definition 3.4. ($s \in \mathbb{N}_0$). For them the definitions, remarks and propositions up to 3.11. can be introduced or stated with the modification that we say s times continuously differentiable r-tensor fields instead of smooth ones, we take $\mathrm{C}^s(M)$ (the set of s times continuously differentiable maps $M \longrightarrow \mathbb{R}$) instead of D($M$) and we consider an $(s+1)$ times continuously differentiable map $F : M \rightarrowtail N$ instead of a smooth one.

3.17. Let $\mathbf{A}$ be a one dimensional real vector space. Then

$$T^r(M)_{\mathbf{A}} := \bigcup_{x \in M} \{x\} \times \frac{\overset{r}{\otimes} T_x(M)}{\mathbf{A}}$$

is called the *r-tangent bundle of cotype* **A** of M. The *symmetric* and *antisymmetric r-*tangent bundles of cotype **A** are defined similarly.

The map

$$\Pi_M : T^r(M)_{\mathbf{A}} \longrightarrow M, \qquad (x, t) \mapsto x$$

is the *canonical projection* of the r-tangent bundle of cotype **A**.

All these spaces are equipped with a topology and a smooth structure in a similar way as it is done for $\mathbf{A} := \mathbb{R}$.

If $F : M \rightarrowtail N$ is a continuously differentiable map then we put

$$\mathrm{D}F_{\mathbf{A}} : T^r(M)_{\mathbf{A}} \rightarrowtail T^r(N)_{\mathbf{A}}, \qquad (x, t) \mapsto (F(x), (\overset{r}{\bigotimes} \mathrm{D}F[x])_{\mathbf{A}} t)$$

$$(x \in \mathrm{Dom}\, F)$$

(see IV. 2.10.).

3.18. We can introduce the *r-tensor fields of cotype* **A** by an evident modification of Definition 3.4.

Then all the subsequent remarks, definitions and propositions up to 3.11. can be introduced or stated with evident modifications.

4. Cotensor fields

M, $\mathrm{D}(M)$, m, r and p have the same meaning as in the previous section.

4.1. Definition.

$$(T^*)^r(M) := \bigcup_{x \in M} \{x\} \times \overset{r}{\otimes}\, T_x(M)^*$$

is called the *r-cotangent bundle* of M (see IV. 1.12.). The map

$$(\Pi^*)_M : (T^*)^r(M) \longrightarrow M, \qquad (x, c) \mapsto x$$

is the *canonical projection* of the r-cotangent bundle.

The *symmetric* and *antisymmetric* r-cotangent bundles $(S^*)^r(M)$ and $(A^*)^r(M)$ are defined similarly, with $\overset{r}{\vee}$ and $\overset{r}{\wedge}$ instead of $\overset{r}{\otimes}$.

Remarks. (i) We have the equalities (see IV. 1.12. and IV. 1.17.)

$$(T^*)^0(M) = (S^*)^0(M) = (A^*)^0(M) = M \times \mathbb{R},$$

$$(T^*)^1(M) = (S^*)^1(M) = (A^*)^1(M) = T^*(M).$$

(ii) The canonical projection is denoted by the same symbol for all r, as in the case of r-tangent spaces. Hopefully, this notation does not lead to confusion.

(iii) $(S^*)^r(M)$ and $(A^*)^r(M)$ are subsets of $(T^*)^r(M)$ (see IV. 1.16.)

(iv) If M is an affine space over $\mathbf{M}$ then $(T^*)^r(M) = M \times \overset{r}{\otimes} \mathbf{M}^*$, $(S^*)^r(M) = M \times \overset{r}{\vee} \mathbf{M}^*$, $(A^*)^r(M) = M \times \overset{r}{\wedge} \mathbf{M}^*$.

4.2. A basis of $T_x(M)^*$ determines a basis of $\overset{r}{\otimes} T_x(M)^*$, $\overset{r}{\vee} T_x(M)^*$ and $\overset{r}{\wedge} T_x(M)^*$ (see IV. 1.7. and IV. 1.19.). Thus if (U, u) is a chart in $M, x \in U$, then u induces a basis in these tensor products (see 2.19.).

The r-cotangent bundles for $r \geq 2$ are equipped with a topology and a smooth structure in a similar way as it was done in 2.16. for $r = 1$. Then all the canonical projections become smooth maps.

4.3. Definition. A smooth map $C: M \rightarrowtail (T^*)^r(M)$ for which $(\Pi^*)_M \circ C \subset \mathrm{id}_M$ holds is called an *r-cotensor field*. If its range is contained in $(S^*)^r(M)$ or $(A^*)^r(M)$, it is called *symmetric*, or *antisymmetric*, respectively.

In the case $r = 1$ we use the name *covector field.*

Everywhere defined cotensor fields are called *global.*

The set of global (symmetric, antisymmetric), r-cotensor fields is denoted by $\mathrm{T}^r\mathrm{F}(M)^*$ ($\mathrm{S}^r\mathrm{F}(M)^*$, $\mathrm{A}^r\mathrm{F}(M)^*$, respectively).

Remarks. (i) The value of C at $x \in \mathrm{Dom}\, C$ is a pair whose first member is x itself and the second one is an element of $\overset{r}{\otimes} T_x(M)^*$; we shall use the notation

$$C(x) = (x, C[x])$$

and we warn the reader as we did it in 2.14.

(ii) 0-cotensor fields can be identified with smooth functions $M \rightarrowtail \mathbb{R}$, hence $\mathrm{T}^0\mathrm{F}(M)^* = \mathrm{D}(M)$.

(iii) An r-cotensor field in an affine space M can be identified with a smooth map $M \rightarrowtail \overset{r}{\otimes} \mathbf{M}^*$.

4.4. Definition.

(i) Let C_1 and C_2 be r-cotensor fields. Their *sum* is the r-cotensor field

$$(C_1 + C_2)(x) := (x, C_1[x] + C_2[x]) \qquad (x \in \mathrm{Dom}\, C_1 \cap \mathrm{Dom}\, C_2).$$

(ii) Let D and C be a p-cotensor field and an r-cotensor field. Their *tensor product* is the $(p+r)$-cotensor field

$$(D \otimes C)(x) := (x, D[x] \otimes C[x]) \qquad (x \in \mathrm{Dom}\, D \cap \mathrm{Dom}\, C).$$

Remark. If f is a 0-cotensor field, i.e. a smooth real valued function, we write fC instead of $f \otimes C$ for all cotensor fields C.

4.5. Proposition. The global (symmetric, antisymmetric) cotensor fields form a module over $\mathrm{D}(M)$ by the operations given in the previous definition. In particular, they are real vector spaces because $\mathbb{R}$ can be considered as the subset of $\mathrm{D}(M)$ consisting of the constant functions.

4.6. Proposition. The modules $T^rF(M)^*$, $S^rF(M)^*$ and $A^rF(M)^*$ can be identified with the duals of the modules $T^rF(M)$, $S^rF(M)$ and $A^rF(M)$, respectively, by

$$(C|T)(x) := (C[x]|T[x]) \qquad (x \in M).$$

Here $\overset{r}{\otimes} T_x(M)^*$ is considered as the dual of $\overset{r}{\otimes} T_x(M)$ (see IV. 1.5. (v)).

Remark. As a consequence of Proposition 3.7., a global r-cotensor field can be viewed as a certain map $T^r(M) \longrightarrow \mathbb{R}$.

4.7. Proposition. The module $T^rF(M)^*$ can be identified with the module of r-fold $D(M)$-linear maps $\overset{n}{\times} V(M) \longrightarrow D(M)$; similarly, the modules $S^rF(M)^*$ and $A^rF(M)^*$ can be identified with the modules of such maps that are symmetric and antisymmetric, respectively, by

$$C(X_1, \ldots, X_r)(x) := C[x](X_1[x], \ldots, X_r[x]) \qquad (x \in M).$$

Here $\overset{r}{\otimes} T_x(M)^*$ is considered as the set of r-fold linear maps on $T_x(M)$ (see IV. 1.11.) ([2], Theorem III. 7.6.).

4.8. Proposition. Let N be another manifold and let $F: M \rightarrowtail N$ be a smooth map. The *pull-back* of the r-cotensor field C in N by F is the r-cotensor field F^*C in M defined by

$$(F^*C)(x) := \left(x, \overset{r}{\otimes} DF[x]^*C[F(x)]\right) \qquad (x \in F^{-1}[\mathrm{Dom}\, C]).$$

Remarks. (i) For $r=0$ the definition gives that

$$F^*(g) = g \circ F \qquad (g \in D(N)).$$

In particular, if c is a constant function on N then $F^*(c)$ is the same constant function on M.

(ii) If $C[F(x)] \in \overset{r}{\otimes} T_{F(x)}(N)^*$ is considered as an r-fold linear map on $T_{F(x)}(N)$ then

$$(F^*C)[x] = C[F(x)] \circ \left(\overset{r}{\times} DF[x]\right).$$

(iii) If M is a submanifold of N and $i: M \longrightarrow N$ is the canonical injection then we use the notation

$$C|_M := i^*C.$$

4.9. Proposition. (i) If C is a symmetric or antisymmetric cotensor field then F^*C is symmetric or antisymmetric, respectively.

(ii) If the global r-cotensor field C is considered as a map $T^r(M) \longrightarrow \mathbb{R}$ (see 4.6. and 3.7.) then

$$F^*C = C \circ \mathrm{D}F$$

(see 3.3.).

4.10. Proposition. Let $F: M \rightarrowtail N$ be a smooth map. Then

(i) $F^*(C_1 + C_2) = F^*C_1 + F^*C_2$

for all r-cotensor fields C_1 and C_2,

(ii) $F(D \otimes C) = (F^*D) \otimes (F^*C)$

for all p-cotensor fields D and r-cotensor fields C.

In particular, $F^*: \mathrm{T}^r\mathrm{F}(N)^* \longrightarrow \mathrm{T}^r\mathrm{F}(M)^*$ is an $\mathbb{R}$-linear map if $\mathrm{Dom}\, F = M$.

If G is another smooth map composable with F then

$$(F \circ G)^* = G^* \circ F^* .$$

Furthermore, if F is a diffeomorphism,

$$(F^{-1})^* = (F^*)^{-1} .$$

4.11. Proposition. If $F: M \longrightarrow N$ is a diffeomorphism and $C \in T^rF(N)^*$ is considered as in Proposition 4.7. then

$$(F^*C)(\mathbf{X}_1, \ldots, \mathbf{X}_r) = C(F_*(\mathbf{X}_1), \ldots, F_*(\mathbf{X}_r)) \circ F$$

for all $\mathbf{X}_1, \ldots, \mathbf{X}_r \in \mathrm{V}(M)$.

4.12. (i) If $f: M \rightarrowtail \mathbb{R}$ is a smooth function then (see 2.18.)

$$x \mapsto \mathrm{d}f(x) := (x, \mathrm{d}f[x])$$

is a covector field defined on $\mathrm{Dom}\, f$, called the *differential* of f.

In particular, if (U, u) is a chart in M

$$x \mapsto \mathrm{d}u^k(x) := (x, \mathrm{d}u^k[x]) \qquad (k = 1, \ldots, m)$$

are covector fields defined on U.

(ii) Let $\mathbf{K}$ be a covector field and let (U, u) be a chart in M, $U \subset \mathrm{Dom}\, \mathbf{K}$. Then Proposition 2.19. yields that for $x \in U$ there are real numbers $K_k(u(x))$ $(k = 1, \ldots, m)$ such that

$$\mathbf{K}(x) = \left(x, \sum_{k=1}^{m} K_k(u(x))\, \mathrm{d}u^k[x]\right) = \sum_{k=1}^{m} K_k(u(x))\, \mathrm{d}u^k(x) ,$$

in other words there are smooth functions $K_k: \mathrm{Ran}\, u \longrightarrow \mathbb{R}$ $(k = 1, \ldots, m)$ such that

$$\mathbf{K}|_U = (K_k \circ u)\, \mathrm{d}u^k$$

where we used the so called summation convention.

The right side of the above equality is called the *local representative* of $\mathbf{K}$ in (U, u) and K_k $(k = 1, \ldots, m)$ are the *local components* of $\mathbf{K}$ with respect to (U, u).

If (U', u') is another chart in M meeting (U, u), the local components of **K** with respect to $(U \cap U', u'|_{U \cap U'})$ are

$$K'_k = K_i \frac{\partial u^i}{\partial u'^k} \qquad (k = 1, \ldots, m)$$

(iii) The local components of a covector field **K** with respect to a chart (U, u) are exactly the components of the covector field $(u^{-1})^*(\mathbf{K}|_U)$ in $\mathbb{R}^m$, i.e.

$$((u^{-1})^*(\mathbf{K}|_U))\,[\xi] = (K_1(\xi), K_2(\xi), \ldots, K_m(\xi))$$
$$(\xi \in \operatorname{Ran} u)\,.$$

(iv) Similar observations can be made for other cotensor fields as well. For instance, the local representative of a 2-cotensor field C in (U, u), $U \subset \operatorname{Dom} C$, is

$$C|_U = (C_{ik} \circ u)\, \mathrm{d}u^i \otimes \mathrm{d}u^k\,,$$

where $C_{ik} : \operatorname{Ran} u \longrightarrow \mathbb{R}$ $(i, k = 1, \ldots, m)$ are smooth functions.

(v) Let $F : M \rightarrowtail N$ be a smooth map and suppose (U, u) is a chart in $\operatorname{Dom} F$, (V, v) is a chart in N.

Then $(F^* \mathrm{d}v^k)\,[x] = \mathrm{d}v^k[F(x)] \circ \mathrm{D}F[x] = \mathrm{d}(v^k \circ F)\,[x]$, hence if $\bar{F}$ denotes the local representative of F with respect to the given charts, i.e. $\bar{F} = v \circ F \circ u^{-1}$, and $\bar{F}^k = v^k \circ F \circ u^{-1}$, then

$$F^*\, \mathrm{d}v^k = \mathrm{d}(\bar{F}^k \circ u) = (\partial_j \bar{F}^k \circ u)\, \mathrm{d}u^j \qquad (k = 1, \ldots, m)\,.$$

As a consequence, if $(K_k \circ v)\, \mathrm{d}v^k$ is the local representative of a covector field **K** in (V, v) then the local representative of $F^*\mathbf{K}$ in (U, u) is

$$(K_k \circ (\bar{F} \circ u))\, \mathrm{d}(\bar{F}^k \circ u) = (((K_k \circ \bar{F})\partial_j \bar{F}^k) \circ u)\, \mathrm{d}u^j\,.$$

Similar formulae can be derived for other r-cotensor fields as well. For instance, if $(C_{ik} \circ v)\, \mathrm{d}v^i \otimes \mathrm{d}v^k$ is the local representative of a 2-cotensor field in (V, v) then the local representative of F^*C in (U, u) is

$$(C_{ik} \circ (\bar{F} \circ u))\, \mathrm{d}(\bar{F}^i \circ u) \otimes \mathrm{d}(\bar{F}^k \circ u) = (((C_{ik} \circ \bar{F})\partial_j \bar{F}^i \partial_l \bar{F}^k) \circ u)\, \mathrm{d}u^j \otimes \mathrm{d}u^l\,.$$

4.13. We can define s times continuously differentiable r-cotensor fields by an evident modification of Definition 4.3. $(s \in \mathbb{N}_0)$. The corresponding definitions, remarks and propositions for them can be introduced or stated with the modification that we say s times continuously differentiable r-cotensor fields and r-tensor fields (see 3.16.) instead of smooth ones, we take $\mathrm{C}^s(M)$ instead of $\mathrm{D}(M)$ and we consider an $(s+1)$ times continuously differentiable map $F : M \rightarrowtail N$ instead of a smooth one.

4.14. Let **A** be a one dimensional real vector space. Then

$$(\mathrm{T}^*)^r(M)_{\mathbf{A}} := \bigcup_{x \in M} \{x\} \times \frac{\overset{r}{\otimes}\, \mathrm{T}_x(M)^*}{\mathbf{A}^*}$$

is called the *r-cotangent bundle of type* $\mathbf{A}$ of M. The *symmetric* and *antisymmetric* r-cotangent bundle of type $\mathbf{A}$ are defined similarly.

The map

$$\Pi_M:(\mathrm{T}^*)^r(M)_{\mathbf{A}} \longrightarrow M, \qquad (x,c)\mapsto x$$

is the *canonical projection* of the *r-cotangent bundle of type* $\mathbf{A}$.

All these spaces are equipped with a topology and a smooth structure in a similar way as it is done for $\mathbf{A}:=\mathbb{R}$.

4.15. We can introduce the *r-cotensor fields of type* $\mathbf{A}$ by an evident modification of Definition 4.3. Note that if C is an r-cotensor field of type $\overset{r}{\otimes}\mathbf{A}$ then $C[x]$ can be considered either as an $\mathbb{R}$-valued r-fold linear map on $\frac{T_x(M)}{\mathbf{A}}$ or as an $\overset{r}{\otimes}\mathbf{A}$-valued r-fold linear map on $T_x(M)$.

Then all the subsequent remarks, definitions and propositions can be introduced or stated with evident modifications: everywhere one has to suffix the term "of type $\overset{r}{\otimes}\mathbf{A}$" to r-tensor fields, vector fields and r-cotensor fields.

Moreover, Proposition 4.7. can be rewritten in a convenient form, too:

Proposition. The module of global (symmetric, antisymmetric) r-cotensor fields of type $\overset{r}{\otimes}\mathbf{A}$ can be identified with the module of r-fold (symmetric, antisymmetric) $\mathrm{D}(M)$-linear maps $\overset{r}{\times}\mathrm{V}(M)\longrightarrow \mathrm{D}\left(M,\overset{r}{\otimes}\mathbf{A}\right)$, where the last symbol denotes the module over D(M) of the $\overset{r}{\otimes}\mathbf{A}$-valued smooth functions defined on M.

5. Differential forms

M, $\mathrm{D}(M)$, m, r and p have the same meaning as in the previous two sections.

5.1. As we know, there are at least two convenient possibilities to consider antisymmetric tensor products as tensor products (see IV. 1.16. and IV. 1.17.). If we choose the possibility $\wedge$ instead of $\oslash$ in Definitions 4.1. and 4.3., we call a global antisymmetric r-cotensor field a *differential form of degree* r or an *r-form* on M.

The r-forms constitute a module over $\mathrm{D}(M)$ which will be denoted by $\mathbf{A}^r(M)$. $\mathbf{A}^0(M)=\mathrm{D}(M)$ and all differential forms of degree $r>m$ are zero.

$\mathbf{A}^r(M)$ is considered mostly as the module of $\mathrm{D}(M)$-valued r-fold $\mathrm{D}(M)$-linear antisymmetric maps on the module $\mathrm{V}(M)$ of global vector fields (see Proposition 4.7.).

The *exterior product*

$$\alpha \wedge \beta := \frac{m!}{r!\,p!} A(\alpha \otimes \beta)$$

(see IV. 1.16. and IV. 1.17.) of $\alpha \in \mathbf{A}^r(M)$ and $\beta \in \mathbf{A}^p(M)$ has the property

$$\beta \wedge \alpha = (-1)^{rp} \alpha \wedge \beta.$$

If (U, u) is a chart in M then the local representative of $\alpha \in \mathbf{A}^r(M)$ is

$$\alpha|_U = \frac{1}{r!} (\alpha_{i_1 \dots i_r} \circ u) du^{i_1} \wedge \dots \wedge du^{i_r}$$

where $\alpha_{i_1 \dots i_r}$: Ran $u \longrightarrow \mathbb{R}$ $(i_1, \dots, i_r = 1, \dots, m)$ are smooth functions, antisymmetrically given in the indices.

5.2. Definition. Let $\alpha \in \mathbf{A}^r(M)$, $r \neq 0$. Then the element $d\alpha$ of $\mathbf{A}^{r+1}(M)$ defined by

$$d\alpha(\mathbf{X}_1, \dots, \mathbf{X}_{r+1}) := \sum_{i=1}^{r+1} (-1)^{i-1} \mathbf{X}_i(\alpha(\mathbf{X}_1, \dots, \not{\mathbf{X}}_i, \dots, \mathbf{X}_{r+1})) +$$

$$+ \sum_{\substack{i<j \\ j=2,\dots,r+1}} (-1)^{i+j} \alpha([\mathbf{X}_i, \mathbf{X}_j], \mathbf{X}_1, \dots, \not{\mathbf{X}}_i, \dots, \not{\mathbf{X}}_j, \dots, \mathbf{X}_{r+1})$$

$$(\mathbf{X}_1, \dots, \mathbf{X}_{r+1} \in \mathrm{V}(M))$$

(where the sign | means that the corresponding entry should be omitted) is called the *exterior differential* of α. The map $d: \mathbf{A}^r(M) \longrightarrow \mathbf{A}^{r+1}(M)$ is the *exterior differentiation*.

The exterior differential of $f \in \mathbf{A}^0(M) = \mathrm{D}(M)$ is defined to be the differential df of f (see 4.12. (i)).

An r-form α is called *closed* if $d\alpha = 0$; it is *exact* if there is an $(r-1)$-form β such that $\alpha = d\beta$.

Proposition. The exterior differentiation has the following properties

(i) it is $\mathbb{R}$-linear,

(ii) $d(\beta \wedge \alpha) = (d\beta) \wedge \alpha + (-1)^p \beta \wedge d\alpha \qquad (\beta \in \mathbf{A}^p(M),\ \alpha \in \mathbf{A}^r(M))$,

(iii) $d(d\alpha) = 0 \quad$ for all $\alpha \in \mathbf{A}^r(M)$.

([2], Proposition IV. 2.4.).

Remarks. (i) If $\alpha \in \mathbf{A}^1(M)$, i.e. α is a covector field, then

$$d\alpha(\mathbf{X}, \mathbf{Y}) = \mathbf{X}(\alpha(\mathbf{Y})) - \mathbf{Y}(\alpha(\mathbf{X})) - \alpha([\mathbf{X}, \mathbf{Y}])$$

$$(\mathbf{X}, \mathbf{Y} \in \mathrm{V}(M)).$$

(ii) If $\alpha \in \mathbf{A}^2(M)$ then

$$\mathrm{d}\alpha(\mathbf{X}, \mathbf{Y}, \mathbf{Z}) = \mathbf{X}(\alpha(\mathbf{Y}, \mathbf{Z})) + \mathbf{Y}(\alpha(\mathbf{Z}, \mathbf{X})) + \mathbf{Z}(\alpha(\mathbf{X}, \mathbf{Y})) -$$
$$- \alpha([\mathbf{X}, \mathbf{Y}], \mathbf{Z}) - \alpha([\mathbf{Y}, \mathbf{Z}], \mathbf{X}) - \alpha([\mathbf{Z}, \mathbf{X}], \mathbf{Y})$$
$$(\mathbf{X}, \mathbf{Y}, \mathbf{Z} \in \mathrm{V}(M)).$$

(iii) Let (U, u) be a chart in M. Then $(\mathrm{d}\alpha)|_U = \mathrm{d}(\alpha|_U)$ for all r-forms α. Hence, if

$$\alpha|_U = \frac{1}{r!}(\alpha_{i_1 \dots i_r} \circ u)\mathrm{d}u^{i_1} \wedge \dots \wedge \mathrm{d}u^{i_r}$$

then Proposition 5.2. (ii) and (iii) yield that

$$\mathrm{d}\alpha|_U = \frac{1}{r!}\mathrm{d}(\alpha_{i_1 \dots i_r} \circ u) \wedge \mathrm{d}u^{i_1} \wedge \dots \wedge \mathrm{d}u^{i_r} =$$

$$= \frac{1}{r+1!}\sum_{k=0}^{r}(-1)^k((\partial_{i_k}\alpha_{i_0 \dots |_k \dots i_r}) \circ u)\mathrm{d}u^{i_0} \wedge \dots \wedge \mathrm{d}u^{i_r}$$

where the sign | means that the corresponding index is to be omitted.

In particular, if $r=1$,

$$\mathrm{d}\alpha|_U = \frac{1}{2}((\partial_i\alpha_j - \partial_j\alpha_i) \circ u)\mathrm{d}u^i \wedge \mathrm{d}u^j,$$

if $r=2$,

$$\mathrm{d}\alpha|_U = \frac{1}{6}((\partial_i\alpha_{jk} + \partial_j\alpha_{ki} + \partial_k\alpha_{ij}) \circ u)\mathrm{d}u^i \wedge \mathrm{d}u^j \wedge \mathrm{d}u^k.$$

5.3. Equality (iii) of Proposition 5.2. shows that exact differential forms are closed. The converse is not true in general but we have the so called *Poincaré lemma*:

Proposition. If M is homeomorphic with $\mathbb{R}^m$ then all closed differential forms on M are exact ([2], Theorem IV. 2.11.). A closed one-form on M is exact if M is simply connected.

5.4. Proposition. Let N be another manifold and let $F: M \longrightarrow N$ be a smooth map. Then

$$F^* \circ \mathrm{d} = \mathrm{d} \circ F^*.$$

(Note that the same symbol d denotes the exterior differentiation both on M and N) ([2], Proposition IV. 2.7.).

5.5. Definition. Let $\mathbf{X}$ be an element of $\mathrm{V}(M)$. The map

$$i_{\mathbf{X}}: \mathbf{A}^{r+1}(M) \longrightarrow \mathbf{A}^r(M), \qquad \alpha \mapsto \alpha(\mathbf{X}, \cdot, \dots, \cdot)$$

is called the *contraction* by $\mathbf{X}$. For $f \in \mathbf{A}^0(M) = \mathrm{D}(M)$ we put

$$i_{\mathbf{X}}(f) := 0.$$

Proposition. The contraction by $\mathbf{X}$ has the properties
(i) it is $\mathbb{R}$-linear,
(ii) $i_{\mathbf{X}}(\beta \wedge \alpha) = (i_{\mathbf{X}}\beta) \wedge \alpha + (-1)^p \beta \wedge i_{\mathbf{X}}\alpha \qquad (\beta \in \mathbf{A}^p(M), \alpha \in \mathbf{A}^r(M))$.
Moreover, the assignment $\mathbf{X} \mapsto i_{\mathbf{X}}$ is $\mathrm{D}(M)$-linear.

Remark. If (U, u) is a chart in M and

$$\mathbf{X}|_U = (X^k \circ u)\frac{\partial}{\partial u^k},$$

$$\alpha|_U = \frac{1}{r!}(\alpha_{i_1 \ldots i_r} \circ u)\mathrm{d}u^{i_1} \wedge \ldots \wedge \mathrm{d}u^{i_r}$$

then

$$(i_{\mathbf{X}}\alpha)|_U = \frac{1}{(r-1)!}((X^k \alpha_{k i_1 \ldots i_{r-1}}) \circ u)\mathrm{d}u^{i_1} \wedge \ldots \wedge \mathrm{d}u^{i_{r-1}}$$

5.6. Proposition. Let N be another manifold and let $F: M \longrightarrow N$ be a smooth map. If $\mathbf{Y} \in \mathrm{V}(N)$ is F-related to $\mathbf{X} \in \mathrm{V}(M)$ then

$$F^* \circ i_{\mathbf{Y}} = i_{\mathbf{X}} \circ F^*.$$

In particular, if F is a diffeomorphism,

$$F^* \circ i_{F_*(\mathbf{X})} = i_{\mathbf{X}} \circ F^*.$$

([2], Example IV. 1.8.).

5.7. Definition. Let $\mathbf{X}$ be an element of $\mathrm{V}(M)$. The map

$$L_{\mathbf{X}} := i_{\mathbf{X}} \circ \mathrm{d} + \mathrm{d} \circ i_{\mathbf{X}} : \mathbf{A}^r(M) \longrightarrow \mathbf{A}^r(M)$$

is called the *Lie differentiation* with respect to $\mathbf{X}$.

Proposition. The Lie differentiation with respect to $\mathbf{X} \in \mathrm{V}(M)$ has the properties
(i) it is $\mathbb{R}$-linear,
(ii) $L_{\mathbf{X}}(\beta \wedge \alpha) = (L_{\mathbf{X}}\beta) \wedge \alpha + \beta \wedge L_{\mathbf{X}}\alpha \qquad (\beta \in \mathbf{A}^p(M),\ \alpha \in \mathbf{A}^r(M))$,
(iii) $L_{\mathbf{X}}(f) = \mathbf{X}(f) \qquad (f \in \mathrm{D}(M))$.

Remark. The local representative of $L_{\mathbf{X}}\alpha$ expressed by the local representative of $\mathbf{X}$ and α is rather complicated. An exception is the important case when α is an m-form; its local representative in the chart (U, u) is

$$\alpha|_U = (\bar{\alpha} \circ u)\mathrm{d}u^1 \wedge \mathrm{d}u^2 \wedge \ldots \wedge \mathrm{d}u^m$$

where $\bar{\alpha}: \operatorname{Ran} u \longrightarrow \mathbb{R}$ is a smooth function. If

$$\mathbf{X}|_U = (X^k \circ u)\frac{\partial}{\partial u^k}$$

then

$$(L_{\mathbf{X}}\alpha)|_U = ((X^k \partial_k \bar{\alpha} + \bar{\alpha}\partial_k X^k) \circ u)\mathrm{d}u^1 \wedge \mathrm{d}u^2 \wedge \ldots \wedge \mathrm{d}u^m.$$

5.8. Proposition. Let N be another manifold and let $F: M \longrightarrow N$ be a smooth map. If $\mathbf{Y} \in \mathrm{V}(N)$ is F-related to $\mathbf{X} \in \mathrm{V}(M)$ then

$$F^* \circ L_{\mathbf{Y}} = L_{\mathbf{X}} \circ F^*.$$

In particular, if F is a diffeomorphism,

$$F^* \circ L_{F_*(\mathbf{X})} = L_{\mathbf{X}} \circ F^*.$$

5.9. Proposition. Let $\mathbf{X}, \mathbf{Y} \in \mathrm{V}(M)$. Then
(i) $i_{[\mathbf{X},\mathbf{Y}]} = L_{\mathbf{X}} \circ i_{\mathbf{Y}} - i_{\mathbf{Y}} \circ L_{\mathbf{X}}$
(ii) $L_{[\mathbf{X},\mathbf{Y}]} = L_{\mathbf{X}} \circ L_{\mathbf{Y}} - L_{\mathbf{Y}} \circ L_{\mathbf{X}}$.
([2], Proposition IV. 3.4.).

5.10. It is evident how the differential forms of type $\mathbf{A}$ are defined for a one dimensional real vector space $\mathbf{A}$.

On the base of 4.15. the exterior derivative of an r-form of type $\mathbf{A}$ can be defined and it is an $(r+1)$-form of type $\mathbf{A}$. Similarly, the contraction of an $(r+1)$-form of type $\mathbf{A}$ by a vector field is an r-form of type $\mathbf{A}$. In the same way, if α is an r-form of type $\mathbf{A}$ and $\mathbf{X} \in \mathrm{V}(M)$, then $L_{\mathbf{X}}\alpha$ is an r-form of type $\mathbf{A}$.

6. One parameter flows

In this section M denotes a smooth manifold whose topology is of Hausdorff type.

6.1. Let A be a one dimensional real affine space over the vector space $\mathbf{A}$. If $r: A \rightarrowtail M$ is a continuously differentiable map then $\mathrm{D}r[a]: \mathbf{A} \longrightarrow T_{r(a)}(M)$ is a linear map for $a \in \mathrm{Dom}\, r$ (see 2.9. and 2.8. (iii)) and thus it can be viewed as an element of $\dfrac{T_{r(a)}(M)}{\mathbf{A}}$ (see IV. 2.7. (vi)).

Then

$$\dot{r}: A \rightarrowtail T(M)_{\mathbf{A}}, \qquad a \mapsto (r(a), \mathrm{D}r[a]) \qquad (a \in \mathrm{Dom}\, r)$$

is a continuous map.

6.2. Definition. Let $\mathbf{X}$ be a continuously differentiable vector field of cotype $\mathbf{A}$. A *solution* of the differential equation

$$\dot{x} = \mathbf{X} \circ x$$

is a continuously differentiable map $r: A \rightarrowtail M$ such that
(i) $\mathrm{Dom}\, r$ is connected,
(ii) $\mathrm{Ran}\, r \subset \mathrm{Dom}\, \mathbf{X}$,
(iii) $\dot{r}(a) = \mathbf{X}(r(a)) \quad (a \in \mathrm{Dom}\, r)$.

The range of a solution is called an *integral curve* of $\mathbf{X}$. An integral curve is *maximal* if it is not contained properly in an integral curve.

Suppose (a_0, x_0) is in $A \times \mathrm{Dom}\,\mathbf{X}$. A *solution of the initial value problem*

$$\dot{x} = \mathbf{X} \circ x, \qquad x(a_0) = x_0$$

is a map r such that

(i) r is a solution of the corresponding differential equation,

(ii) $a_0 \in \mathrm{Dom}\, r$, $r(a_0) = x_0$.

6.3. Introducing the function $\mathbf{A} \rightarrowtail M$, $\mathbf{a} \mapsto r(a_0 + \mathbf{a})$ ($\mathbf{a} \in \mathrm{Dom}\, r - a_0$) we can replace the above problem by the initial value problem

$$\dot{x} = \mathbf{X} \circ x, \qquad x(\mathbf{0}) = x_0,$$

whose solution is a continuously differentiable map $r: \mathbf{A} \rightarrowtail M$ such that $\mathbf{0} \in \mathrm{Dom}\, r$, $r(\mathbf{0}) = x_0$, $\mathrm{Ran}\, r \subset \mathrm{Dom}\,\mathbf{X}$, $\dot{r}(\mathbf{a}) = \mathbf{X}(r(\mathbf{a}))$ for $\mathbf{a} \in \mathrm{Dom}\, r$.

Moreover, if $\mathbf{b}$ is a non-zero element of $\mathbf{A}$ then $\mathbf{b} \otimes \mathbf{X}$ is a continuously differentiable vector field (of type $\mathbb{R}$) and we easily reduce the above initial value problem to a similar initial value problem posed for a function $\mathbb{R} \rightarrowtail M$.

Since the local form of the differential equations in question is an ordinary differential equation in $\mathbb{R}^{\dim M}$, we have the following results which we formulate for global smooth vector fields; similar assertions are true for not necessarily global and only s times continuously differentiable vector fields as well ($s \in \mathbb{N}$).

6.4. Definition. A *one parameter flow* on M is a pair (S, F) where

(i) S is an open subset of $\mathbb{R} \times M$ containing $\{0\} \times M$,

(ii) $F: S \longrightarrow M$, $(t, x) \mapsto F_t(x)$ is a smooth map

such that for all $x \in M$

1) $I_x := \{t \in \mathbb{R} : (t, x) \in S\}$ is a (necessarily open) interval,

2) $F_0(x) = x$,

3) if (t, x), $(s+t, x)$ and $(s, F_t(x))$ are in S then

$$F_{s+t}(x) = F_s(F_t(x)).$$

6.5. Proposition. Let $\mathbf{X}$ be a global vector field on M. Then there is a unique one parameter flow $(S(\mathbf{X}), \mathrm{Exp}\,\mathbf{X})$ such that $t \mapsto (\mathrm{Exp}\,\mathbf{X})_t(x_0)$ is the unique maximal solution of the initial value problem

$$\dot{x} = \mathbf{X} \circ x, \quad x(0) = x_0,$$

for all $x_0 \in M$.

Conversely, if (S, F) is a one parameter flow then

$$\mathbf{X}(x) := \left.\frac{\mathrm{d}F_t(x)}{\mathrm{d}t}\right|_{t=0} \qquad (x \in M)$$

(the right hand side stands for the derivative at zero of the map $\mathbb{R} \rightarrowtail M$, $t \mapsto F_t(x)$, see 6.1.) is a vector field such that $S \subset S(\mathbf{X})$ and $F = (\mathrm{Exp}\,\mathbf{X})|_S$ ([2], Corollary V. 1.3.).

6.6. Remarks. (i) We have for all $s, t \in \mathbb{R}$

$$(\text{Exp } s\mathbf{X})_t = (\text{Exp } \mathbf{X})_{st}.$$

(ii) Let (S, F) be a one parameter flow. Put

$$U_t := \{x \in M \,\vdots\, (t, x) \in S\} \qquad (t \in \mathbb{R}).$$

U_t is an open subset of M,

$$U_t \subset U_s \quad \text{if} \quad t \geq s \geq 0,$$

$$U_t \subset U_s \quad \text{if} \quad t \leq s \leq 0.$$

Define $t_+ := \sup\{t > 0 \,\vdots\, U_t \neq \emptyset\}$,

$t_- := \inf\{t < 0 \,\vdots\, U_t \neq \emptyset\}$.

It can be shown that for $t \in]t_-, t_+[$, $F_t : U_t \longrightarrow U_{-t}$, $x \mapsto F_t(x)$ is a diffeomorphism, and $F_t^{-1} = F_{-t}$.

(ii) We can distinguish two important special types of one parameter flows.

1) There is a positive real number T such that $]-T, T[\times M \subset S$. Then $F_t : M \longrightarrow M$ is a diffeomorphism for all $t \in]-T, T[$ and

$$F_{s+t} = F_s \circ F_t \qquad (s, t, s+t \in]-T, T[).$$

2) $S = \mathbb{R} \times M$. Then F_t is a diffeomorphism of M for all $t \in \mathbb{R}$ and the above equality holds for all $s, t \in \mathbb{R}$.

(iii) To show that there is a one parameter flow which has none of the above properties and that there is a one parameter flow which has the first property but has not the second one and that there is a one parameter flow which has the second property, we give the following examples:

0) $M := \mathbb{R}^+$, $S := \{(t, x) \in \mathbb{R} \times M \,\vdots\, x^2 > |t|\}$, $F_t(x) := \sqrt{x^2 - t}$;

1) $M := \mathbb{R}^+ \setminus]0, 1]$, S and F are given by the previous formulae;

2) $M := \mathbb{R}$, $F_t(x) := e^t x$, $((t, x) \in \mathbb{R} \times \mathbb{R})$.

(iv) A vector field is called *semi-complete* or *complete* if the corresponding one parameter flow has property 1 or 2 given in (ii), respectively.

6.7. Definition. A *local Lie representation of* $\mathbb{R}$ *on* M is a pair (T, R) where

(i) T is a positive real number,

(ii) $R :]-T, T[\longrightarrow \text{Diff}(M)$, $t \mapsto R_t$ is a map

such that

1) $R_{s+t} = R_s \circ R_t$, $\quad (s, t, s+t \in]-T, T[)$,

2) $]-T, T[\times M \longrightarrow M$, $(t, x) \mapsto R_t(x)$ is smooth.

Remark. A local Lie representation of $\mathbb{R}$ on M is a one parameter flow of first type given in 6.6. (ii). Hence there is a unique vector field $\mathbf{X}$ such that $R_t = (\text{Exp } \mathbf{X})_t$ for $t \in]-T, T[$. We call $\mathbf{X}$ the *infinitesimal generator* of R. Observe that $(\text{Exp } \mathbf{X})_t$ can exist even for $t \notin]-T, T[$ but it is not necessarily a diffeomorphism of M, it can be a little diffeomorphism.

6.8. The following results can be formulated for arbitrary one parameter flows in a suitable way; however, they are simpler for local Lie representations of $\mathbb{R}$ and we need them only in this special case.

6.9. Proposition. Let (T, R) be a local Lie representation of $\mathbb{R}$ and let $\mathbf{X}$ be its infinitesimal generator. Then

$$L_{\mathbf{X}}\alpha = \lim_{t\to 0}\frac{R_t^*\alpha-\alpha}{t} = \lim_{t\to 0}\frac{\alpha-R_{-t}^*\alpha}{t}$$

for all r-forms α $(r=0, 1, \dots)$; the limits are to be taken pointwise. ([2], Proposition V. 2.5.).

6.10. Proposition. Use the previous notations. Then α is invariant for R, i.e. $R_t^*\alpha=\alpha$ $(t\in]-T, T[)$ if and only if $L_{\mathbf{X}}\alpha=0$ ([2], Lemma VIII. 2.1.).

6.11. Proposition. Let (T, R) be a local Lie representation of $\mathbb{R}$ and let $\mathbf{X}$ be its infinitesimal generator. Then

$$[\mathbf{X}, \mathbf{Y}] = \lim_{t\to 0}\frac{(R_{-t})_*\mathbf{Y}-\mathbf{Y}}{t} = \lim_{t\to 0}\frac{\mathbf{Y}-(R_t)_*\mathbf{Y}}{t}$$

for all global vector fields $\mathbf{Y}$; the limits are to be taken pointwise ([2], Proposition V. 2.6.).

6.12. Proposition. Let (T, R) and (T, S) be local Lie representations of $\mathbb{R}$ on the manifolds M and N, and suppose $\mathbf{X}$ and $\mathbf{Y}$ are their infinitesimal generators, respectively. Let $F: M \longrightarrow N$ be a smooth map. The equality

$$F\circ R_t=S_t\circ F \qquad (t\in]-T,T[)$$

holds if and only if $\mathbf{Y}$ is F-related to $\mathbf{X}$, i.e.

$$\mathrm{D}F\circ\mathbf{X}=\mathbf{Y}\circ F.$$

6.13. Proposition. Let (T, R) and (T, S) be local Lie representations of $\mathbb{R}$ on the same manifold M, and suppose $\mathbf{X}$ and $\mathbf{Y}$ are their infinitesimal generators, respectively. The local Lie representations of $\mathbb{R}$ commute, i.e.

$$R_t\circ S_s=S_s\circ R_t \qquad (s, t\in]-T, T[)$$

if and only if the infinitesimal generators commute, i.e.

$$[\mathbf{X}, \mathbf{Y}]=0$$

([2], Proposition V. 2.7.).

6.14. Proposition. Let $\mathbf{X}$ be a vector field on M. If r is a solution of the differential equation $\dot{x}=\mathbf{X}\circ x$ then $f\circ r:\mathbb{R}\rightarrowtail\mathbb{R}$ is a smooth function for all $f\in\mathrm{D}(M)$ and

$$\mathrm{D}(f\circ r)=\mathbf{X}(f)\circ r.$$

If $\mathbf{X}$ is the infinitesimal generator of the local Lie representation (T, R) of $\mathbb{R}$ then the above equality implies

$$\frac{\mathrm{d}f(R_t(x))}{\mathrm{d}t} = (\mathbf{X}(f))(R_t(x)) \qquad (t \in]-T, T[, x \in M).$$

6.15. Proposition. Let N be another manifold and suppose $F: M \longrightarrow N$ is a diffeomorphism. If C is an integral curve of $\mathbf{X} \in \mathrm{V}(M)$ then $F[C]$ is an integral curve of $F_* \mathbf{X} \in \mathrm{V}(N)$.

Proof. It suffices to show that if r is a solution of the differential equation $\dot{x} = \mathbf{X} \circ x$ then $F \circ r$ is a solution of $\dot{y} = (F_* \mathbf{X}) \circ y$, and this is quite evident.

7. Foliations

M denotes a manifold of Hausdorff type.

7.1. Let $D_x^k(M)$ denote the set of k dimensional linear subspaces of $T_x(M)$ $(x \in M, k = 0, 1, \dots m)$. We put

$$D^k(M) := \bigcup_{x \in M} \{x\} \times D_x^k(M)$$

$$\Gamma : D^k(M) \longrightarrow M, \qquad (x, S) \mapsto x.$$

Definition. A *k dimensional tangent subspace field* is a map $E: M \rightarrowtail D^k(M)$ for which $\Gamma \circ E \subset \mathrm{id}_M$ holds, and which is smooth in the sense that for all charts (U, u) there exist vector fields $\mathbf{X}_1, \dots, \mathbf{X}_k$ defined on U such that $\mathbf{X}_1[x], \dots, \mathbf{X}_k[x]$ span $E[x]$. (Here we use a notation similar to that introduced in Remark 3.4.(i).)

It is *involutive* if for all vector fields $\mathbf{X}$ and $\mathbf{Y}$ having common domain U and $\mathbf{X}[x], \mathbf{Y}[x] \in E[x]$ $(x \in U)$, it follows that $[\mathbf{X}, \mathbf{Y}][x] \in E[x]$ $(x \in U)$.

7.2. Definition. A submanifold Q of M is called an *integral manifold* of a k dimensional tangent subspace field E if $Q \subset \mathrm{Dom}\, E$ and

$$T_x(Q) = E[x] \quad \text{for all} \quad x \in Q.$$

An integral manifold of E is *maximal* if it is connected and is not contained properly in another integral manifold.

E is called a k dimensional *foliation* if *for all* $x \in \mathrm{Dom}\, E$ there exists an integral manifold containing x.

7.3. Proposition. If E is a foliation, for all $x \in \mathrm{Dom}\, E$ there exists a unique maximal integral manifold of E containing x. ([11], Theorem II. 5.5.).

7.4. Proposition. A tangent subspace field E is a foliation if and only if it is involutive ([11]), Theorem II. 5.4.).

7.5. Definition. Two foliations D and K are *transversal* if $D[x]$ and $K[x]$ are complementary subspaces, i.e. $D[x] \cap K[x] = \{0\}$ and $D[x]$ and $K[x]$ span $T_x(M)$ for all $x \in \operatorname{Dom} D \cap \operatorname{Dom} K$.

7.6. Proposition. Let D and K be transversal foliations of dimension d and k, respectively, given on the open subset G of M. Then ([2], Lemma V. 3.7.) every point of G has a neighbourhood U such that a chart (U, u) can be realized with the properties

$$u = (u_D, u_K) : U \longrightarrow \mathbb{R}^d \times \mathbb{R}^k,$$

$$\{u_D^{-1}[\xi] : \xi \in \operatorname{Ran} u_D\}$$

is the set of maximal integral manifolds of $K|_U$,

$$\{u_K^{-1}[\eta] : \eta \in \operatorname{Ran} u_K\}$$

is the set of maximal integral manifolds of $D|_U$.

If ϕ is an integral manifold of K meeting U then $(\phi \cap U, u_K|_\phi)$ is a chart of ϕ.

7.7. Let D and K be transversal foliations given on an open subset G of M. Assume the notation I_D and I_K for the set of maximal integral manifolds of D and K, respectively. Suppose that an element of I_D and an element of I_K intersect each other at most in a single point.

Let $\phi \in I_K$ and let H_ϕ be an open subset of ϕ. Then we put

$$\hat{H}_\phi := \{Q \in I_D : Q \cap \phi \in H_\phi\}$$

and we equip I_D with the topology generated by such subsets as ϕ runs through I_K and H_ϕ runs through the open subsets of ϕ.

Let (V_ϕ, v_ϕ) be a chart in $\phi \in I_K$. Then we define a chart $(\hat{V}_\phi, \hat{v}_\phi)$ in I_D by

$$\hat{v}_\phi(Q) := v_\phi(\phi \cap Q) \qquad (Q \in \hat{V}_\phi).$$

Proposition. The charts given above form an atlas of I_D as ϕ runs through I_K and (V_ϕ, v_ϕ) runs through the charts of ϕ. This atlas turns I_D into a manifold whose dimension equals the dimension of K.

Proof. Let $M_D(x)$ denote the maximal integral manifold of D passing through $x \in G$. For $\phi, \psi \in I_K$ assume the notation

$$Z_{\phi\psi} : \phi \rightarrowtail \psi, \qquad x \mapsto \psi \cap M_D(x).$$

If $(\hat{V}_\phi, \hat{v}_\phi)$ and $(\hat{W}_\phi, \hat{w}_\phi)$ are two charts in I_D then

$$\hat{v}_\phi \circ \hat{w}_\psi^{-1} = v_\phi \circ Z_{\phi\psi} \circ w_\psi^{-1},$$

hence we prove compatibility of the charts if we prove that $Z_{\phi\psi}$ is smooth.

If ϕ and ψ are such that for a chart (U, u) described in 7.6. both intersect U in a non-void set then the restriction of $Z_{\phi\psi}$ to a convenient open subset is evidently smooth.

In general, since $M_D(x)$ is connected, we can give an arc γ connecting x and $\psi \cap M_D(x)$. Ran γ is a compact set, hence it can be covered with finitely many charts described in 7.6. As a consequence, the restriction of $Z_{\phi\psi}$ to a convenient open subset can be realized as a composition of finitely many evidently smooth map.

Obviously, a similar assertion is true for I_K.

7.8. Proposition. Let D and K be as before. Then

$$G \longrightarrow I_D, \qquad x \mapsto M_D(x),$$

$$G \longrightarrow I_K, \qquad x \mapsto M_K(x)$$

are smooth maps, and

$$S: G \longrightarrow I_D \times I_K, \qquad x \mapsto (M_D(x), M_K(x))$$

is a little diffeormorphism ($M_D(x)$ and $M_K(x)$ are the maximal integral manifolds of D and K passing through x).

Proof. We prove only the assertion concerning S which is evidently an injective map.

Let (U, u) be a chart in G described in 7.6. and let $(\hat{V}_\phi, \hat{v}_\phi)$ and $(\hat{Z}_Q, \hat{z}_Q)$ be charts in I_D and I_K, respectively, such that $V_\phi \cap U \neq \emptyset, Z_Q \cap U \neq \emptyset$ and $v_\phi = u|_{V_\phi}, z_Q = u|_{Z_Q}$. Then

$$(v_\phi, z_Q) \circ S \circ (u_D, u_K)^{-1} \subset \mathrm{id}_{\mathbb{R}^d \times \mathbb{R}^k}.$$

Since such charts form atlases of the manifolds in question, S is clearly smooth. Its inverse is smooth by reason of similar arguments.

7.9. Proposition. Let K and K' be two foliations transversal to the foliation D. Then the topologies and the smooth structures on I_D defined by K and K' according to 7.7. are the same.

Proof. Let I_D denote the set I_D equipped with the smooth structure defined by K and assume the notation I'_D for the set I_D equipped with the smooth structure defined by K'. If S and S' are the little diffeomorphisms corresponding to K and K', respectively, then $S' \circ S^{-1}: I_D \times I_K \longrightarrow I'_D \times I_{K'}$ is a little diffeomorphism which has the form $(Q, \phi) \mapsto (Q, F(Q, \phi))$.

Fix a $\phi \in I_K$. Then

$$I_D \rightarrowtail I'_D, \qquad Q \mapsto (Q, \phi) \mapsto (Q, F(Q, \phi)) \mapsto Q$$

is evidently a little diffeomorphism.

7.10. Proposition. Let D and K be transversal foliations given on the open subset G of M. Then the tangent space of I_D at $M_D(x)$ can be identified with $K[x]$. We consider an element $\mathbf{v}$ of $K[x]$ as a tangent vector (i.e. a derivation) of I_D at $M_D(x)$ by the formula

$$f \mapsto \mathbf{v}(f \circ M_D) \qquad (f \in C^1(M_D(x)))$$

where x is an element of G and $M_D: M \rightarrowtail I_D$ is the smooth map given in 7.8.

Proof. Evidently, the given correspondence $K[x] \longrightarrow T_{M_D(x)}(I_D)$ is linear. Now we have to show only that it is injective, for $K[x]$ and $T_{M_D(x)}(I_D)$ are of equal dimension.

Suppose $\mathbf{v} \in K[x]$ and $\mathbf{v}(f \circ M_D) = 0$ for all $f \in C^1(M_D(x))$. Take a chart (U, u) of M having the form given in 7.6., such that $x \in U$. Then $(M_K(x) \cap U, u_K|_{M_K(x)})$ is a chart of $M_K(x)$ giving rise to a chart of I_D according to 7.7. and

$$\widehat{u_K|_{M_K(x)}} \circ M_D = u_K(M_K(x) \cap M_D(\,\cdot\,)).$$

We easily find that if $\mathbf{v}$ applied to the components of the map above results in zero then $\mathbf{v}$ is necessarily zero.

Remarks. (i) An element $\mathbf{v}$ of $T_x(M)$ $(x \in G)$ is decomposed uniquely in the form $\mathbf{v}\|_D + \mathbf{v}\|_K$ where $\mathbf{v}\|_D \in K[x]$, $\mathbf{v}\|_K \in D[x]$ ($\mathbf{v}\|_D$ is the "projection of $\mathbf{v}$ onto $K[x]$ parallel to $D[x]$"). Making the identification $K[x] = T_{M_D(x)}(I_D)$, we obtain

$$\mathrm{D}M_D[x]\mathbf{v} = \mathbf{v}\|_D .$$

(ii) The identification $K[x] = T_{M_D(x)}(I_D)$ is not unique, since it may happen that $x \neq y$, $K[x] \neq K[y]$ but $M_D(x) = M_D(y)$.

(iii) Similarly, the tangent space of I_K at $M_K(x)$ can be identified with $D[x]$.

(iv) Though the identifications above are not unique, the tangent space of $I_D \times I_K$ at $(M_D(x), M_K(x))$ can be uniquely identified with $K[x] \times D[x]$ $(x \in G)$.

7.11. The previous characterization of tangent spaces is sometimes unconvenient because the tangent space of I_D at $M_D(x)$ involves the other foliation K. We have seen that the smooth structure of I_D is independent of the transversal foliation. Hence we should like to have a characterization of tangent spaces of I_D only in terms of D.

7.12. Definition. A foliation is called *secable* if there is a foliation transversal to it.

Proposition. Let D be a secable foliation. Then the tangent space of I_D at $M_D(x)$ can be identified with $T_x(M)/D[x]$ by considering an element $\mathbf{r}$ of this quotient vector space as a tangent vector of I_D at $M_D(x)$ (i.e. as a derivation at $M_D(x)$) $(x \in \operatorname{Dom} D)$ with the aid of the formula

$$f \mapsto \mathbf{v}(f \circ M_D) \qquad (f \in C^1(M_D(x)))$$

$$(\mathbf{v} \in T_x(M),\ \Gamma_x(\mathbf{v}) = \mathbf{r})$$

where $\Gamma_x : T_x(M) \longrightarrow T_x(M)/D[x]$ is the canonical surjection.

Proof. First of all, the given correspondence is well defined because if $\mathbf{v} \in D[x] \subset T_x(M)$ then $\mathbf{v}(f \circ M_D) = 0$ for all $f \in C^1(M_D(x))$. This can be seen by the use of charts having the form given in 7.6. Further we can follow a line similar to that of the proof of Proposition 7.10.

7.13. Proposition. Let D be a secable foliation. Suppose N is a manifold and $F: M \longrightarrow N$ is a smooth map constant on the maximal integral manifolds of D. Then there is a unique smooth map $F_D: I_D \longrightarrow N$ such that $F = F_D \circ M_D$.

As a consequence, if $R: M \longrightarrow M$ is a smooth map that maps maximal integral manifolds of D into maximal integral manifolds, then there is a unique smooth map $R_D: I_D \longrightarrow I_D$ such that $R_D \circ M_D = M_D \circ R$. If $Z: M \longrightarrow M$ is another such smooth map then $(Z \circ R)_D = Z_D \circ R_D$.

8. Integration

M is an m dimensional oriented manifold whose topology is of Hausdorff type and satisfies the second axiom of countability.

A is a one dimensional oriented real vector space.

8.1. Let λ denote the Lebesque measure on $\mathbb{R}^m$. If $f: \mathbb{R}^m \longrightarrow \mathbf{A}$ is a continuous function which is non-negative or has compact support then we define the product of f and λ as the **A**-valued measure given by

$$(f\lambda)(E) := \begin{cases} \int_E f \, d\lambda & \text{if } f \text{ is Lebesgue integrable on } E, \\ \infty & \text{if } f \text{ is not Lebesgue integrable on } E \end{cases}$$

(E is a Borel subset of $\mathbb{R}^m$).

8.2. Since M is oriented, $T_x(M)$ and the one dimensional vector space $\overset{m}{\wedge} \dfrac{T_x(M)}{\mathbf{A}}$ are oriented as well. Hence it makes sense to say that an m-form of type $\overset{m}{\otimes} \mathbf{A}$ takes non-negative values.

Definition. An m-form of type $\overset{m}{\otimes} \mathbf{A}$ which is non-negative or has compact support is called a *measure form* of type $\overset{m}{\otimes} \mathbf{A}$ on M. A *volume form* is a positive measure form.

8.3. A measure form μ of type $\overset{m}{\otimes} \mathbf{A}$ induces an $\overset{m}{\otimes} \mathbf{A}$-valued Radon measure on M, denoted also by μ, in the following way.

Let (U, u) be a positively oriented chart in M and suppose

$$\mu|_U = (\bar{\mu} \circ u) \otimes (du^1 \wedge du^2 \wedge \ldots \wedge du^m).$$

If B is an open subset of U we put

$$\mu(B) := (\bar{\mu}\lambda)(u[B]),$$

or in a more familiar form,

$$\int_B \mu := \int_{u[B]} \bar{\mu}\, d\lambda$$

where we allow that the integral be infinite.

One can easily show that the definition is independent of the positively oriented chart (U, u). Furthermore, since M is second countable, there exists an atlas of M consisting of denumerable charts, hence we can fortwith extend the definition of the measure for all open subsets of M and thus for all Borel subsets as well.

8.4. Proposition. Let N be another oriented manifold and suppose $F: M \longrightarrow N$ is an orientation preserving diffeomorphism. If μ is a measure form on N then

$$\int_{F[B]} \mu = \int_B F^* \mu$$

for all Borel subsets B of M ([2], Corollary IV. 4.3.).

8.5. Proposition. Let $F: M \longrightarrow M$ be an orientation preserving diffeomorphism. The measure induced by the measure form μ is invariant under F if and only if $F^*\mu = \mu$.

(A measure is invariant under F if B and $\overset{-1}{F}(B)$ has the same measure for all Borel subsets B.)

8.6. Proposition. Let (T, R) be a local Lie representation of $\mathbb{R}$ on M with infinitesimal generator $\mathbf{X}$. The measure induced by a measure form μ is invariant under R_t for all $t \in]-T, T[$ if and only if $L_{\mathbf{X}}\mu = 0$ ([2], Proposition VIII. 2.2.).

9. Covariant derivations

9.1. Definition. A *covariant derivation* (or *connection*) on a manifold M is a map

$$\nabla : \mathrm{V}(M) \times \mathrm{T}^r\mathrm{F}(M) \longrightarrow \mathrm{T}^r\mathrm{F}(M), \qquad (\mathbf{X}, T) \mapsto \nabla_{\mathbf{X}} T$$

$(r = 0, 1, \ldots,)$ such that

(i) $\mathbf{X} \mapsto \nabla_{\mathbf{X}}$ is $\mathrm{D}(M)$-linear,

and for all $\mathbf{X} \in \mathrm{V}(M)$

(ii) $\nabla_{\mathbf{X}} : \mathrm{T}^r\mathrm{F}(M) \longrightarrow \mathrm{T}^r\mathrm{F}(M)$ is $\mathbb{R}$-linear,

(iii) $\nabla_{\mathbf{X}}(T \otimes S) = (\nabla_{\mathbf{X}} T) \otimes S + T \otimes \nabla_{\mathbf{X}} S$ for all tensor fields T and S,

(iv) $\nabla_{\mathbf{X}} f = \mathbf{X}(f)$ for $f \in \mathrm{T}^0\mathrm{F}(M) = \mathrm{D}(M)$.

Remarks. (i) According to (iii) of the definition above, it suffices to give a covariant derivation on vector fields, i.e. a map $(\mathbf{X}, \mathbf{Y}) \mapsto \nabla_{\mathbf{X}} \mathbf{Y}$.

(ii) If $\mathbf{X}$ is a vector field of cotype $\mathbf{A}$ and $\mathbf{Y}$ is a vector field of cotype $\mathbf{B}$ where $\mathbf{A}$ and $\mathbf{B}$ are one dimensional real vector spaces, then we put

$$\nabla_{\mathbf{X}} \mathbf{Y} := \frac{\nabla_{\mathbf{a} \otimes \mathbf{X}} (\mathbf{b} \otimes \mathbf{Y})}{\mathbf{a} \otimes \mathbf{b}} \qquad (\mathbf{a} \in \mathbf{A} \setminus \{0\},\ \mathbf{b} \in \mathbf{B} \setminus \{0\})$$

which is a vector field of cotype $\mathbf{A} \otimes \mathbf{B}$.

9.2. Proposition. Let M be an affine space over $\mathbf{M}$. Then vector fields can be identified with smooth maps $M \longrightarrow \mathbf{M}$ (see 3.4. (iii)) and

$$(\nabla_{\mathbf{X}}\mathbf{Y})[x] := D\mathbf{Y}[x]\mathbf{X}[x] \qquad (x \in M,\ \mathbf{X}, \mathbf{Y} \in \mathrm{V}(M))$$

defines a covariant derivation.

Definition. This covariant derivation is called the *canonical covariant derivation* on the affine space M.

9.3. (i) Let (U, u) be a chart in the manifold M and suppose ∇ is a covariant derivation on M. One easily shows that $\nabla|_U$ defined by

$$(\nabla|_U)_{\mathbf{X}|_U}\mathbf{Y}|_U := (\nabla_{\mathbf{X}}\mathbf{Y})|_U \qquad (\mathbf{X}, \mathbf{Y} \in \mathrm{V}(M))$$

is a covariant derivation on U. Then

$$(\nabla|_U)_{\frac{\partial}{\partial u^i}} \frac{\partial}{\partial u^j} = (\Gamma^k_{ij} \circ u)\frac{\partial}{\partial u^k};$$

Γ^k_{ij}: Ran $u \longrightarrow \mathbb{R}$ $(i, j, k = 1, \ldots, \dim M)$ are called the *Cristoffel symbols* of ∇ corresponding to (U, u).

(ii) If

$$\mathbf{X}|_U = (X^k \circ u)\frac{\partial}{\partial u^k}, \qquad \mathbf{Y}|_U = (Y^k \circ u)\frac{\partial}{\partial u^k}$$

then

$$(\nabla_{\mathbf{X}}\mathbf{Y})|_U = [[X^i(\partial_i Y^k + Y^j \Gamma^k_{ij})] \circ u]\frac{\partial}{\partial u^k}.$$

(iii) If (U', u') is another chart, $U \cap U' \neq \emptyset$ and Γ'^s_{pr} $(s, p, r = 1, \ldots, \dim M)$ are the Christoffel symbols of ∇ corresponding to (U', u') then

$$\Gamma'^s_{pr} = \frac{\partial u^i}{\partial u'^p}\frac{\partial u^j}{\partial u'^r}\frac{\partial u'^s}{\partial u^k}\Gamma^k_{ij} + \frac{\partial^2 u^j}{\partial u'^p \partial u'^r}\frac{\partial u'^s}{\partial u^j}.$$

9.4. Definition. Let C be a one dimensional submanifold of M and let i: $C \longrightarrow M$ denote the inclusion map. A smooth map

$$\mathbf{X}: C \longrightarrow T(M) \quad \text{for which} \quad \Pi_M \circ \mathbf{X} = i \quad \text{holds}$$

is called a *vector field along* C.

The set of vector fields along C is denoted by $\mathrm{V}(C, M)$. ∎

Evidently, a vector field on C is a vector field along C, in other words, $\mathrm{V}(C) \subset \mathrm{V}(C, M)$.

$\mathrm{V}(C, M)$ is a module over the smooth functions $C \longrightarrow \mathbb{R}$.

9.5. Proposition.

(i) Let $\mathbf{X} \in \mathrm{V}(C, M)$. Then there exists an $\hat{\mathbf{X}} \in \mathrm{V}(M)$ such that $\mathbf{X} = \hat{\mathbf{X}}|_C$.

(ii) Let $\dot{\mathbf{C}}$ be a nowhere zero element of $\mathrm{V}(C)$. Then

$$\nabla_{\dot{\mathbf{C}}}\mathbf{X} := (\nabla_{\hat{\dot{\mathbf{C}}}}\hat{\mathbf{X}})|_C \qquad (\mathbf{X} \in \mathrm{V}(C, M))$$

is a well defined element of V(C, M), i.e. it is independent of the extensions $\hat{\mathbf{C}}$ and $\hat{\mathbf{X}}$ of $\dot{\mathbf{C}}$ and $\mathbf{X}$, ([4], Lemma I. 5.1.).

9.6. Definition. A *parametrization* of the one dimensional submanifold C is a little diffeomorphism $p:\mathbb{R}\rightarrowtail C$ (i.e. a parametrization is the inverse of a coordinatization). A smooth map

$$Z:\mathbb{R}\rightarrowtail T(M) \qquad \text{for which} \qquad \Pi_M \circ Z = p$$

holds is called a *parametrized vector field along* C, and we put

$$\nabla_{\dot{p}} Z := [\nabla_{\dot{p}\circ p^{-1}}(Z\circ p^{-1})]\circ p$$

which is a parametrized vector field along C.

Remark. More generally, a parametrization of C is a little diffeomorphism from a one dimensional affine space I over $\mathbf{I}$ into C. If Z is a parametrized vector field of cotype $\mathbf{I}$ along C then $\nabla_{\dot{p}} Z$ can be defined by combining Definition 9.6. and Remark 9.1. (ii) and it will be a parametrized vector field of cotype $\mathbf{I}\otimes\mathbf{I}$ along C.

9.7. Definition. A one dimensional submanifold C is called *geodesic* with respect to the covariant derivation ∇ if $\nabla_{\dot{\mathbf{C}}}\dot{\mathbf{C}}=0$ for some nowhere zero $\dot{\mathbf{C}}\in$ V(C).

9.8. Let p be a parametrization of C and let (U, u) be a chart in M, (Ran p)$\cap U \neq \emptyset$. Then introducing

$$c := u\circ i\circ p, \quad c^k := \mathrm{pr}^k\circ c, \quad J := p^{-1}[U]$$

we get that

$$\nabla_{\dot{p}} Z|_J = [\dot{Z}^k + (\Gamma^k_{ij}\circ c)Z^i Z^j]\frac{\partial}{\partial u^k}\circ p$$

where Z is a parametrized vector field along C and

$$Z|_J = Z^k\frac{\partial}{\partial u^k}\circ p\,.$$

In particular, C is geodesic if and only if

$$\ddot{c}^k + (\Gamma^k_{ij}\circ c)\dot{c}^i\dot{c}^j = 0 \qquad (k=1, \ldots, \dim M)$$

for some parametrization.

9.9. Suppose M is an affine space and ∇ is the canonical covariant derivation on M, and $p:I\rightarrowtail C$ is a parametrization of the one dimensional submanifold C, where I is a one dimensional affine space over $\mathbf{I}$. Then by VI. 2.5. we have

$$\dot{p}:I\rightarrowtail\frac{\mathbf{M}}{\mathbf{I}}, \qquad \ddot{p}:I\rightarrowtail\frac{\mathbf{M}}{\mathbf{I}\otimes\mathbf{I}}.$$

If $Z: I \rightarrowtail \frac{\mathbf{M}}{\mathbf{I}}$ is a parametrized vector field of cotype $\mathbf{I}$ along C then

$$\nabla_{\dot{p}} Z = \dot{Z} : I \rightarrowtail \frac{\mathbf{M}}{\mathbf{I}\otimes\mathbf{I}},$$

in particular, $\nabla_{\dot{p}}\dot{p} = \ddot{p}$ (see 3.4. (iii)).

Consider now M as a manifold; then (see 6.1.)

$$\dot{p}: I \rightarrowtail M \times \frac{\mathbf{M}}{\mathbf{I}}, \qquad t \longrightarrow (p(t), \dot{p}(t)),$$

$$\ddot{p}: I \rightarrowtail \left(M \times \frac{\mathbf{M}}{\mathbf{I}}\right) \times \left(\frac{\mathbf{M}}{\mathbf{I}} \times \frac{\mathbf{M}}{\mathbf{I}\otimes\mathbf{I}}\right), \qquad t \longrightarrow (p(t), \dot{p}(t), \dot{p}(t), \ddot{p}(t)),$$

$$\nabla_{\dot{p}}\dot{p}: I \rightarrowtail M \times \frac{\mathbf{M}}{\mathbf{I}\otimes\mathbf{I}}, \qquad t \longrightarrow (p(t), \ddot{p}(t)).$$

Hence we see that $\ddot{p}$ is the non trivial part of both $\ddot{p}$ and $\nabla_{\dot{p}}\dot{p}$. It is very important that $\ddot{p}$ *and* $\nabla_{\dot{p}}\dot{p}$ are essentially different objects and they coincide only if we make the usual identifications. However, if we are forced to reject these identifications—e.g. because of the use of curvilinear coordinates—we must know whether $\ddot{p}$ represents $\ddot{p}$ or $\nabla_{\dot{p}}\dot{p}$.

10. Pseudo-Riemannian manifolds

10.1. Definition. A triplet $(V, \mathbf{B}, b)$ is called a *pseudo-Riemannian (Riemannian, Einsteinian) manifold* if

(i) V is a smooth manifold,

(ii) $\mathbf{B}$ is a one dimensional real vector space,

(iii) b is a 2-contensor field of type $\mathbf{B}\otimes\mathbf{B}$ such that

$$(T_x(V), \mathbf{B}, b[x])$$

is a pseudo-Euclidean (Euclidean, Minkowskian) vector space for all $x \in V$. ∎

Note that here $b[x]$ is considered as a $\mathbf{B}\otimes\mathbf{B}$ valued bilinear map on $T_x(V)$ (see 4.15.).

In the sequel $(V, \mathbf{B}, b)$ *denotes a pseudo-Riemannian manifold*, and we assume that

the topology of V is of Hausdorff type and satisfies the second axiom of countability,

V is oriented, $\mathbf{B}$ is oriented,

$n := \dim V$.

10.2. Proposition. The module of global vector fields of cotype $\mathbf{B}\otimes\mathbf{B}$ can be identified with the module of global covector fields by

$$(\mathbf{P}|\mathbf{X}) := \frac{b(\mathbf{h}\otimes\mathbf{P}, \mathbf{X})}{\mathbf{h}}$$

where $\mathbf{P}$ is a global vector field of cotype $\mathbf{B}\otimes\mathbf{B}$, $\mathbf{X} \in \mathrm{V}(V)$ and $\mathbf{h}$ is a non-zero element of $\mathbf{B}\otimes\mathbf{B}$.

As a consequence, we can give a 2-tensor field of type $\mathbf{B}^*\otimes\mathbf{B}^*$ by

$$b^{(*)}(\mathbf{C}, \mathbf{D}) := \frac{b(\mathbf{h}\otimes\mathbf{C}, \mathbf{h}\otimes\mathbf{D})}{\mathbf{h}\otimes\mathbf{h}}$$

where $\mathbf{C}$ and $\mathbf{D}$ are global covector fields, $\mathbf{0} \neq \mathbf{h} \in \mathbf{B}\otimes\mathbf{B}$.

Similarly, the module of global vector fields of cotype $\mathbf{B}$ can be identified with the module of global covector fields of type $\mathbf{B}^*$.

10.3. Since V is oriented, $T_x(V)$ is oriented as well for all $x \in V$. Hence we can give the canonical volume form $\mu[x]$ on $T_x(V)$ which is a $\overset{n}{\otimes}$ $\mathbf{B}$ valued antisymmetric m-fold linear map on $T_x(V)$ (see V. 2.15.).

Definition. The n-form of type $\overset{n}{\otimes}$ $\mathbf{B}$, $x \mapsto (x, \mu[x])$ is called the *canonical volume form* on V and is denoted by $\sqrt[\otimes]{|b|}$. ∎

The local representative of $\sqrt[\otimes]{|b|}$ in a positively oriented chart (U, u) is

$$\sqrt[\otimes]{|\det\{(b_{ij})_{i,j=1,\ldots,n}\} \circ u|} \otimes (\mathrm{d}u^1 \wedge \mathrm{d}u^2 \wedge \ldots \wedge \mathrm{d}u^n)$$

where

$$(b_{ij} \circ u) \otimes (\mathrm{d}u^i \otimes \mathrm{d}u^j)$$

is the local representative of b; note that b_{ij}: Ran $u \longrightarrow \mathbf{B}\otimes\mathbf{B}$ (see V. 2.15.).

10.4. Proposition. On the pseudo-Riemannian manifold $(V, \mathbf{B}, b)$ there is a unique covariant derivation ∇ with the properties

(i) $\nabla_{\mathbf{X}}\mathbf{Y} - \nabla_{\mathbf{Y}}\mathbf{X} = [\mathbf{X}, \mathbf{Y}]$,

(ii) $\nabla_{\mathbf{X}} b^{(*)} = 0$

for all vector fields $\mathbf{X}$ and $\mathbf{Y}$ ([4], Theorem I. 9.1.).

Remark. A pseudo-Euclidean affine space $(V, \mathbf{B}, b)$ is evidently a pseudo-Riemannian manifold. The corresponding covariant derivation is the canonical covariant derivation on V, independently of the bilinear form b. (cf. 9.2. and 3.14. (iii)).

10.5. Proposition. If b_{ik} and b^{ik} are the local components of b and $b^{(*)}$ with respect to a chart (U, u), then the Christoffel symbols of ∇ corresponding to (U, u) are

$$\Gamma^k_{ij} = \frac{1}{2} b^{km} \otimes (\partial_i b_{jm} + \partial_j b_{mi} - \partial_m b_{ij})$$

([11], p. 174.).

10.6. Proposition. Every point of a pseudo-Riemannian manifold has a neighbourhood in which arbitrary two distinct points can be connected by a unique geodesic, i.e. for arbitrary two distinct points in the neighbourhood there is a unique connected geodesic containing the two points ([4], Theorem I. 6.2.).

10.7. Proposition. For all points x of V there is a chart (U, u) such that $x \in U$ and

(i) $\left\{\frac{\partial}{\partial u^k}[x] : k=1, \ldots, n\right\}$ is a $b[x]$-orthogonal basis in $T_x(V)$, normed to some $\mathbf{a} \in \mathbf{B}$ (see 4. 2.2.), i.e. the local components of b satisfy the relations

$$b_{ik}(u(x))=0 \quad \text{if} \quad i \neq k$$

$$b_{kk}(u(x)) \quad \text{is either} \quad \mathbf{a}\otimes\mathbf{a} \quad \text{or} \quad -\mathbf{a}\otimes\mathbf{a}$$

$$(i, k=1, \ldots, n);$$

(ii) the first order derivative of b_{ik} at $u(x)$ is zero, i.e.

$$\partial_j b_{ik}(u(x))=0,$$

or, equivalently,

$$\Gamma^k_{ij}(u(x))=0 \quad \text{for all} \quad i, j, k=1, \ldots, n.$$

([11], Theorem II. 10.5.).

10.8. The pull-back of b to a submanifold of V is not necessarily a pseudo-Riemannian cotensor field; it is, for instance, if b is a Riemannian cotensor field.

If C is a one dimensional oriented submanifold of V and $(C, \mathbf{B}, b|_C)$ is a pseudo-Riemannian manifold (notation: 4.8. (iii)) then $\sqrt[\otimes]{|b|_C|}$ is a measure form on C. If C is connected and x and y are distinct points of C then we can integrate $\sqrt[\otimes]{|b|_C|}$ on the interval of C between x and y; the result is called the *distance between x and y along C*.

If C is a geodesic, we speak of the *geodesic distance* between x and y.

10.9. Let $(M, \mathbf{I}, g)$ be an Einsteinian manifold. According to 10.1., $g[x]$ is considered as an $\mathbf{I}\otimes\mathbf{I}$-valued bilinear map on $T_x(M)$ $(x \in M)$, in other words, g is considered as a D(M)-bilinear map on V(M) taking values in D$(M, \mathbf{I}\otimes\mathbf{I})$ (see 4.15.); transferred to a D(M)-valued bilinear map on the vector fields of cotype $\mathbf{I}$, it will be denoted by $(\ \dagger\ )$. Furthermore, we remind of the notation $(\ |\ )$ introduced in 10.2.

Suppose $r: \mathbf{I} \rightarrowtail M$ is a smooth map such that $(\dot{r} \dagger \dot{r}) = -1$. Then $C := \operatorname{Ran} r$ is a one dimensional submanifold of M. Put

$$\mathbf{u} := \dot{r}, \qquad \mathbf{a} := \nabla_{\dot{r}} \dot{r}.$$

Definition. A parametrized vector field Z along C is called of *Fermi-Walker type* if

$$\nabla_{\dot{r}} Z = (\mathbf{a}|Z)\mathbf{u} - [\mathbf{u}|Z)\otimes\mathbf{a}$$

where we used a notation defined by $m\otimes[\mathbf{u}|\mathbf{Z}) := (m\otimes\mathbf{u}|\mathbf{Z}) \quad (m \in \mathbf{I}^*)$.

An evident modification leads us to the definition of *vector fields of Fermi-Walker type along C*.

If $x_1, x_2 \in C$, $\mathbf{v}_1 \in T_{x_1}(M)$, $\mathbf{v}_2 \in T_{x_2}(M)$ and there is a (necessarily unique, see below) Fermi-Walker vector field $\mathbf{X}$ along C such that $\mathbf{X}[x_1] = \mathbf{v}_1$, $\mathbf{X}[x_2] = \mathbf{v}_2$, then we say that $\mathbf{v}_2$ is the *Fermi-Walker translation* of $\mathbf{v}_1$.

Proposition. If $x_0 \in C$, $\mathbf{v}_0 \in T_{x_0}(M)$, then there is a unique $\mathbf{X} \in \mathrm{V}(C, M)$ of Fermi-Walker type such that $\mathbf{X}[x_0] = \mathbf{v}_0$.

Proof. The defining equality of Fermi-Walker vector fields turns locally into a first order differential equation, whose solution is unique under a given initial value.

Remark. A vector field of Fermi-Walker type can be interpreted as a translation of a vector along C without rotation (see Exercise 4. of Chapter II. and Section 9.8. of [5]).

10.10. Definition. An Einsteinian manifold $(M, \mathbf{I}, g)$ is called *arrow orientable* if there is a vector field $\mathbf{X}$ on M such that $g \circ (\mathbf{X}, \mathbf{X})$ is a negative constant. If such a vector field is given the Einsteinian manifold is called *arrow oriented.*

A little diffeomorphism F between arrow oriented Einsteinian manifolds whose arrow orientations are given by $\mathbf{X}$ and $\mathbf{X}'$, respectively, is called *arrow preserving* or *arrow reversing* if

$$g'(F_*\mathbf{X}, \mathbf{X}') < 0 \quad \text{or} \quad g'(F_*\mathbf{X}, \mathbf{X}') > 0, \quad \text{respectively.}$$

(notation: 3.9.).

Remarks. (i) If $(M, \mathbf{I}, g)$ is an arrow oriented Einsteinian manifold then $(T_x(M), \mathbf{I}, g[x])$ is an arrow oriented Minkowskian vector space for all $x \in M$; its arrow orientation is given by the arrow class of $\mathbf{X}[x]$.

(ii) If $(M, \mathbf{I}, g)$ is a Minkowskian affine space which is arrow oriented in the above-given sense then one can easily see that $\mathbf{X}[x]$ gives the same arrow orientation to $T_x(M) = \mathbf{M}$ for all $x \in M$, hence $(M, \mathbf{I}, g)$ is arrow oriented in the sense of V. 4.8. Conversely, an arrow orientation of $(M, \mathbf{I}, g)$ in the sense of V. 4.8. determines an arrow orientation on $(M, \mathbf{I}, g)$, e.g. by a constant map.

VIII. LIE GROUPS

1. Fundamentals

1.1. Definition. A *Lie group* is a group G endowed with a smooth structure such that the maps

$$G \times G \longrightarrow G, \qquad (g, h) \mapsto gh,$$

$$G \longrightarrow G, \qquad g \mapsto g^{-1}$$

are smooth.

A *Lie group homomorphism (isomorphism)* from a Lie group into another is a smooth group homomorphism (isomorphism).

1.2. Proposition. Lie groups are of Hausdorff type and regular ([11], Theorem I. 6.1.).

1.3. Proposition. Let G be a connected Lie group and let U be a neighbourhood of the identity of G. Then

$$G = \bigcup_{n \in \mathbb{N}} U^n$$

where $U^n := \{g_1 g_2 \ldots g_n : g_i \in U, i = 1, \ldots, n\}$ ([11], Theorem I. 6.4.).

1.4. Proposition. The connected component of the identity of a Lie group is an open normal subgroup ([11], Theorem I. 6.3.).

1.5. Definition. Let G be a Lie group, $g \in G$. Then the maps

$$L_g : G \longrightarrow G, \qquad h \mapsto gh,$$

$$R_g : G \longrightarrow G, \qquad h \mapsto hg,$$

$$I_g := L_g \circ R_{g^{-1}}$$

are called *left translation by g, right translation by g, inner automorphism by g*, respectively.

Proposition. The maps above are diffeomorphisms of G and they satisfy the relations

$$L_{gh} = L_g \circ L_h, \qquad L_{g^{-1}} = L_g^{-1},$$

$$R_{gh} = R_h \circ R_g, \qquad R_{g^{-1}} = R_g^{-1},$$

$$L_g \circ R_h = R_h \circ L_g,$$

$$I_{gh} = I_g \circ I_h, \qquad I_{g^{-1}} = I_g^{-1}$$

for all $g, h \in G$.

1.6. Definition. Let G be a Lie group. A vector field $\mathbf{X}$ on G is called *left* or *right invariant* if

$$(L_g)_* \mathbf{X} \quad \text{or} \quad (R_g)_* \mathbf{X} = \mathbf{X}, \quad \text{respectively,}$$

for all $g \in G$ (see VII. 3.9.).

Remark. $\mathbf{X}$ is left invariant if and only if $DL_g[h]\,\mathbf{X}[h] = \mathbf{X}[gh]$ for all $g, h \in G$ which implies that

$$\mathbf{X}[g] = DL_g[e]\,\mathbf{X}[e] \qquad (g \in G)$$

where e is the identity of G.

A similar observation can be made for right invariant vector fields as well.

1.7. Definition. A *Lie algebra* is a real vector space $\mathbf{G}$ endowed with a so called *commutator* or *Lie product*

$$[\ ,\ ]: \mathbf{G} \times \mathbf{G} \to \mathbf{G}$$

which

(i) is bilinear and antisymmetric,

(ii) satisfies the *Jacobi identity*

$$[[\mathbf{x}, \mathbf{y}], \mathbf{z}] + [[\mathbf{y}, \mathbf{z}], \mathbf{x}] + [[\mathbf{z}, \mathbf{x}], \mathbf{y}] = 0 \qquad (\mathbf{x}, \mathbf{y}, \mathbf{z} \in \mathbf{G}).$$

A *Lie algebra homomorphism (isomorphism)* from a Lie algebra into another is a linear map (bijection) which preserves the commutator.

1.8. Proposition. Endowed with the usual $\mathbb{R}$-linear structure and commutator (see VII. 3.13.), the set of left invariant vector fields (and the set of right invariant vector fields) of a Lie group forms a Lie algebra.

Definition. The Lie algebra of left invariant vector fields of a Lie group G is called the *Lie algebra of* G and is denoted by $\mathbf{L}(G)$.

1.9. Proposition. Let e denote the identity of the Lie group G. Then

$$\mathbf{L}(G) \longrightarrow T_e(G), \qquad \mathbf{X} \mapsto \mathbf{X}[e]$$

is a linear bijection whose inverse is

$$T_e(G) \longrightarrow \mathbf{L}(G), \qquad \mathbf{v} \mapsto (g \mapsto (g, DL_g[e]\mathbf{v})).$$

As a consequence, $\dim \mathbf{L}(G) = \dim G$.

Remarks. (i) With the aid of this bijection $T_e(G)$ can be equipped with a commutator and it is sometimes convenient to consider $T_e(G)$ as the Lie algebra of G.

(ii) A similar linear bijection exists from the Lie algebra of right invariant vector fields onto $T_e(G)$. Hence we have a linear bijection between the Lie algebras of the right invariant vector fields and left invariant vector fields. One can show that this bijection is a Lie algebra anti-isomorphism, i.e. reverses commutators.

(iii) Let J denote the diffeomorphism $G \longrightarrow G$, $g \mapsto g^{-1}$. Then $J_*(\mathbf{X})$ is a right invariant vector field if $\mathbf{X}$ is left invariant. Moreover, $\mathbf{X} \mapsto J_*(\mathbf{X})$ is a Lie algebra isomorphism.

2. The exponential mapping

In this section G is an n dimensional Lie group, e is the identity of G, and $\mathsf{L}(G)$ is the Lie algebra of G.

2.1. Proposition. (i) All left invariant vector fields on G are complete.

(ii) $\mathsf{L}(G) \longrightarrow G$, $\mathbf{X} \mapsto \exp \mathbf{X} := (\mathrm{Exp}\,\mathbf{X})_1\,(e)$ (see VII. 6.5.) is a smooth map, and

$$\exp \mathbf{0} = e, \qquad \exp(-\mathbf{X}) = (\exp \mathbf{X})^{-1},$$

$$\exp(t+s)\mathbf{X} = (\exp t\mathbf{X})(\exp s\mathbf{X})$$

for all $t, s \in \mathbb{R}$, $\mathbf{X} \in \mathsf{L}(G)$.

(iii) exp is a diffeomorphism between a neighbourhood of $\mathbf{0} \in \mathsf{L}(G)$ and a neighbourhood of $e \in G$ ([11], Theorem I. 10.4.).

Definition. exp is called the *exponential mapping* of G.

If we have to distinguish the exponential mapping of G from the exponential mapping of another Lie group, we write $\exp_G$.

2.2. Proposition. For all $\mathbf{X} \in \mathsf{L}(G)$ the map $\mathbb{R} \longrightarrow G$, $t \mapsto \exp t\mathbf{X}$ is a smooth group homomorphism whose range is

(i) the identity of G if and only if $\mathbf{X} = \mathbf{0}$,

(ii) diffeomorphic to $\mathbb{R}$ if and only if $t \mapsto \exp t\mathbf{X}$ is injective,

(iii) diffeomorphic to $\mathbb{T}$ in the other cases. ([11], Corollary I. 10.1.).

Definition $\{\exp t\mathbf{X} \,\vdots\, t \in \mathbb{R}\}$ is called the one parameter subgroup generated by $\mathbf{X}$.

Remark. According to Remark VII. 6.6. (i) $\exp t\mathbf{X} = (\mathrm{Exp}\,t\mathbf{X})_1(e) = (\mathrm{Exp}\,\mathbf{X})_t(e)$, thus the one parameter subgroup generated by $\mathbf{X}$ is determined by the initial value problem

$$\dot{x} = \mathbf{X} \circ x, \qquad x(0) = e\,.$$

2.3. Definition. Let U be a neighbourhood of e in the neighbourhood given by (iii) of 2.1. If $L: \mathbf{L}(G) \longrightarrow \mathbb{R}^n$ is a linear bijection then $(U, L \circ \exp^{-1}|_U)$ is called a *canonical chart* of G and $L \circ \exp^{-1}|_U$ is a *canonical coordinatization.*∎

Evidently, the image of a one parameter subgroup by a canonical coordinatization is a straight line segment in $\mathbb{R}^n$, and the group multiplication in it becomes the vector addition.

2.4. Proposition. Let $\mathbf{X}_1, \ldots, \mathbf{X}_n$ be a basis of $\mathbf{L}(G)$. Then

$$\sum_{i=1}^{n} t_i \mathbf{X}_i \mapsto (\exp t_1 \mathbf{X}_1) \ldots (\exp t_n \mathbf{X}_n)$$

is a diffeomorphism between a neighbourhood of $\mathbf{0} \in \mathbf{L}(G)$ and a neighbourhood of $e \in G$ ([11], Theorem I. 10.5.).

Remark. A simple and important consequence of this result is that in a neighbourhood of $e \in G$ each element can be realized as a product of elements taken from one parameter subgroups generated by the elements of an arbitrary basis of $\mathbf{L}(G)$. If G is connected the same is true for all elements of G (see 1.3.).

2.5. Proposition. Let $\mathbf{X}, \mathbf{Y} \in \mathbf{L}(G)$. Then for all $t \in \mathbb{R}$

(i) $(\exp t\mathbf{X})(\exp t\mathbf{Y}) = \exp\left(t\mathbf{X} + t\mathbf{Y} + \frac{1}{2} t^2[\mathbf{X}, \mathbf{Y}] + \mathrm{Ordo}\,(t^3)\right)$

(ii) $(\exp t\mathbf{X})(\exp t\mathbf{Y})(\exp -t\mathbf{X}) = \exp\,(t\mathbf{Y} + t^2[\mathbf{X}, \mathbf{Y}] + \mathrm{Ordo}\,(t^3))$

(iii) $(\exp t\mathbf{X})(\exp t\mathbf{Y})(\exp -t\mathbf{X})(\exp -t\mathbf{Y}) = \exp\,(t^2[\mathbf{X}, \mathbf{Y}] + \mathrm{Ordo}\,(t^3))$

(iv) $\exp t(\mathbf{X}+\mathbf{Y}) = (\exp t\mathbf{X})(\exp t\mathbf{Y})(\exp \mathrm{Ordo}\,(t^2))$

(v) $\exp t^2[\mathbf{X}, \mathbf{Y}] = (\exp t\mathbf{X})(\exp t\mathbf{Y})(\exp -t\mathbf{X})(\exp -t\mathbf{Y})(\exp \mathrm{Ordo}\,(t^3))$

([11], Lemma II. 6.1.).

3. Lie subgroups and Lie group homomorphisms

3.1. Definition. A Lie group H is a Lie *subgroup* of a Lie group G if H is a subgroup of G and the canonical embedding $i: H \longrightarrow G$ is an immersion, i.e. $Di[h]$ is injective for all $h \in H$.

3.2. Proposition. If H is a Lie subgroup of a Lie group G then H is a closed subgroup of G.

Conversely, if H is a closed subgroup of a Lie group G then there is a unique smooth structure on H such that H is a Lie subgroup of G ([11], Theorem I. 10.12. and Theorem II. 6.3.).

3.3. Proposition. If H is a Lie subgroup of the Lie group G then $\mathbf{L}(H)$ is a Lie subalgebra of $\mathbf{L}(G)$.

For every Lie subalgebra $\mathbf{A}$ of $\mathbf{L}(G)$ there is a unique connected Lie subgroup H of G such that $\mathbf{L}(H) = \mathbf{A}$ ([11], Theorem I. 10.2. and Theorem II. 6.2.).

3.4. Definition. Let G and H be Lie groups. A *local Lie group homomorphism* from G into H is a pair (U, F) where

(i) U is a neighbourhood of the identity of G,

(ii) $F: U \longrightarrow H$ is a smooth map such that

$$F(g_1 g_2) = F(g_1)\, F(g_2),$$

$$F(g^{-1}) = F(g)^{-1}$$

whenever $g_1, g_2, g_1 g_2 \in U$ and $g, g^{-1} \in U$.

A local Lie group homomorphism (U, F) is called a *local Lie group isomorphism* if F is a little diffeomorphism.

Remark. Evidently, a Lie group homomorphism $F: G \longrightarrow H$ can be considered as a local Lie group homomorphism (G, F).

3.5. Proposition. Suppose (U, F) is a local Lie group homomorphism (isomorphism) from G into H. Then

$$F_*: \mathbf{L}(G) \longrightarrow \mathbf{L}(H), \qquad \mathbf{X} \mapsto (h \mapsto (h, DL_h[e]\, DF[e]\mathbf{X}[e]))$$

is a Lie algebra homomorphism (isomorphism) (e denotes the identity of both G and H), and

$$F \circ \exp_G = \exp_H \circ F_*$$

([11], Theorem I. 10.8. and [5] Lemma II. 1.12.).

Remarks. (i) If we identify $\mathbf{L}(G)$ with $T_e(G)$ and $\mathbf{L}(H)$ with $T_e(H)$ (as usually, the identities of both groups are denoted by the same symbol) then F_* becomes simply $DF[e]$.

(ii) One can easily show that $F_*(\mathbf{X})$ is F-related to $\mathbf{X}$ (see VII. 3.8.); note that though F is not necessarily a diffeomorphism, the group structure allows us to define F_* on the left invariant vector fields.

3.6. Proposition. Let G and H be Lie groups. Their Lie algebras are isomorphic if and only if there is a local Lie group isomorphism between G and H ([11], Theorem I. 10.9., [5] Theorem II. 1.11.).

3.7. Definition. Let G be a connected Lie group. A *universal covering group* of G is a Lie group $\hat{G}$ having the properties

(i) $\hat{G}$ is simply connected,

(ii) there is a Lie group surjection $\phi: \hat{G} \longrightarrow G$ (called the *covering map*),

(iii) there is a neighbourhood U of the identity of $\hat{G}$ such that $(U, \phi|_U)$ is a local Lie group isomorphism.

Proposition. For every Lie group there exists a universal covering group ([7], Theorem 8.13.).

4. Some special Lie groups

In this section V denotes an n dimensional real affine space over the vector space $\mathbf{V}$, and $I := \mathrm{id}_{\mathbf{V}}$.

4.1. The Lie algebra gL(**V**). The set of linear maps $\mathbf{V} \longrightarrow \mathbf{V}$ form an n^2 dimensional real vector space which we denote by L(**V**).

L(**V**) is an n^2 dimensional manifold whose tangent space at every point is identified with L(**V**) (cf. VII. 2.8. (iii)). A convenient global chart on L(**V**) can be given by a matrix representation of the elements of L(**V**) corresponding to a basis of **V**.

The "multiplication"

$$\mathrm{L}(\mathbf{V}) \times \mathrm{L}(\mathbf{V}) \longrightarrow \mathrm{L}(\mathbf{V}), \qquad (A, B) \mapsto A \circ B =: AB$$

is a smooth map. Moreover, if $A \in \mathrm{L}(\mathbf{V})$ is given, the left and right multiplication by A

$$L_A : \mathrm{L}(\mathbf{V}) \longrightarrow \mathrm{L}(\mathbf{V}), \qquad B \mapsto AB,$$

$$R_A : \mathrm{L}(\mathbf{V}) \longrightarrow \mathrm{L}(\mathbf{V}), \qquad B \mapsto BA$$

are linear maps and thus they are smooth and

$$\mathrm{D}L_A = L_A, \qquad \mathrm{D}R_A = R_A .$$

Endowed with the commutator

$$[A, B] := AB - BA \qquad (A, B \in \mathrm{L}(\mathbf{V}))$$

L(**V**) becomes a Lie algebra which is denoted by gL(**V**).

4.2. The Lie group GL(**V**). The set of invertible elements of L(**V**) is an open submanifold which we denote by GL(**V**). The multiplication inherited from L(**V**) turns GL(**V**) into a group whose identity is I.

The inversion $\mathrm{GL}(\mathbf{V}) \longrightarrow \mathrm{GL}(\mathbf{V})$, $F \mapsto F^{-1}$ is a smooth map, hence GL(**V**) is a Lie group. We identify the tangent space of $\mathrm{GL}(\mathbf{V}) \subset \mathrm{L}(\mathbf{V})$ at I by L(**V**).

Proposition. The Lie algebra gL(**V**) can be considered as the Lie algebra of GL(**V**).

Proof. If **X** is a left invariant vector field on GL(**V**) and $\mathbf{X}[I] = A$ then (see Remark 1.6.)

$$\mathbf{X}[F] = \mathrm{D}L_F[I]A = FA = R_A F \qquad (F \in \mathrm{GL}(\mathbf{V})).$$

As a consequence, if **X** and **Y** are left invariant vector fields, $\mathbf{X}[I] = A$, $\mathbf{Y}[I] = B$, then

$$[\mathbf{X}, \mathbf{Y}][F] = \mathrm{D}\mathbf{Y}[F]\mathbf{X}[F] - \mathrm{D}\mathbf{X}[F]\mathbf{Y}[F] = F(AB - BA) \qquad (F \in \mathrm{GL}(\mathbf{V}))$$

(see VII. 3.14. (ii) and recall that $\mathrm{GL}(\mathbf{V}) \subset \mathrm{L}(\mathbf{V})$). Apply Remark 1.9. (i) to get the desired result.

4.3. The exponential mapping $\mathrm{gL}(\mathbf{V}) \longrightarrow \mathrm{GL}(\mathbf{V})$. For $A \in \mathrm{gL}(\mathbf{V})$ we put $A^0 := = I$ (even if $A = 0$) and

$$e^A := \sum_{n \in \mathbb{N}_0} \frac{A^n}{n!}$$

It is well known that this series converges absolutely and its sum is in $\mathrm{GL}(\mathbf{V})$.

Let $\mathbf{X}$ be a left invariant field on $\mathrm{GL}(\mathbf{V})$, $\mathbf{X}[I] = A$. Then $\mathbb{R} \longrightarrow \mathrm{GL}(\mathbf{V})$, $t \mapsto \exp t\mathbf{X}$ is the solution of the initial value problem (see Remark 2.2.)

$$\dot{F}(t) = \mathbf{X}[F(t)] = F(t)A \qquad (t \in \mathbb{R}),$$

$$F(0) = I$$

It is well known that its solution is $t \mapsto e^{tA}$. Consequently, $e^{tA} = \exp t\mathbf{X}$ $(t \in \mathbb{R})$, i.e. if $\mathrm{gL}(\mathbf{V})$ is identified with the Lie algebra of $\mathrm{GL}(\mathbf{V})$ then the map defined by the series above coincides with the exponential mapping defined in 2.1.

4.4. The Lie algebra $\mathrm{gA}(\mathbf{V})$. The set $\mathrm{A}(\mathbf{V})$ of affine maps $\mathbf{V} \longrightarrow \mathbf{V}$ can be identified with the set $\mathbf{V} \times \mathrm{L}(\mathbf{V})$ by

$$(\mathbf{a}, A)\mathbf{x} := A\mathbf{x} + \mathbf{a} \qquad ((\mathbf{a}, A) \in \mathbf{V} \times \mathrm{L}(\mathbf{V}), \quad \mathbf{x} \in \mathbf{V}).$$

Hence $\mathrm{A}(\mathbf{V})$ is an $(n+n^2)$ dimensional real vector space; as such, it is an $(n+n^2)$ dimensional manifold whose tangent space at every point is identified with $\mathrm{A}(\mathbf{V})$.

The composition of such affine maps gives a "multiplication" on $\mathrm{A}(\mathbf{V})$:

$$((\mathbf{a}, A), (\mathbf{b}, B)) \mapsto (\mathbf{a}, A) \circ (\mathbf{b}, B) = (\mathbf{a} + A\mathbf{b}, AB)$$

which is a smooth map. Moreover, if $(\mathbf{a}, A) \in \mathrm{A}(\mathbf{V})$ is given, the left and right multiplications by $(\mathbf{a}, A)$

$$L_{(\mathbf{a}, A)} : \mathrm{A}(\mathbf{V}) \longrightarrow \mathrm{A}(\mathbf{V}), \qquad (\mathbf{b}, B) \mapsto (\mathbf{a}, A) \circ (\mathbf{b}, B),$$

$$R_{(\mathbf{a}, A)} : \mathrm{A}(\mathbf{V}) \longrightarrow \mathrm{A}(\mathbf{V}), \qquad (\mathbf{b}, B) \mapsto (\mathbf{b}, B) \circ (\mathbf{a}, A)$$

are an affine and a linear map, respectively, thus they are smooth and

$$\mathrm{D}L_{(\mathbf{a}, A)} = L_{(\mathbf{0}, A)}, \qquad \mathrm{D}R_{(\mathbf{a}, A)} = R_{(\mathbf{a}, A)}.$$

Endowed with the commutator

$$[(\mathbf{a}, A), (\mathbf{b}, B)] := (A\mathbf{b} - B\mathbf{a}, AB - BA)$$

$$((\mathbf{a}, A), (\mathbf{b}, B) \in \mathrm{A}(\mathbf{V}))$$

$\mathrm{A}(\mathbf{V}) = \mathbf{V} \times \mathrm{L}(\mathbf{V})$ becomes a Lie algebra which is denoted by $\mathrm{gA}(\mathbf{V})$.

Note that we can write

$$[(\mathbf{a}, A), (\mathbf{b}, B)] = \mathrm{D}(\mathbf{a}, A) \circ (\mathbf{b}, B) - \mathrm{D}(\mathbf{b}, B) \circ (\mathbf{a}, A)$$

where $\mathrm{D}(\mathbf{a}, A)$ is the derivative of the affine map $(\mathbf{a}, A)$ and equals A.

4.5. The Lie group GA(V). The set of invertible elements of A(V) is an open submanifold which as a set, can be identified with $\mathbf{V} \times GL(\mathbf{V})$. The multiplication inherited from $\mathbf{V} \times L(\mathbf{V})$ turns $\mathbf{V} \times GL(\mathbf{V})$ into a group which is denoted by GA(V) and whose identity is $(\mathbf{0}, I)$.

One often says that GA(V) is the *semidirect product* of the additive group of **V** and GL(V).

The inversion $GA(\mathbf{V}) \to GA(\mathbf{V})$, $(\mathbf{f}, F) \mapsto (\mathbf{f}, F)^{-1} = (-F^{-1}\mathbf{f},\ F^{-1})$ is a smooth map, hence GA(V) is a Lie group. We identify the tangent space of $GA(\mathbf{V}) \subset$ $\subset A(\mathbf{V})$ at $(\mathbf{0}, I)$ by $A(\mathbf{V}) = \mathbf{V} \times L(\mathbf{V})$.

Proposition. The Lie algebra gA(V) can be considered as the Lie algebra of GA(V).

Proof. If **X** is a left invariant vector field on GA(V) and $\mathbf{X}[\mathbf{0}, I] = (\mathbf{a}, A)$ then (see Remark 1.6.)

$$\mathbf{X}[\mathbf{f}, F] = DL_{(\mathbf{f}, F)}[\mathbf{0}, I]\,(\mathbf{a}, A) = (F\mathbf{a}, FA) = R_{(\mathbf{a}, A)}D(\mathbf{f}, F)$$

for all $(\mathbf{f}, F) \in GA(\mathbf{V})$. As a consequence, if **X** and **Y** are left invariant vector fields, $\mathbf{X}[\mathbf{0}, I] = (\mathbf{a}, A)$, $\mathbf{Y}[\mathbf{0}, I] = (\mathbf{b}, B)$, then

$$[\mathbf{X}, \mathbf{Y}][\mathbf{f}, F] = (F(A\mathbf{b} - B\mathbf{a}), F(AB - BA)) \qquad ((\mathbf{f}, F) \in GA(\mathbf{V}))$$

(see VII. 3.14 (iii) and recall that $GA(\mathbf{V}) \subset A(\mathbf{V})$). Apply Remark 1.9. (i) to get the desired result.

4.6. The exponential mapping $gA(\mathbf{V}) \to GA(\mathbf{V})$.

Take gA(V) as the Lie algebra of GA(V). Then $\exp(\mathbf{a}, A)$ for $(\mathbf{a}, A) \in gA(\mathbf{V})$ is determined by the initial value problem posed for the right invariant vector field corresponding to $(\mathbf{a}, A)$:

$$(\dot{\mathbf{f}}(t), \dot{F}(t)) = (F(t)\mathbf{a}, F(t)A) \qquad (t \in \mathbb{R}),$$

$$(\mathbf{f}(0), F(0)) = (\mathbf{0}, I).$$

We easily deduce that

$$F(t) = e^{tA} \qquad (t \in \mathbb{R})$$

and consequently,

$$\mathbf{f}(t) = \sum_{n \in \mathbb{N}_0} \frac{t^{n+1} A^n}{(n+1)!}\mathbf{a} \qquad (t \in \mathbb{R}).$$

Thus we find that

$$\exp(\mathbf{a}, A) = I + \sum_{n \in \mathbb{N}_0} \frac{A^n}{(n+1)!} \circ (\mathbf{a}, A).$$

We call attention to that A is the derivative of the affine map $(\mathbf{a}, A)$. Two special cases are worth being mentioned:

$$\exp(\mathbf{0}, A) = (\mathbf{0}, e^A), \qquad \exp(\mathbf{a}, 0) = (\mathbf{a}, I).$$

Moreover, observe that the initial value problem determining exp$(\mathbf{a}, A)$ is equivalent to the following one:

$$(\dot{\mathbf{f}}(t), \dot{F}(t)) = (\mathbf{a}, A) \circ (\mathbf{f}(t), F(t)) \qquad (t \in \mathbb{R}),$$

$$(\mathbf{f}(0), F(0)) = (\mathbf{0}, I).$$

4.7. The Lie algebra gA(V). The set of affine maps $V \longrightarrow \mathbf{V}$ is an $n+n^2$ dimensional real vector space which we denote by A(V, **V**). As such, it is an $n+n^2$ dimensional manifold whose tangent space at every point is A(V, **V**).

Note that the elements of A(V, **V**) cannot be composed, hence there is no multiplication on A(V, **V**), in contradistinction to A(**V**).

However, if $A \in$ L(**V**) the we can define the multiplication

$$M_A : \mathrm{A}(V, \mathbf{V}) \longrightarrow \mathrm{A}(V, \mathbf{V}), \qquad X \mapsto A \circ X$$

which is a linear map.

Equipped with the commutator

$$[X, Y] := \mathrm{D}X \circ Y - \mathrm{D}Y \circ X \qquad (X, Y \in \mathrm{A}(V, \mathbf{V}))$$

A(V, **V**) becomes a Lie algebra which is denoted by gA(V).

Proposition. Let $o \in V$. Then

$$w_o : \mathrm{gA}(V) \longrightarrow \mathrm{gA}(\mathbf{V}), \qquad X \mapsto X \circ O_o^{-1}$$

(notation: VI. 1.1. (iii)) is a Lie algebra isomorphism (which shows in particular that dim A(V, **V**) $= n+n^2$). If we identify the elements of gA(**V**) with the elements of **V** × gL(**V**) then

$$w_0(X) = (X(o), \mathrm{D}X) \qquad (X \in \mathrm{gA}(V)),$$

$$(w_o^{-1}(\mathbf{a}, A))(x) = A(x-o) + \mathbf{a} \qquad ((\mathbf{a}, A) \in \mathrm{gA}(\mathbf{V}),\ x \in V).$$

If o' is another element of V then

$$(w_{o'} \circ w_o^{-1})(\mathbf{a}, A) = (\mathbf{a} + A(o'-o), A) \qquad ((\mathbf{a}, A) \in \mathrm{gA}(\mathbf{V})).$$

4.8. The Lie group GA(V). The set A(V) of affine maps $V \longrightarrow V$, endowed with the subtraction given in VI. 1.14., becomes an affine space over A(V, **V**).

The "multiplication"

$$\mathrm{A}(V) \times \mathrm{A}(V) \longrightarrow \mathrm{A}(V), \qquad (S, T) \mapsto S \circ T$$

is a smooth map. Moreover, if $S \in$ A(V) is given, the left and right multiplications by S

$$L_S : \mathrm{A}(V) \longrightarrow \mathrm{A}(V), \qquad T \mapsto S \circ T,$$

$$R_S : \mathrm{A}(V) \longrightarrow \mathrm{A}(V), \qquad T \mapsto T \circ S$$

are affine maps and

$$N_S : \mathrm{A}(V, \mathbf{V}) \longrightarrow \mathrm{A}(V, \mathbf{V}), \qquad X \mapsto X \circ S$$

is a linear map for which we have

$$DL_S = M_{DS}, \qquad DR_S = N_S.$$

The set of invertible elements of $A(V)$ is an open submanifold which we denote by $GA(V)$. The multiplication inherited from $A(V)$ turns $GA(V)$ into a group whose identity is id_V.

The inversion $GA(V) \longrightarrow GA(V)$, $S \mapsto S^{-1}$ is a smooth map which becomes clear from the following proposition, hence $GA(V)$ is a Lie group. We identify the tangent space of $GA(V) \subset A(V)$ at id_V by $A(V, \mathbf{V})$.

(i) **Proposition.** Let $o \in V$. Then

$$z_o : GA(V) \longrightarrow GA(\mathbf{V}), \qquad S \mapsto O_o \circ S \circ O_o^{-1}$$

(notation: VI. 1.1. (iii)) is a group isomorphism and a diffeomorphism (in other words, it is a Lie group isomorphism). If we identify the elements of $GA(\mathbf{V})$ with the elements of $\mathbf{V} \times GL(\mathbf{V})$ then

$$z_o(S) = (S(o) - o, DS) \qquad (S \in GA(V)),$$

$$(z_o^{-1}(\mathbf{f}, F))(x) = o + F(x - o) + \mathbf{f} \qquad ((\mathbf{f}, F) \in GA(\mathbf{V}),\ x \in V).$$

If o' is another element of V then

$$(z_{o'} \circ z_o^{-1})(\mathbf{f}, F) = (\mathbf{f} + (F - I)(o' - o), F) \qquad ((\mathbf{f}, F) \in GA(\mathbf{V})).$$

(ii) **Proposition.** The Lie algebra $gA(V)$ can be considered as the Lie algebra of $GA(V)$.

Proof. If $\mathbf{X}$ is a left invariant vector field on $GA(V)$ and $\mathbf{X}[\mathrm{id}_V] = X$, then

$$\mathbf{X}[S] = DL_S[\mathrm{id}_V]X = M_{DS}X = DS \circ X \qquad (S \in GA(V)).$$

Note that for given $S \in GA(V)$, $\mathbf{X}[S]$ is an affine map; $D\mathbf{X}[S]$ will denote its derivative. On the other hand, we need the derivative of the map $(S \mapsto \mathbf{X}[S])$; we denote it by $D(S \mapsto \mathbf{X}[S])$.

Since our previous result says that $GA(V) \longrightarrow A(V, \mathbf{V})$, $S \mapsto \mathbf{X}[S]$ is the restriction of the affine map $A(V) \longrightarrow A(V, \mathbf{V})$, $S \mapsto DS \circ X$, we have that

$$D(S \mapsto \mathbf{X}[S])Z = DZ \circ \mathbf{X} \qquad (S \in GA(V),\ Z \in A(V, \mathbf{V})).$$

Consequently, if $\mathbf{X}$ and $\mathbf{Y}$ are left invariant vector fields, $\mathbf{X}[\mathrm{id}_V] = X$, $\mathbf{Y}[\mathrm{id}_V] = Y$, then

$$\begin{aligned}[\mathbf{X}, \mathbf{Y}][S] &= D\mathbf{Y}[S]\mathbf{X}[S] - D\mathbf{X}[S]\mathbf{Y}[S] = \\ &= D\mathbf{X}[S] \circ Y - D\mathbf{Y}[S] \circ X = \\ &= DS \circ (DX \circ Y - DY \circ X) \qquad (S \in GA(V)).\end{aligned}$$

4.9. The exponential mapping $\mathrm{gA}(V) \to \mathrm{GA}(V)$.

According to the previous considerations and Proposition 3.5. we have for all $o \in V$

$$\exp_{\mathrm{GA}(\mathbf{V})}(w_o(X)) = z_0(\exp_{\mathrm{GA}(V)} X) \qquad (X \in \mathrm{gA}(V)).$$

Thus we conclude from 4.6. that the exponential of X is determined by the initial value problem

$$\dot{S}(t) = X \circ S(t) \qquad (t \in \mathbb{R})$$

$$S(0) = \mathrm{id}_V,$$

and

$$\exp X = \mathrm{id}_V + \sum_{n \in \mathbb{N}_0} \frac{(\mathrm{D}X)^n}{(n+1)!} \circ X.$$

4.10. Subgroups of $\mathrm{GA}(V)$. (i) If $\mathbf{a} \in \mathbf{V}$ then the translation by $\mathbf{a}$,

$$T_{\mathbf{a}} : V \to V, \qquad x \mapsto x + \mathbf{a}$$

is an affine bijection whose derivative is $\mathrm{id}_{\mathbf{V}}$.

$$\mathrm{Tn}(\mathbf{V}) := \{T_{\mathbf{a}} : \mathbf{a} \in \mathbf{V}\}$$

is a commutative normal subgroup of $\mathrm{GA}(V)$:

$$LT_{\mathbf{a}}L^{-1} = T_{\mathrm{D}L\mathbf{a}} \qquad (\mathbf{a} \in \mathbf{V},\ L \in \mathrm{GA}(V)).$$

The Lie algebra of $\mathrm{Tn}(\mathbf{V})$ is $\mathbf{V}$ equipped with the zero commutator. Considered as the set of constant functions $V \to \mathbf{V}$, $\mathbf{V}$ is a Lie subalgebra (and even an ideal) of $\mathrm{gA}(V)$:

$$[X, \mathbf{a}] = \mathrm{D}X\mathbf{a} \qquad (\mathbf{a} \in \mathbf{V},\ X \in \mathrm{gA}(V)).$$

(ii) The map

$$\mathrm{GA}(V) \to \mathrm{GL}(\mathbf{V}), \qquad L \mapsto \mathrm{D}L$$

is a Lie group homomorphism whose kernel is $\mathrm{Tn}(\mathbf{V})$.

The corresponding Lie algebra homomorphism (see 3.5.) is

$$\mathrm{gA}(V) \to \mathrm{gL}(\mathbf{V}), \qquad X \mapsto \mathrm{D}X.$$

Consequently,

$$\mathrm{D}(\exp X) = \exp \mathrm{D}X.$$

(iii) Let K be a Lie subgroup of $\mathrm{GL}(\mathbf{V})$. The Lie algebra of K is a subalgebra of $\mathrm{gL}(\mathbf{V})$. K is a closed subgroup and thus

$$H := \{L \in \mathrm{GA}(V) : \mathrm{D}L \in K\}$$

is a closed subgroup, whence a Lie subgroup of $\mathrm{GA}(V)$.

We easily derive that

$$\mathrm{L}(H) = \{A \in \mathrm{gA}(V) : \mathrm{D}A \in \mathrm{L}(K)\}.$$

4.11. The pseudo-orthogonal group O(b). Suppose ($\mathbf{V}, \mathbf{B}, b$) is a pseudo-Euclidean vector space and assume the notations O(b) and $\mathbf{A}(b)$ introduced in V. 2.10.

(i) **Proposition.** O(b) is an $\frac{n(n-1)}{2}$ dimensional Lie subgroup of GL($\mathbf{V}$).

Proof. O(b) is clearly a closed subgroup of GL($\mathbf{V}$) and so it is a Lie group.

Take an arbitrary b-orthogonal basis $\{\mathbf{e}_k : k=1, \ldots, n\}$. It is evident that $F \in \mathrm{GL}(\mathbf{V})$ is in O(b) if and only if

$$b(F\mathbf{e}_i, F\mathbf{e}_k) = b(\mathbf{e}_i, \mathbf{e}_k) \qquad (i, j = 1, \ldots, n).$$

Thus we have $n + \binom{n}{2}$ equations for determining O(b) which yields that the submanifold O(b) is $\frac{n(n-1)}{2}$ dimensional.

(ii) **Proposition.** The Lie algebra of O(b) is $\mathbf{A}(b)$.

Proof. The Lie algebra of O(b) is a Lie subalgebra of gL($\mathbf{V}$). Recall that gL($\mathbf{V}$) is the tangent space of GL($\mathbf{V}$) at I (see 4.2.), hence we look for the Lie algebra of O(b) as a linear subspace of the tangent space of GL($\mathbf{V}$) at I.

Let $\mathbb{R} \rightarrowtail \mathrm{O}(b)$, $t \mapsto F(t)$ be a smooth curve, $F(0) = I$, $\mathrm{D}F[0] =: A$.

Differentiating the map $t \mapsto F(t)^* F(t) = I$ at zero we get

$$A^* + A = 0.$$

Since the vector space $\mathbf{A}(b)$ is of $\frac{n(n-1)}{2}$ dimension, we obtain the desired result.

Remark. In a similar way we get that the tangent space of O(b) at F is

$$\{A \in \mathrm{gL}(\mathbf{V}) : AF^{-1} \in \mathbf{A}(b)\}.$$

4.12. The three dimensional rotation group. Suppose ($\mathbf{E}, \mathbf{D}, \gamma$) is a three dimensional Euclidean vector space (see Section V. 3.), $I := \mathrm{id}_{\mathbf{E}}$.

Let us accept the notations

$$\mathrm{SO}(\gamma) := \mathrm{O}^+(\gamma) := \{R \in \mathrm{O}(\gamma) : \det R = +1\},$$

$$\mathrm{O}^-(\gamma) := \{R \in \mathrm{O}(\gamma) : \det R = -1\}.$$

We know from the previous results that O(γ) is a three dimensional Lie group, its Lie algebra is $\mathbf{A}(\gamma)$ and $\mathrm{O}(\gamma) = \mathrm{O}^+(\gamma) \cup \mathrm{O}^-(\gamma)$ (see V. 2.12.).

(i) **Proposition.** For every $R \in \mathrm{SO}(\gamma)$ there is a non-zero $\mathbf{e} \in \mathbf{E}$ such that $R\mathbf{e} = \mathbf{e}$.

Proof. Proposition IV. 1.23. results in

$$\det(R - I) = \det(R - R^* R) = \det(I - R^*) \det R = -\det(R - I)$$

which implies $\det(R - I) = 0$. Consequently, $R - I$ is not injective, there is a non-zero $\mathbf{e} \in \mathbf{E}$ such that $(R - I)\mathbf{e} = 0$.

(ii) Assume the notations

$$N:=\{R\in \mathrm{SO}(\gamma) : R\mathbf{x}\neq -\mathbf{x} \text{ for all } \mathbf{x}\in \mathbf{E}\},$$

$$P:=\{\mathbf{r}\in \frac{\mathbf{E}}{\mathbf{D}} : |\mathbf{r}| <\pi\}.$$

Proposition. For every $R\in N$ there exists a unique $\mathbf{r}\in P$ such that $\mathbf{0}\in P$ corresponds to $I\in N$ and $\mathbf{r}\neq \mathbf{0}$ for $R\neq I$ and

$$R = \frac{1-\cos|\mathbf{r}|}{|\mathbf{r}|^2}\mathbf{r}\otimes\mathbf{r}+\cos|\mathbf{r}|I - \frac{\sin|\mathbf{r}|}{|\mathbf{r}|}j^{-1}(\mathbf{r})$$

where j is given in V. 3.6. and $\mathbf{r}\otimes\mathbf{r}$ is considered as an element of $\mathbf{E}\otimes\mathbf{E}^*$(see V. 2.6.), i.e. a linear map $\mathbf{E}\longrightarrow\mathbf{E}$, $(\mathbf{r}\otimes\mathbf{r})\mathbf{x}=\mathbf{r}\otimes\frac{\gamma(\mathbf{d}\otimes\mathbf{r},\mathbf{x})}{\mathbf{d}}$ for $\mathbf{x}\in E$, $\mathbf{0}\neq\mathbf{d}\in\mathbf{D}$.

Proof. Let $R\in N$, $R\neq I$. Take a non-zero $\mathbf{e}\in\mathbf{E}$ such that $R\mathbf{e}=\mathbf{e}$, and a non-zero $\mathbf{x}\in\mathbf{E}$ which is γ-orthogonal to $\mathbf{e}$. Then

$$\mathbf{r}:=\frac{\mathbf{e}}{|\mathbf{e}|}\arccos\frac{\gamma(\mathbf{x},R\mathbf{x})}{|\mathbf{x}|^2}.$$

where $|\mathbf{e}|:=\sqrt[\otimes]{\gamma(\mathbf{e},\mathbf{e})}\in\mathbf{D}$. ∎

We say that R is a rotation of angle $|\mathbf{r}|$ around the axis $\mathbf{r}\otimes\mathbf{D}$.

(iii) **Proposition.** $\phi: N\longrightarrow P$, $R\mapsto\mathbf{r}$ is a diffeomorphism.

Proof. ϕ is evidently bijective, and ϕ^{-1} is evidently smooth. Moreover, the derivative of ϕ^{-1} is nowhere zero, hence ϕ^{-1} and thus ϕ are diffeomorphisms.

(iv) Let R be an element of $\mathrm{SO}(\gamma)$ such that $R\mathbf{x}=-\mathbf{x}$ for some non-zero $\mathbf{x}\in\mathbf{E}$. Then we can find two elements, $\mathbf{r}$ and $-\mathbf{r}$ of $\frac{\mathbf{E}}{\mathbf{D}}$ such that R is realized by the formula given in (ii). Thus we see that $\mathrm{SO}(\gamma)$ is homeomorphic to a set which is obtained from the closure of P by sticking together the diametrical points of the boundary of P.

Proposition. $\mathrm{SO}(\gamma)$ is compact, connected but not simply connected.

(v) **Proposition.** $\mathrm{O}(\gamma)$ has two connected components, $\mathrm{O}^+(\gamma)$ and $\mathrm{O}^-(\gamma)$.

Proof. $\mathrm{O}^-(\gamma)=(-I)O^+(\gamma)$, hence $\mathrm{O}^-(\gamma)$ is connected as well. Since det is continuous, $\mathrm{O}(\gamma)$ is disconnected.

(vi) **Proposition.** Let $A\in\mathbf{A}(\gamma)$. Then for an $\mathbf{x}\in\mathbf{E}$

$$(\exp tA)\mathbf{x}=\mathbf{x}\qquad (t\in\mathbb{R})$$

is equivalent to the fact that $\mathbf{x}$ is in $\operatorname{Ker} A$. ∎

Consequently, the elements of a one parameter subgroup are rotations around the same axis.

(vii) The bijection $j^{-1}:\frac{\mathbf{E}}{\mathbf{D}}\longrightarrow\mathbf{A}(\gamma)$ (see V. 3.7.) transfers P into

$$K:=\{A\in\mathbf{A}(\gamma) : |A|<\pi\}$$

and the previous results show that $\phi\circ j=\exp|_K$.

Proposition. The exponential mapping $\mathbf{A}(\gamma)\longrightarrow\mathrm{SO}(\gamma)$ is surjective, even $\exp[\bar{K}]=\mathrm{SO}(\gamma)$. Every element of $\mathrm{SO}(\gamma)$ is in a one parameter subgroup.

LIST OF SYMBOLS

■	marks the end of either a proposition or a proof or a definition, if it is necessary

1. Basic notations

$:=$ $=:$	defining equalities; the symbol on the side of the colon is defined to equal the other one
$\mathbb{N}$	the set of natural numbers
$\mathbb{N}_0$	the set of non-negative integers
$\mathbb{Z}$	the set of integers
$\mathbb{R}$	the set of real numbers
$\mathbb{R}^+$	the set of positive real numbers
$\mathbb{R}_0^+$	the set of non-negative real numbers
$]a, b[$	the open interval between $a, b \in \mathbb{R}$
$[a, b]$	the closed interval between $a, b \in \mathbb{R}$
$\mathbb{C}$	the set of complex numbers
$\mathbb{T}$	the set of unit complex numbers
$\overset{n}{\times} X := X^n$	the n-fold Cartesian product of the set X with itself ($n \in \mathbb{N}$)
$\operatorname{Dom} f$	the *domain* of the map f
$\operatorname{Ran} f$	the *range* of the map f
$f: X \longrightarrow Y$	f is a map with $\operatorname{Dom} f = X$, $\operatorname{Ran} f \subset Y$
$f: X \rightarrowtail Y$	f is a map with $\operatorname{Dom} f \subset X$, $\operatorname{Ran} f \subset Y$
$\mapsto$	the symbol showing a mapping rule
$f\|_A$	the *restriction* of the map f to $A \cap \operatorname{Dom} f$
$f \subset g$	the map g is an *extension* of f, i.e. $\operatorname{Dom} f \subset \operatorname{Dom} g$ and $g\|_{\operatorname{Dom} f} = f$
$\overset{-1}{f}$	the *total inverse* of the map $f: X \rightarrowtail Y$, i.e. if $H \subset Y$, $$\overset{-1}{f}(H) := \{x \in \operatorname{Dom} f : f(x) \in H\}$$

$f \circ g$	the *composition* of the maps $g: X \rightarrowtail Y$, $f: Y \rightarrowtail Z$, $\mathrm{Dom}\,(f \circ g) := \overset{-1}{g}\,(\mathrm{Dom}\, f) \cap \mathrm{Dom}\, g$, $(f \circ g)(x) := f(g(x))$
$\underset{i \in I}{\times} f_i$	the *product* of the maps $f_i: X_i \longrightarrow Y_i \quad (i \in I)$: $\underset{i \in I}{\times} f_i: \left(\underset{i \in I}{\times} X_i\right) \longrightarrow \left(\underset{i \in I}{\times} Y_i\right), \quad (x_i)_{i \in I} \mapsto (f_i(x_i))_{i \in I}$
$\overset{n}{\times} f$	the n-fold product of the map f with itself $(n \in \mathbb{N})$
$(f_i)_{i \in I}$	the *joint* of the maps $f_i: X \longrightarrow Y_i \quad (i \in I)$: $(f_i)_{i \in I}: X \longrightarrow \underset{i \in I}{\times} Y_i, \quad x \mapsto (f_i(x))_{i \in I}$
$\mathrm{Ker}\, A$	$:= \overset{-1}{A}\,\{0\}$, A is a linear map
pr_X	the canonical projection $X \times Y \longrightarrow X$
pr_k	the canonical projection $\underset{i \in I}{\times} X_i \longrightarrow X_k$
pr^k	$\mathbb{R}^n \longrightarrow \mathbb{R}$, the projection on the k-th coordinate
id_X	the identity map $X \longrightarrow X$, $x \mapsto x$

2. Other notations

$*$	marks duals of vector spaces (modules) and transposes of linear maps
$\ast$	marks adjoints of linear maps, V. 2.
$\star$	marks pull-backs, VII. 4.
$()_{\star}$	VII. 3.8.
$\otimes$	IV. 1.2.
⊚	IV. 1.8.
$=\!=$	IV. 2.2.
$\sqrt[\otimes]{}$	IV. 4.
$(\,\mid\,)$	the bilinear form of duality
(⫲)	V. 2.1., II. 1.5.
(⫲)	V. 2.7.
(⫳)	II. 1.5.
$[\,\mid\,)$	II. 1.5.
[,]	commutator, VII. 3.13., VIII. 1.7.
ar	I. 7.4., I. 7.8., II. 7.4., II. 7.7.
$\mathbf{A}(\gamma)$	VIII. 4.12.
$\mathbf{A}(g)$	II. 7.2.
$\mathrm{A}^r\mathrm{F}(M)$	VII. 3.4.
$\mathrm{A}^r\mathrm{F}(M)^*$	VII. 4.3.
$A^r(M)$	VII. 3.1.

LIST OF SYMBOLS

BIBLIOGRAPHY

[1] **Dombrowski, H. D.–Horneffer, K.:** Die differentialgeometrie des Galileischen Relativitätsprinzip, Math. Ż. 86 (1964) 291–311.

[2] **Godbillon, C.:** Géometrie differentielle et mécanique analytique, Coll. méthodes, Hermann, Paris. 1969.

[3] **de Groot, S. Z.–Suttorp, L. G.:** Foundations of Electrodynamics, North-Holland, Amsterdam 1972.

[4] **Helgason, S.:** Differential Geometry and Symmetric Spaces, Acad. Press, New York–London, 1962.

[5] **Møller, C.:** The theory of relativity, Second edition, Clarendon Press, Oxford, 1972.

[6] **Petrow, A. S.:** Einstein-Räume, Akademie-Verlag, Berlin, 1964.

[7] **Sagle, A. A.–Walde, R. F.:** Introduction to Lie Groups and Lie Algebras, Acad. Press, New York–London, 1973.

[8] **Schaefer, H. H.:** Topological Vector Spaces, The Macmillan Company, New York–London, 1966.

[9] **Schwartz, L.:** Analyse mathématique, Hermann, Paris, 1967.

[10] **Segal, I. E.:** Mathematical Problems of Relativistic Physics, Am. Math. Soc. Providence, Rhode Island, 1963.

[11] **Sulanke, R.–Wintgen, P.:** Differentialgeometrie und Faserbündel, VEB Deutscher Verlag der Wissenschaften, Berlin, 1972.

[12] **Varadarajan, V. S.:** Geometry of Quantum Theory, Vol. II. Van Nostrand Rein. Co. New York, 1970.

[13] **Wightman, A. S.:** Revised notes for lectures at the French Summer School of Theoretical Physics, Cargêse, Corsica, July 1964 (or, Les problèmes Mathématique de la Théorie Quantique des Champes (CNRS, Paris 1959)).

SUBJECT INDEX

270.—